Hydrology-Shaped Plant Communities: Diversity and Ecological Function

Hydrology-Shaped Plant Communities: Diversity and Ecological Function

Editors

Alenka Gaberščik
Igor Zelnik

MDPI • Basel • Beijing • Wuhan • Barcelona • Belgrade • Manchester • Tokyo • Cluj • Tianjin

Editors
Alenka Gaberščik
University of Ljubljana
Slovenia

Igor Zelnik
University of Ljubljana
Slovenia

Editorial Office
MDPI
St. Alban-Anlage 66
4052 Basel, Switzerland

This is a reprint of articles from the Special Issue published online in the open access journal *Water* (ISSN 2073-4441) (available at: https://www.mdpi.com/journal/water/special_issues/hydrology-shaped_plant).

For citation purposes, cite each article independently as indicated on the article page online and as indicated below:

LastName, A.A.; LastName, B.B.; LastName, C.C. Article Title. *Journal Name* **Year**, *Volume Number*, Page Range.

ISBN 978-3-0365-2968-4 (Hbk)
ISBN 978-3-0365-2969-1 (PDF)

Cover image courtesy of Alenka Gaberščik and Matej Holcar

Contents

About the Editors

Alenka Gaberščik (Associate Professor, Ph.D.) obtained her Ph.D. in Biology from the University of Ljubljana in 1991. She is Associate Professor at the Chair for Ecology and Environment conservation at the University of Ljubljana and the head of the Plant Ecology research group. Her research interests are various aspects of the ecology of plants, especially macrophytes and agricultural plants as well as ecosystem processes. She has published over 110 scientific articles, 15 book chapters, edited a monograph on the world's famous intermittent Lake Cerknica and co-edited a book Macrophytes of the River Danube Basin. She has organized several international symposia on UV radiation and plants, Macrophytes and Wetland ecology.

Igor Zelnik (Assistant Professor, Ph.D.) obtained his Ph.D. in Biology from the University of Ljubljana in 2005. He is Assistant Professor at the Chair for Ecology and Environment conservation at the University of Ljubljana. He gives lectures in General Ecology and Ecology of Aquatic Ecosystems. His research interests are various aspects of the Ecology from wetland and aquatic plants, to diatoms and benthic invertebrates. A great part of his research work includes publications about biodiversity of aquatic ecosystems, ecology of different types of wetlands, vegetation ecology. He has published more than 30 articles and several book chapters.

Preface to "Hydrology-Shaped Plant Communities: Diversity and Ecological Function"

Water is a crucial source for enabling the survival of plants and other organisms. However, different species show different needs for water as well as other relationships to it in their habitats. Flooded and water-saturated soils prevent the thriving of certain species while favoring others. The hydrological regime and movement of water are important factors that shape plant communities in rivers, lakes and wetlands. In 2021 we celebrated the 50th anniversary of the Ramsar Convention, which was established to protect wetlands in a wider sense. Despite the efforts of many individuals and organizations these habitats and the biodiversity they host are facing a continuous decline in their quality and quantity due to the endless human needs for electricity, water supply, food production, tourism and recreation. Since the plant species and communities living in these ecosystems are threatened due to changes in their habitats, including in hydrological conditions, every attempt to provide new knowledge on this topic is warmly welcome.

In this Special Issue authors present a wide range of ecosystems where water level, time of inundation or water movement influence the distribution of plants and shape the composition of plant communities. From the wet meadows and floodplain forests, where the water on the surface of soil generates the conditions that the majority of the plant species cannot withstand, to the rivers, where plants face currents that hinder their development. Besides the long list of studied ecosystems and plant communities there is also a considerable number of countries where the influence of hydrology on plant communities has been studied: Germany, Czech Republic, Poland, Lithuania, Austria, Slovakia, Hungary, Slovenia, Croatia, Serbia, Romania, Bulgaria and China.

We are grateful to the publisher, who invited us to edit the present Special Issue—Hydrology-Shaped Plant Communities: Diversity and Ecological Function—and in particular to Ms. Hailey Wu, Section Managing Editor of Water. We would like to thank to all authors who contributed their original articles and made this Special Issue valuable.

Alenka Gaberščik, Igor Zelnik
Editors

water

MDPI

Editorial

Hydrology-Shaped Plant Communities: Diversity and Ecological Function

Alenka Gaberščik and Igor Zelnik *

Department of Biology, Biotechnical Faculty, University of Ljubljana, Jamnikarjeva 101, 1000 Ljubljana, Slovenia; alenka.gaberscik@bf.uni-lj.si
* Correspondence: igor.zelnik@bf.uni-lj.si

Citation: Gaberščik, A.; Zelnik, I. Hydrology-Shaped Plant Communities: Diversity and Ecological Function. *Water* **2021**, *13*, 3525. https://doi.org/10.3390/w13243525

Received: 23 November 2021
Accepted: 7 December 2021
Published: 9 December 2021

Publisher's Note: MDPI stays neutral with regard to jurisdictional claims in published maps and institutional affiliations.

1. Introduction

Water is not only the essence of life and medium of organisms but also the environmental factor that shapes biocenoses, including plant communities. These communities are affected by the availability of water for single plant species or by hydrological regimes that may cause disturbance by physical force or stress due to high water levels and droughts. In many ecosystems, the hydrological regime presents major determinant that affects plant species composition, succession, productivity, and stability of plant communities. The hydrological regime is characterized by changes in the levels and volumes of water in rivers, lakes, reservoirs, and wetlands in time and space as well as by the currents and waves. According to Keddy [1], wetlands comprise a variety of ecosystems marked by permanent or temporal presence of water above the ground, namely floodplain forests and mangals, marshes with prevailing helophytes rooted in waterlogged soils, bogs dominated by peat mosses overgrowing thick layer of peat, fens with prevailing sedges and grasses and shallow peat layer, wet meadows dominated by herbaceous plants on occasionally flooded soils, and shallow water bodies colonized by truly aquatic plants. The structure of specific wetland depends on the climate, landscape morphology, and hydrological regime. The following aspects of a hydrological regime may affect diversity and functioning of plant communities: the presence of water in time (permanent, seasonal, and occasional), the frequency and the duration of flooding, the volume and depth of water, the variability in changes over time, and the predictability of a water regime [2]. The functional traits of plant species colonizing different wetlands or water bodies are adjusted to specific hydrological regimes and define community characteristics and their colonization and competition potential.

This Special Issue includes a variety of papers reporting about long-term researches of ecosystems where the hydrological regime presents an important factor in shaping plant communities, such as lake, intermittent lake, kettle-holes, lagoon, rivers, floodplain forests, and wet meadows. To show these effects of a hydrological regime along with other environmental factors, some authors concentrated on single species, on a group of plants, or on plant communities. The studied plant species range from fragile stonewort to giant oak-tree, but most of the authors focused on herbaceous vascular plants. Most of these ecosystems were studied from their banks or by wading in their shallow parts, but some of them were monitored by boat or even from the air.

Beside the widening specific knowledge on the certain ecosystem, especially the knowledge on interactions between plant communities and hydrological regime, some contributions published in this book may have important implications from the viewpoint of global warming and climate changes that affect the amount and distribution of precipitation and thus the hydrological regimes of many ecosystems. Aquatic ecosystems monitored for several years or even long-term studies provide valuable results of the changes with time, which we urgently need for more reliable forecasts of future trends.

2. Main Messages of This Special Issue

The study of Šumberová et al. [3] investigated the diversity of vascular plants, charophytes, and diatoms in relation to a range of environmental factors in a large hypertrophic fishpond in order to point out the main drivers affecting these communities and to improve the understanding and the basis for protection of threatened aquatic ecosystems. The advantage of this study is that the authors summarized the results of several growing seasons. Even though the life histories of the studied groups differ significantly, the outcomes show that they responded in the same way to specific locally important environmental factors, such as water depth and transparency, water level fluctuations, nutrient contents, sediment type, littoral morphology, and disturbance intensity. A complex data analysis revealed that, due to the synergistic effects of a hydrological regime, climatic conditions, and management measures, the overall conditions in the littoral zone of a fishpond were similar such as in temporary wetlands. Due to the wide range of tested variables and extreme climatic conditions during the experimental period, this study presents a basis for the prediction of directions of community changes in hypertrophic water bodies under ongoing climate change.

Cyperus fuscus is a plant species that colonizes wet areas, especially disturbed places such as temporary ponds and ditches. The paper by Kúr et al. [4] discussed the ecology and genetics of *C. fuscus*, which serves as a model for ephemeral wetland plant research and their conservation. The author presumed that plant community diversity may exert a direct effect on within-population genetic diversity, which may vary in different species regarding their competitive ability. The study showed that *C. fuscus* thrives in a broad range of vegetation types, with an optimal distribution in the communities of the *Isoëto-Nanojuncetea* class. As a thermophilous species, it is expected to spread to previously cooler habitats due to climate change. This possibly holds true also for the great share of co-habitating thermophilous species. Even though *C. fuscus* was considered basiphilous species, it seems to be a generalist regarding soil reaction since it was found at locations with lower calcium availability. Another important result of this paper is that the relevés with *C. fuscus* included relatively high number of alien species, but no negative effect on *C. fuscus* cover was detected. The relationship between vegetation diversity in natural habitats and genetic diversity is slightly negative, while anthropogenically modified habitats exerted a strong negative effect on the genetic diversity of this species. These outcomes have important implications for conservation, since only the preservation of natural habitats ensures the maintenance of the adaptive potential of the species.

The territory of Slovenia is geomorphologically, geologically, climatically, and edaphically very diverse and belongs to different ecoregions. This diversity reflects the high biotic diversity of terrestrial ecosystems [5]; however, the diversity in aquatic systems is known to the lesser extent. Zelnik et al. [6] studied vascular plant communities in 33 watercourses in different regions of Slovenia. Hydrologically, the most outstanding is the Karst region, where watercourses are marked by extreme water level fluctuations and even intermittence, which affect the presence of vascular plants. The overall analysis showed that species presence and abundance were at most affected by the longitudinal characteristics of the rivers. The species that were the most abundant in these watercourses have wide ecological ranges. The authors determined 87 vascular plant taxa in total and distinguished 25 different plant associations belonging to 5 classes and 9 alliances. The study showed that the majority of plant communities were distributed in different watercourses belonging to different hydro-ecoregions. Only seven communities showed narrow distribution ranges, and three of them were recorded in karst poljes, including the new association between *Mentho aquaticae* and *Oenanthetum fistulosae* from the karst river Mali Obrh. This distribution pattern reveals specific environmental conditions in these karst poljes and a lack of regionality elsewhere, even though this regionality is evident in the case of terrestrial ecosystems [5].

The Danube River is the second-longest river in Europe that is subjected to various human-made alterations, including those related to hydro-power plants. The extensive study of Janauer et al. [7] is a unique compilation of knowledge on macrophyte species, aquatic plant communities, and their habitat preferences in the Danube main channel from 2582 river kilometers (rkm) to 171 rkm. The authors recorded 89 different aquatic macrophyte species along the entire course—among them, neophytes *Elodea nuttallii*, *E. canadensis*, *Vallisneria spiralis*, and *Azolla filiculoides*—and distinguished 15 plant communities, which were classified into five alliances and represented vegetation classes—namely, rooted hydrophytes *Potamogetonetea*, pleustophytes *Lemnetea*, and helophytes *Phragmitetea*. The Danube is far from being a natural river, but some free-running sections have reasonably low human interference. The study revealed that the distribution of macrophyte species and their communities reflect neither the river continuum concept [8] nor a specific regionality [9]. The present study revealed that hydrological regime significantly affected species presence and abundance, since impounded sections of the river were richer in number and abundance of plant species, whereas the free-running sections mainly show the opposite result. Based on the distribution of aquatic macrophytes and communities, four different sections of the river can be distinguished. The main reason for this distribution pattern is human activity generating disturbances (e.g., navigation, hydropower-plants, and pollution). This study not only is a summary of the present knowledge on macrophyte and their communities in the Danube River but also presents a baseline for determining future changes of the aquatic vegetation.

Kettle holes are glacially shaped small lentic waterbodies characterized by a wet–dry-cycle that supports high species diversity. Due to eutrophication, they are a source of greenhouse gasses; thus, effective conservation and management strategies should be established. Pätzig and Düker [10] used manually digitized maps derived by unmanned aerial systems to quantify differences among dominant plant communities of kettle holes between 2015 and 2020, when several extreme weather events took place in order to contribute to the understanding of functions and services of these specific ecosystems. A high variability in an open water area in these kettle holes in 2016, 2018, and 2020 caused a shift in the occurrence and dominance of plant communities. Various dominant plant communities were differently affected by dry and wet intervals with a major increase in terrestrial plants. High correlation with decreased water permanence was observed for nitrophilous perennials and woody plants. Communities with dominant taxa *Carex* spp., *Phragmites australis*, and *Phalaris arundinacea* revealed no significant changes over the entire period, since they colonize kettle holes with different characteristics. The authors claimed that they expected a lower precipitation rate due to climate changes preventing the reset of early succession stages; thus, the succession of woody species will be accelerated under drier conditions, especially in the absence of management. This results in increased homogenization of the landscape, and loss of certain ecosystem functions and services.

Floodplain forests are complex ecosystems that colonize natural levees, low ridges, bottoms, sloughs, and back-swamps of rivers. In the past, they were cleared due to agriculture, dam impoundments, farming, and urbanization and reduced to narrow bands along the rivers. Hardwood floodplain forests serve as natural carbon sinks; therefore, the reforestation of floodplains may increase their carbon storage potential. Shupe et al. [11] studied carbon stocks in floodplain forests of different ages, structures, and hydrological conditions. Trees in these forests are adapted to cope with large water level fluctuations that may result in hypoxic or anoxic conditions during flooding and water shortage during the dry period. Authors report that the carbon stock accumulated in young trees is significantly lower than in trees of old dense or sparse forests. Old forests thriving on low active floodplains, the high active floodplain, the seepage water zone, and tributaries have the same level of carbon storage; therefore, such locations seem suitable for reforestation process. The comparison of different species revealed that *Quercus robur* accumulated much more carbon than any other species in all hydrological conditions, while other species accumulated much less carbon, and their potential varied among different habitats.

Wet meadows are wetlands with specific hydrological conditions. They develop on poorly drained areas such as shallow lake basins and on land in the transition between marshes and upland areas. Wet meadows are usually without standing water, with exception to the periods of seasonal high waters when they are inundated or waterlogged. These specific conditions and fertile soil create favorable habitats for a variety of plant communities. Inappropriate hydrological interventions or management practices may endanger these communities. The contribution of Čarni et al. [12] discussed wet meadow plant communities of the alliance *Trifolion pallidi* that colonize riverside terraces or gentle slopes along watercourses in the southeastern part of the Pannonian Plain. The hydrological gradient along with nutrient availability is a key factor that determines the growth of different species, as shown by Ellenberg indicator values. The Extent database enabled the distinction of four plant associations: *Trifolio pallidi–Alopecuretum pratensis, Ventenato dubii–Trifolietum pallidi, Ranunculo strigulosi–Alopecuretum pratensis*, and *Ornithogalo pyramidale–Trifolietum pallidi*. The main threats to these communities are natural succession and reforestation, intensive agriculture, and changes of hydrological regime; therefore, mowing, extensive agriculture, and maintenance of favorable hydrological regime are crucial for maintaining these highly diverse plant communities.

Charophytes that are present in all continents with the exception of Antarctica are a group of algae that colonize fresh and brackish water bodies, where they can form large stands. The establishment of these stands is affected by nutrient availability, water transparency, wave disturbances, water level fluctuations, substrate type, and water depth [13]. Charophytes are bioindicators of ecological status in lakes and rivers, while their oospores in the sediment can be used for the reconstruction of trophic status or salinity of such ecosystems in the past. Stragauskaitė et al. [14] studied the distribution and seasonal patterns of a charophyte oospore bank and its relationship with abiotic factors and charophyte stands distribution in the largest lagoon in the Baltic Sea. The study revealed that the distance of oospores from the charophyte stands, salinity, and bottom slope were environmental factors defining the distribution of charophyte fructifications. In addition, wave exposure was also an important factor affecting oospore occurrence. The future increase in storm frequency and consequently increased wave action and turbidity may thus cause changes in spatial patterns of charophyte communities.

The contribution of Gaberščik et al. [15] analyzes the long-term response of cosmopolitan species *Phragmites australis* to extreme water level fluctuations. The study was performed in intermittent wetland Lake Cerknica, which is located at the bottom of the Cerkniško polje depression. This intermittence is a consequence of the karst features of the area, such as a high precipitation range and carbonate bedrock, which is mainly limestone. As an outstanding system with extreme water level fluctuations, resulting in long aquatic and relatively short dry phases, the lake is also listed as a Ramsar site. The lake area is colonized by extensive reed stands that thrive in different habitat types. *P. australis* is a highly productive species; however, unfavorable environmental conditions may easily reduce this potential. The authors compared common reed production parameters in plants from riparian and lake (littoral) stands within 13 years. The growth and assimilate allocation parameters in the lake reed were related to water levels for June and July, while temperatures were important at the beginning of the growing season in May. In the case of riparian reed, the water levels and temperatures out of the vegetation season appeared to be more important. The authors concluded that habitats with permanent water presence are more suitable for reeds than those with fluctuating water, which occasionally dry out. However, such an extreme hydrological regime that is becoming a reality not only in wetlands but also in other water bodies colonized by common reed can result in positive and negative feedback loops that may affect reed production. This study is an important contribution to the understanding of the changes in the structure and function of reed-dominated wetlands in the future.

Worldwide, lakes show a significant decrease in aquatic vegetation (cca. 65%) [16] presumably due to water level changes, nutrient input, and global climate change. Water level fluctuations in lakes may be due to natural causes or due to human interventions. To build a basis of sustainable lake restoration and management, Tan et al. [17] examined the relationship between aquatic vegetation coverage and water levels in an artificially regulated shallow lake Longgan Lake located in the middle section of the Yangtze River. Based on long-term data that were obtained using medium-resolution Landsat satellite imagery, they aimed to understand the long-term dynamic of aquatic vegetation. Aquatic vegetation distribution showed a significant negative relationship with a water level. As water depth is the most important factor for aquatic plant growth and development, the expansion and retreat of aquatic vegetation was also related to the lake-bottom topography. The aquatic vegetation was negatively affected by high spring and summer water level. In the seasons with extremely high precipitation rates, the area colonized by aquatic vegetation was low. The authors concluded that the optimal water depths for successful aquatic vegetation growth are 1 m in spring and 3 m in summer, respectively.

3. Conclusions

A hydrological regime is one of the main factors shaping vegetation patterns across a wide range of scales. A hydrological gradient and a fluctuating water level present major driving forces, which determine the distribution of different plant communities as well as plant species composition within them. This was evidenced in various aquatic ecosystems, such as wet-meadows [12], marshes [4], intermittent lake [15], flood-forests [11], ponds [10], lakes [3,17], or rivers [6]. Waterlogged or flooded soil can become an inhospitable environment, which excludes the majority of species, while in the remaining species, it triggers the expression of various traits and mechanisms that enable survival despite oxygen shortage in the root zone. In addition, moving water masses as waves or current present a physical factor preventing successful colonization of certain communities [7,14]. Thus, when various factors shaping the plant communities are assessed [6] according to their importance, a hydrological regime often overrides all other environmental factors, also altering the pattern of regionality, which is generally accepted in biogeography [7].

This Special Issue contributed significantly to the understanding of the responses of plant species and communities that are favorized by specific hydrological regimes. Nowadays, these plant species and communities seem to be very vulnerable and threatened, since favorable hydrological conditions are changing due to anthropogenic alterations as well as by climate change that additionally endanger their long-term survival.

Author Contributions: Writing—original draft preparation, A.G. and I.Z.; writing—review and editing, I.Z. and A.G. All authors have read and agreed to the published version of the manuscript.

Funding: This research received no direct external funding.

Institutional Review Board Statement: Not applicable.

Informed Consent Statement: Not applicable.

Acknowledgments: Thanks to the authors who contributed their articles to this Special Issue as well as to the editors of the journal *Water*. A special thank goes to the anonymous reviewers, who provided their comments and suggestions and helped to improve the quality of the articles.

Conflicts of Interest: The authors declare no conflict of interest.

References

1. Keddy, P.A. *Wetland Ecology: Principles and Conservation*, 2nd ed.; Cambridge University Press: Cambridge, UK, 2010.
2. Boulton, A.J.; Brock, M.A. *Australian Freshwater Ecology: Processes and Management*; Gleneagles Publishing: Adelaide, Australia, 1999; p. 118.
3. Šumberová, K.; Vild, O.; Ducháček, M.; Fabšičová, M.; Potužák, J.; Fránková, M. Drivers of macrophyte and diatom diversity in a shallow hypertrophic lake. *Water* **2021**, *13*, 1569. [CrossRef]
4. Kúr, P.; Píšová, S.; Tremetsberger, K.; Dřevojan, P.; Kącki, Z.; Böckelmann, J.; Bernhardt, K.-G.; Hroudová, Z.; Mesterházy, A.; Šumberová, K. Ecology and genetics of cyperus fuscus in central Europe—A model for ephemeral wetland plant research and conservation. *Water* **2021**, *13*, 1277. [CrossRef]
5. Šilc, U.; Vreš, B.; Čelik, T.; Gregorič, M. Biodiversity of Slovenia. In *The Geography of Slovenia*; Perko, D., Ciglič, R., Zorn, M., Eds.; Springer Nature Switzerland AG: Cham, Switzerland, 2020; pp. 109–124.

6. Zelnik, I.; Kuhar, U.; Holcar, M.; Germ, M.; Gaberščik, A. Distribution of vascular plant communities in Slovenian watercourses. *Water* **2021**, *13*, 1071. [CrossRef]
7. Janauer, G.; Exler, N.; Anačkov, G.; Barta, V.; Berczik, Á.; Boža, P.; Dinka, M.; Georgiev, V.; Germ, M.; Holcar, M.; et al. Distribution of the macrophyte communities in the danube reflects river serial discontinuity. *Water* **2021**, *13*, 918. [CrossRef]
8. Vannote, R.L.; Minshall, G.W.; Cummins, K.W.; Sedell, J.R.; Cushing, C.E. The river continuum concept. *Can. J. Fish. Aquat. Sci.* **1980**, *37*, 130–137. [CrossRef]
9. Illies, J.; Botosaneanu, L. Problèmes et méthodes de la classification et de la zonation écologique des eaux courantes, considerées surtout du point de vue faunistique. *SIL Commun. 1953–1996* **1963**, *12*, 1–57. [CrossRef]
10. Pätzig, M.; Düker, E. Dynamic of dominant plant communities in Kettle Holes (Northeast Germany) during a five-year period of extreme weather conditions. *Water* **2021**, *13*, 688. [CrossRef]
11. Shupe, H.A.; Hartmann, T.; Scholz, M.; Jensen, K.; Ludewig, K. Carbon stocks of hardwood floodplain forests along the Middle Elbe: The influence of forest age, structure, species, and hydrological conditions. *Water* **2021**, *13*, 670. [CrossRef]
12. Čarni, A.; Ćuk, M.; Zelnik, I.; Franjić, J.; Igić, R.; Ilić, M.; Krstonošić, D.; Vukov, D.; Škvorc, Ž. Wet meadow plant communities of the alliance *Trifolion pallidi* on the southeastern margin of the Pannonian Plain. *Water* **2021**, *13*, 381. [CrossRef]
13. Schneider, S.C.; García, A.; Martín-Closas, C.; Chivas, A.R. The role of charophytes (Charales) in past and present environments: An overview. *Aquat. Bot.* **2015**, *120*, 2–6. [CrossRef]
14. Stragauskaitė, V.; Bučas, M.; Martin, G. Distribution of charophyte oospores in the curonian lagoon and their relationship to environmental forcing. *Water* **2021**, *13*, 117. [CrossRef]
15. Gaberščik, A.; Grašič, M.; Abram, D.; Zelnik, I. Water level fluctuations and air temperatures affect common reed habitus and productivity in an intermittent wetland ecosystem. *Water* **2020**, *12*, 2806. [CrossRef]
16. Zhang, Y.; Jeppesen, E.; Liu, X.; Qin, B.; Shi, K.; Zhou, Y.; Thomaz, S.M.; Deng, J. Global loss of aquatic vegetation in lakes. *Earth-Sci. Rev.* **2017**, *173*, 259–265. [CrossRef]
17. Tan, W.; Xing, J.; Yang, S.; Yu, G.; Sun, P.; Jiang, Y. Long term aquatic vegetation dynamics in Longgan Lake using landsat time series and their responses to water level fluctuation. *Water* **2020**, *12*, 2178. [CrossRef]

water

MDPI

Article

Drivers of Macrophyte and Diatom Diversity in a Shallow Hypertrophic Lake

Kateřina Šumberová [1,*], Ondřej Vild [1], Michal Ducháček [2], Martina Fabšičová [1], Jan Potužák [3] and Markéta Fránková [4]

[1] Department of Vegetation Ecology, Institute of Botany of the Czech Academy of Sciences, Lidická 25/27, 602 00 Brno, Czech Republic; ondrej.vild@ibot.cas.cz (O.V.); martina.fabsicova@ibot.cas.cz (M.F.)
[2] Department of Botany, National Museum, Cirkusová 1740, CZ-193 00 Praha 9—Horní Počernice, Czech Republic; michal.duchacek@nm.cz
[3] Vltava River Authority, State Enterprise, Holečkova 3178, 150 00 Praha 5—Smíchov, Czech Republic; jan.potuzak@pvl.cz
[4] Department of Paleoecology, Institute of Botany of the Czech Academy of Sciences, Lidická 25/27, 602 00 Brno, Czech Republic; marketa.frankova@ibot.cas.cz
* Correspondence: katerina.sumberova@ibot.cas.cz

Abstract: We studied macrophyte and diatom assemblages and a range of environmental factors in the large hypertrophic Dehtář fishpond (Southern Bohemia, Czech Republic) over the course of several growing seasons. The spatial diversity of the environment was considered when collecting diatoms and water samples in three distinct parts of the fishpond, where automatic sensor stations continually measuring basic factors were established. Macrophytes were mapped in 30 segments of the fishpond littoral altogether. High species richness and spatiotemporal variability were found in assemblages of these groups of autotrophs. Water level fluctuations, caused by the interaction of fish farming management and climatic extremes, were identified as one of the most important factors shaping the structure and species composition of diatom and macrophyte assemblages. The distance of the sampling sites from large inflows reflected well the spatial variability within the fishpond, with important differences in duration of bottom drainage and exposure to disturbances in different parts of the fishpond. Disturbances caused by intensive wave action are most probably a crucial factor allowing the coexistence of species with different nutrient requirements under the hypertrophic conditions of the Dehtář fishpond. Due to a range of variables tested and climatic extremes encountered, our study may be considered as a basis for predictive model constructions in similar hypertrophic water bodies under a progressing climate change.

Keywords: automatic sensor system; Central Europe; fish farming; freshwater algae; epiphytic diatoms; functional species groups; threatened species; vascular plants; wave action; wetland vegetation

Citation: Šumberová, K.; Vild, O.; Ducháček, M.; Fabšičová, M.; Potužák, J.; Fránková, M. Drivers of Macrophyte and Diatom Diversity in a Shallow Hypertrophic Lake. *Water* **2021**, *13*, 1569. https://doi.org/10.3390/w13111569

Academic Editors: Alenka Gaberščik and Igor Zelnik

Received: 12 April 2021
Accepted: 30 May 2021
Published: 1 June 2021

1. Introduction

Shallow lakes are often considered unique habitats of rich aquatic and wetland biota [1–3]. Increasing attention is paid particularly to natural oligotrophic and mesotrophic lakes, which started to be rare due to overall eutrophication, or, in some regions, acidification [4,5]. On the other hand, heavily eutrophicated water bodies usually become the focus of scientists as the objectives of restoration projects [6,7]. Many are considered sites where a substantial part of biodiversity has already been lost and where mitigation measures are needed in order to turn back this unfavourable trend [1,8]. Many fishponds of these more than 20,000 artificial water bodies occurring in the Czech Republic, a country that lacks natural lakes, belong among aquatic habitats with the status of high eutrophy to hypertrophy [9]. Since their construction, mainly in the Middle Ages [10], most of these fishponds have served for semi-intensive carp breeding and recently to a certain extent also for other purposes such as recreation [9] or as nature reserves for biodiversity protection [11–13]. Fishponds exceeding 1 hectare in size should be classified as a special type of shallow artificial lakes [14,15]. Old fishponds, particularly the larger ones, even resemble natural lakes [16]. The results of the fishpond research thus might be relevant for the interpretation of processes in other types of shallow lakes [16,17]. On the other hand, fishponds have some specific characteristics, especially regular water level fluctuations associated with fish stocking, rearing, and harvesting and other management practices [16–20].

Within the number of fishponds, small water bodies of up to about 10 hectares in size, 1–2.5 m of maximum depth, and small spatial variability predominate [9]. Small fishponds, similarly to other small water bodies, have been identified as habitats playing a crucial role in the protection of local diversity of wetland and aquatic biota [18–25]. They are usually stocked with fish fry [18–20], with a relatively low level of fish-mediated disturbances, and provide suitable conditions for a range of aquatic vascular plants and charophytes, including threatened species [22,26]. This is probably one of the reasons why pond systems involving small fishponds attract more attention than similar systems consisting mainly of large fishponds (see, e.g., data from France [21,22,27], Germany [28–30], Poland [26,29,31], Austria [32,33], Belgium [34], and Slovakia [23]).

Special focus on vascular plant and charophyte assemblages in large fishponds of a size from several dozen up to a few hundred hectares and a maximum depth of several meters is rare (see, e.g., [11,12,35,36] for the exceptions). These fishponds have recently been experiencing progressive eutrophication [37–39] and are considered by most ecologists as habitats of low biological value. It may actually seem that their biodiversity is not worth studying and protecting. Despite this common view, many of the large fishponds are important ornithological sites, providing suitable nesting places and food sources to a variety of bird taxa such as herons (*Ardeidae*), geese and ducks (Anseriformes), and waders (*Charadriidae*) [11,13,40,41]. The few existing plant-ecological studies also show high diversity of macrophyte assemblages in large fishponds, particularly those with fluctuating water levels [11,12,35].

Another argument for the research of large fishponds is a range of features typical of large water bodies, setting them apart from the smaller ones, e.g., large spatial variability of the littoral zone, high maximum water depth and the associated large water volume, and possibly also a higher resilience in relation to climatic and other environmental extremes. The fishponds with so-called biennial management cycles offer suitable habitats to a number of plant species from various functional and ecological groups, which is related to partial summer fishpond drainages [19,20]. Only large fishponds, however, may be exposed to wind-mediated wave action [9], resulting in a high substrate diversity and large areas of sandy or gravelly shores (see, e.g., [42–44] for similar processes in other large water bodies). These habitats have probably preserved threatened biota across various taxonomic groups, as suggested by recent records of rare vascular plants [18,45,46]. These are particularly species of nutrient-poor substrates that vanished from most of their former sites [18,20], and large fishponds with sandy shores are one of their last refugia [45,46]. Nevertheless, our knowledge of wetland vascular plant and charophyte diversity and its drivers in large fishponds is still very incomplete, as only a negligible number of large fishponds was subjected to systematic research. It is remarkable especially if compared to the state of knowledge of macrophyte diversity in natural lakes, studied within many research projects (e.g., [1,3,5,7,42,43,47,48]).

Epiphytic diatoms are related to macrophyte assemblages and thus their research along with vascular plants and charophytes seems to be logical. However, real interdisciplinary research combining data on diatom and macrophyte assemblages is scarce and focuses mainly on spring fens or similar wetland habitats. Its results show that diatoms follow the same environmental gradients as vascular plants and bryophytes [49,50]. We do not know if the macrophyte and epiphytic diatom assemblages exhibit similar trends also in water bodies. Studies of this type from fishponds or natural lakes are, to our knowledge, not available so far.

Papers focused purely on epiphytic diatom assemblages in fishponds exist, but they are even more scarce than studies on fishpond macrophytes, being usually devoted to small fishponds [51–54]. The only study on diatom assemblages in a large fishpond is probably our own previous paper [55]. Although the studies on epiphytic diatom assemblages from other types of aquatic environments such as natural lakes, mires, bogs, fens, springs, rivers, and streams are quite frequent [49,52,56–61], attention paid to the substrate for the growth of epiphytic diatoms, i.e., to the macrophytic species sampled, is surprisingly low in most papers. The macrophytes are usually identified only to the genus or family level [52,53] or are not identified at all [58,62]. Diatoms in samples from different macrophyte species are sometimes analysed jointly for the whole locality [54], or only a few, usually the most common and easy-to-identify macrophytic species, are selected for the study (e.g., *Phragmites australis*, *Typha* spp., *Lemna* spp., *Chara* spp., or *Potamogeton* spp.; [53,57,60]). In such cases, any analysis of similarities in habitat ecology between epiphytic diatoms and host macrophytes is impossible. For instance, we do not know if ecologically specialised macrophyte species do or do not support the diversity of similarly specialised epiphytic diatoms, what the indication potential of both groups of autotrophs within a single water body is, or if the species diversity of epiphytic diatoms and macrophytes follows the same spatiotemporal gradients.

In 2014, we started a multidisciplinary project focused on the functioning and biotic diversity of the Dehtář fishpond, one of the largest fishponds of the Czech Republic. Among others, vascular plant, charophyte, and epiphytic diatom assemblages have been studied in detail. The analyses of diatom assemblages from 2015 together with basic nutrient balances have been already published [16,17,55]. In the present study, we want to analyse data on vascular plant and diatom assemblages along with selected environmental and management-related factors, collected during the whole project (2014–2016) and shortly afterward (2019). The study specifically aims to answer the following questions: (1) What are the main drivers of species and functional diversity of vascular plants, charophytes, and diatoms in a large hypertrophic fishpond exposed to large water level fluctuations and wind-mediated disturbances? (2) Do vascular plants and charophytes follow the same environmental gradients as epiphytic diatoms? (3) What types of interactions have developed between aquatic macrophytes (vascular plants, charophytes) and epiphytic diatoms under the conditions of hypertrophy, drought stress, and intensive mechanical disturbances?

The research presented is, to our knowledge, the first to combine a detailed analysis of vascular plant, charophyte, and diatom diversity and its drivers in a large hypertrophic water body. We suppose that it may facilitate a broader understanding and the protection of threatened aquatic ecosystems and their biota but also encourage similar interdisciplinary research of macrophyte and periphytic algal communities.

2. Materials and Methods

2.1. Study Site

The research was carried out in the Dehtář fishpond (49°0′30.064″ N; 14°18′22.302″ E) which thanks to its size, 238 ha, and volume, 5,167,710 m^3, belongs to the ten largest fishponds in the Czech Republic. It is, therefore, included in the regular monitoring activities of the Vltava River Authority state enterprise. The fishpond is rather deep compared to most other fishponds, with a maximum depth of around 6 m at the dam (segment 30, Figure 1) and average depth of 2.6 m (all the values are given for the 2nd year of bi-annual management cycles, characterised by higher water levels; [55]). It is an ancient fishpond, constructed in the 15th century (it was finalised in 1483; [9]) in South Bohemia (southwestern Czech Republic), 12 km northwest of the city of České Budějovice at an altitude of cca 404 m a.s.l. It is situated in an undulating landscape on the transition between the South Bohemian fishpond basins and the Blanský les Hills. The geological composition of the immediate surroundings of the Dehtář fishpond is predominately made up of various types of unstable sediments (sands, gravels, clays), mainly of tertiary age, and in the southern and eastern parts by gneisses and paragneisses [63]. The predominating soil types of the area are Stagnic Cambisols [64].

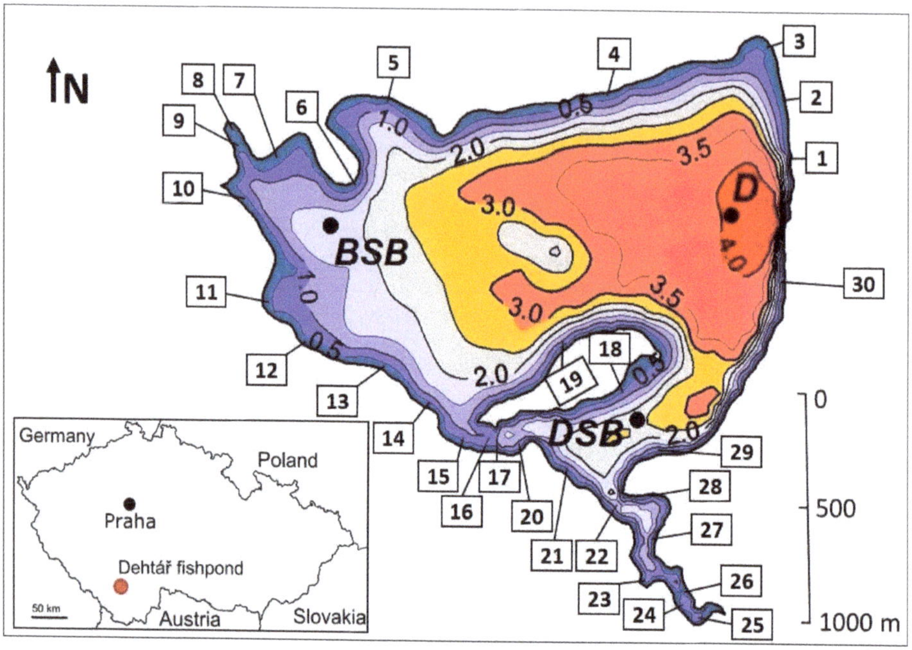

Figure 1. Location of the Dehtář fishpond and its particular study sites. The numbers show the mid-points of segments for the study of the flora of vascular plants and charophytes. The points show the position of high-frequency monitoring stations: D (=FD) = Fishpond dam, BSB = Babický stream bay, DSB = Dehtářský stream bay; samples for water analysis and the analysis of epiphytic diatom assemblages were collected nearby. The two main inflows, Babický stream (BS) and Dehtářský stream (DS), occur at the segments 8 and 25, respectively.

Our earlier studies [16,55] showed that Dehtář is a strongly eutrophic to slightly hypertrophic fishpond (according to the OECD classification of 1982 [65], adjusted for fishpond conditions [66]). Similarly to the majority of Czech fishponds, it is a carp pond, i.e., serving mainly common carp (*Cyprinus carpio*) breeding and to a lesser extent also other fish species with similar habitat requirements as the carp (more details are included in Figures S1–S15). Dehtář is a so-called main fishpond, i.e., the production of marketable fish (fish of consumable size) is its farming purpose. Like the majority of large fishponds in the Czech Republic, Dehtář is managed in bi-annual management cycles. In the first year of the management cycle, the fishpond is stocked with young fish; as the whole fish biomass is, due to a smaller size of fish individuals, much lower than in the second year of the management cycle, the pressure of fish stock on the aquatic environment, particularly on its vegetation, is lower as well [19,20]. In the second year of the management cycle, the fish biomass strongly increases; the whole management cycle ends in October/November by harvesting the fish stock. Before the harvesting, a substantial part of the fishpond is drained (Figure S1), and although it is flooded again after the fish harvesting, the water level is kept lower during the first year of the management cycle than it is in the second year. Therefore, the values of other fishpond parameters (the total area, total volume, and average depth) in the first year should also be lower than in the second year and increase only slowly.

Strips of the fishpond bottom are usually exposed in the first year of the management cycle, offering a suitable habitat to a range of wetland vascular plant species and their communities (Figures S2–S12). Large areas of shallow water support the growth of some amphibious and aquatic plant taxa. During our research period, however, the low water level from the first year of the management cycle in 2015 continued until May 2016, i.e., the second year of the management cycle (Figures S11 and S12). This was a consequence of extremely dry weather in 2015 and in the spring of 2016, enhanced by extraordinarily high temperatures in the summer of 2015 (for basic climatic characteristics of the Dehtář fishpond from the study period compared with the climate of the whole region, see Table A1). This dry period ended in heavy rains in the summer of 2016 ([17]; Table A1, Figure S13), which elevated the water level above its common management value. Moreover, 2014, also the 2nd year of the management cycle, was also characterised by a lower water level due to previous dry years (Figure S11). According to our information from fish farmers [67], a lack of water was always quite frequent in the Dehtář fishpond; for this reason, the exposed bottom used to be sown by clover and other crops in the 1960s–1970s in order to get forage for domestic animals and to increase the fishpond fertility (see also [68]).

Semi-intensive fishpond management practiced on the Dehtář fishpond (and the majority of fishponds throughout the Czech Republic) includes fertilisation with manure, supplementary feeding of fish stock with cereals, and occasional liming of some fishpond parts [68]. Another source of nutrients is the surrounding agricultural landscape and the water brought by the inflows, particularly the large ones, the Babický stream, the Kamenný stream, and the Dehtářský stream.

As already reported elsewhere (e.g., [16,38,39,55]), hypertrophy combined with the relatively high average depth of the Dehtář fishpond contributes to strong fluctuations of oxygen concentrations and other water parameters. Another important factor is strong winds, mainly from the northwest, causing intensive wave action and associated mechanical disturbances in the littoral zone (Figure S14). Thus, despite the cyanobacterial blooms appearing in the last years (Figure S15), Dehtář is popular as a recreational fishpond, particularly among windsurfers [9,55].

Dehtář is a fishpond with several bays separated by natural barriers such as peninsulas and an island. Its surroundings include large areas of arable land and cattle pastures, and small patches of wet to semi-dry meadows, woodlands, and settlement areas. As the natural barriers and diversity of the surrounding landscape may cause differences in some environmental parameters as well as in the biota between the different fishpond parts, for the aim of this study, we divided the fishpond into a total of 30 segments (Figure 1), expecting more or less homogeneous environmental conditions within each of them. The length of the segments varied according to local topography of the fishpond littoral and the surrounding landscape, and their area (=vegetated zone) was also dependent on water level fluctuations determining the extent of exposed pond bottom and shallow water zones (see Table S1 for more details).

Additionally, three parts of the Dehtář fishpond with presumably the largest differences in environmental conditions were selected for the research of the physico-chemical parameters of water and epiphytic diatom assemblages. All these sites have been equipped with automatic, sensor-based stations designed for high-frequency monitoring of several basic environmental parameters; a range of additional parameters was recorded at regular intervals at the stations (for detailed description, see the Methods). The sites are, namely, (1) the fishpond dam—FD, (2) the bay of the Dehtářský stream—DSB, and (3) the bay of the Babický stream—BSB (Figure 1). Epiphytic diatom assemblages were studied in the vegetation zones as close as possible to the three monitored sites, about 200 m from the stations in the Dehtářský stream bay (DSB) and Babický stream bay (BSB) and about 600–650 m from the FD station. The sampling site at FD was situated in a narrow bay at the northeastern corner of the fishpond, naturally preserved against waves (segment 2 near the border with segment 3, and segments 3 and 4; Figures S2 and S3). In contrast, BSB and DSB sampling sites were in broad bays, open and fully exposed to wave action. While the fishpond bottom in the DSB and especially in the BSB exhibits moderate slope and thus remains non-flooded for a relatively long time after the fish harvesting (Figures S11 and S12), the site near the fishpond dam is characterised by a relatively steep slope and very fast water level increase.

Since the 1930s, the Dehtář fishpond has attracted the interest of botanists (e.g., [69,70]). It is also an important ornithological locality, protected as a part of the Natura 2000 network [40,71]. The earlier research of Dehtář was, however, not systematic. Thus, only individual records on plant and animal species are reported in the literature.

2.2. *Methods*

2.2.1. Physico-Chemical Parameters of Water

In 2014, three monitoring sensor-based stations were established in three different parts of the Dehtář fishpond (for the location and description of these sites see Section 2.1. Study Site and Figure 1). Each station was equipped with several sensors. The following parameters were measured: water temperature and concentration of dissolved oxygen (M4016, Fiedler AMS) at all stations, global radiation, rainfall, speed and direction of wind (M4016-A-G3, Fiedler AMS) at the BSB station, and photosynthetically active radiation (PhAR; LI-COR sensors) at the FD station (further details on the stations and parameters measured are included in Figures S16–S22). At the end of 2015, the station at FD was further equipped with a sensor for water level detection. After 2016 (end of the targeted research project and of the substantial part of funding), only the station at FD was kept, and some sensors from the other stations were moved there.

In addition to the measurements provided by the sensor system, basic physico-chemical parameters (temperature, concentration and saturation of oxygen, pH, conductivity, and water transparency) were measured by the multiparametric probe YSI 6600 V2, and samples for water chemistry were taken at bi-weekly intervals near the stations (some of the parameters were taken only at FD).

All samples were analysed for total (TP) and soluble reactive phosphorus (SRP); total (TN), ammonia (N-NH4), and nitrate (N-NO3) nitrogen; total (TC), total organic (TOC), and dissolved organic carbon (DOC); dissolved ions of sodium (Na), magnesium (Mg), calcium (Ca), and potassium (K); and total iron (Fe) according to standard methods certified and intercalibrated in the laboratories of the Vltava River Authority, state enterprise. Moreover, the concentration of chlorophyll-a (Chl-a), and alkalinity (AT) were also analysed in epilimnetic samples (integrated samples from 0–1 m water layer). Water transparency (WT) was measured simultaneously. An overview of the physico-chemical characteristics obtained for the growing seasons (i.e., IV–VIII) of 2014–2016 and 2019 is given in Table S2.

2.2.2. Other Environmental Factors, Fishpond Management Data

A range of environmental factors, related to vascular plants, was recorded for each segment directly in the field. These were, e.g., type of substrate, depth of mud, occurrence of stones, or organic detritus (see Table S1 for the complete list of factors). Additionally, the length of segments, their area in individual years, the proportion of the landscape types surrounding each of the segments, and other characteristics (Table S1) were measured in the national internet mapping application (www.mapy.cz, accessed on 15 November 2020). Fishpond management data on the fish stock composition and biomass, amounts of manure, lime, and supplementary feeding applied were provided by the Hluboká nad Vltavou Fish Farm Cz, s.r.o. However, these data were used only for interpretations of some parameters studied, as they usually could not be linked to particular segments of the fishpond (see Table S1 for more details).

2.2.3. Vascular Plant Data Recording

Vascular plant and charophyte species were mapped in the field separately for each of the 30 segments using a three degree scale: 1—rare (up to ten individuals in a segment), 2—frequent throughout the segment but not dominant, and 3—dominant, forming its own stands. The data were collected throughout the growing season of a particular year in monthly intervals (IV–IX 2014, III–IX 2015, III–IX 2016). Segments 1–19 were mapped during each visit, while segments 20–30, mostly difficult to access, were visited twice a year in 2014 and 2015 and only once in 2016. In 2019 (after the end of the research project), only segments 1–19 were mapped during three sampling events in May, June, and August. Taxonomically critical taxa and/or difficult to identify taxa were collected into a herbarium and revised by experts (see the Acknowledgements). The herbarium vouchers are deposited in the herbaria PR and BRNU (the abbreviations follow Thiers [72]).

2.2.4. Diatom Sampling and Processing

The sampling was performed during three one-day sampling events (16 June 2015, 23 June 2016, and 19 June 2019). Macrophyte species, water depth, distance of a sample from the bank, and disturbance category were noted for each of the samples directly in the field (for more details see Table S3 with the list of samples).

Epiphyton was collected with the whole substrate, which means that the plants were cut with scissors above the bottom and rinsed in water to wash away nonepiphytic species. The samples for light microscopy analysis (LM) were placed into polyethylene bags, kept in a refrigerator, and transported to the laboratory. For Environmental Scanning Electron Microscopy (ESEM), up to 10 cm long sections of stalks and leaves of *Phalaris arundinacea* and *Phragmites australis* in 2016 and *Limosella aquatica* in 2019 were cut, rinsed in water, attached to a polystyrene matrix in order to keep diatom assemblages in the state best corresponding to natural conditions, placed into plastic containers containing water from the locality, and transported to the laboratory in a cooling thermo box.

In the laboratory, the samples for LM were elaborated as follows: diatoms were squeezed out of the sampled plants, treated in hydrogen peroxide [73], and mounted in Naphrax. In each sample, at least 400 diatom valves were identified and enumerated on random transects at 1000× magnification using an Olympus BX51 light microscope. The identification literature used included Krammer and Lange-Bertalot [74–77], Krammer [78–80], Lange-Bertalot [81], Hofmann et al. [82], and Lange-Bertalot et al. [83].

In order to get better insight into the structure of epiphytic diatom assemblages collected on sites with contrasting environmental conditions and/or different macrophyte taxa, observation using ESEM was employed. Sample observation and photography in 2016 took place at the Institute of Scientific Instruments of the Czech Academy of Sciences in Brno. The Low Temperature Method for sample stabilisation in ESEM was used. This method was already successfully applied to diatom assemblages [84]. Observations were performed on an ESEM QUANTA 650 FEG at a temperature of −20 °C. Sample observation and photography in 2019 were conducted at the Institute of Botany in Brno on a table scanning electron microscope Phenom ProX using a temperature-controlled sample holder at a temperature of −8 °C.

2.2.5. Data Processing and Analyses

Data from the automatic sensor stations, recorded at 10 minute intervals, were downloaded from the hosting server and stored regularly. They were checked for possible inconsistencies caused by station failures (e.g., during extreme storms) or by removal of biofilms from some of the sensors. The potentially erroneous data were removed. In this paper, we analysed data on the following factors: air temperature, precipitation, wind speed, water temperature, oxygen concentration, and photosyntetically active radiation (PhAR), while the data from the sensors for global radiation and water level fluctuation data recording were used only in descriptions of site conditions. After the prescreening, only the data measured by the sensors at the depth of 0.3 m were selected and further elaborated. We used absolute values of wind speed, air temperature, oxygen concentrations, water temperature, and PhAR and displayed them as line graphs in order to compare the course and fluctuations of these parameters between years. We also prepared short graphs with a detailed visualisation of wind speed, oxygen concentrations, and PhAR at the time of diatom sampling, as a several-day course of these factors may potentially impact fast-changing diatom communities [85].

Additionally, for air temperature and precipitation, we calculated monthly means and sums, respectively. For the months that were missing or incomplete in our own data, we used publicly available data from the nearest weather station of the Czech Hydrometeorological Institute (CHMI) in České Budějovice-Rožnov [86]. In order to highlight climatic extremes during the study period, we also used CHMI data [86], where long-term means for the whole South-Bohemian region are available. To get better insight into the frequency of wave action in individual years, we classified each wind speed record in one of the classes defined according to the Beaufort wind force scale (https://cs.wikipedia.org/wiki/Beaufortova_stupnice, accessed on 1 March 2021) and calculated the proportion of winds of each class in each year (Table 1).

Table 1. Frequency of selected wind events based on the data on wind speed from an automatic sensor station (10 min measurement interval). Absolute counts of each event in each year, along with their share in total numbers of records, are given.

Wind Events	2014 (12.5–31.8)	2015 (1.4–31.8)	2016 (23.5–31.8)	2019 (1.4–31.8)
no wind (<0.2 m/s)	548 (3.42%)	1602 (7.3%)	749 (5.17%)	8415 (38.20%)
5.5–7.9 m/s	892 (5.56%)	1516 (6.91%)	548 (3.78%)	80 (0.36%)
8.0–13.9 m/s	190 (1.18%)	395 (1.80%)	120 (0.83%)	2 (0.01%)
>13.9 m/s	3 (0.02%)	27 (0.12%)	3 (0.02%)	0
rest (>0.02 and <5.5 m/s, %)	89.82	81.87	90.20	61.43
mean (m/s)	2.36	2.50	1.99	0.95
max. (m/s)	16.64	19.18	17.39	11.57

All other data were computerised and prepared for statistical analyses (vascular plants and charophytes, diatoms, environmental and management-related data from the field, data from the maps, and average monthly chlorophyl-a concentrations as a proxy of nutrient concentrations and water transparency) and/or transformed into the tables for direct presentations (all physico-chemical water parameters, vascular plant, charophyte, and diatom abundances and frequencies). Vascular plants were classified according to their origin [87] and threat status (Grulich [88], with the updates by Grulich & Chobot [89]). Additionally, the classification of vascular plants and charophytes into functional groups was performed according to Francová et al. [19,20], with an extension to the species not presented in the cited papers (Table S4). An adjusted version of this classification was also used in macrophytes collected for diatom analyses (Table S3). In order to get a better overview of the terrestrial species imported to the fishpond, the ecological classification was performed in parallel, based on our own field experience and species ecological characteristics summarised within the PLADIAS database (www.pladias.cz, accessed on 20 November 2020; [90]). The nomenclature of vascular plants follows Danihelka et al. [91], with the exception of *Spergularia kurkae*, where we accepted the concept of Kúr et al. [33]. The nomenclature, threat status, and origin of charophytes are based on the charophyte overview by Caisová and Gąbka [92]. All the classification categories for each species are given in Table S4.

Based on the diatom relation to the host plant surface (according to Fránková et al. [55]), five groups, according to diatom functional types (FT1–FT5) were used: FT1—planktonic taxa represented by centric diatoms; FT2—typically periphytic taxa adhering to the surface directly by a mucous film or with a mucilaginous stalk; FT3—facultatively periphytic araphid, passively moving diatoms able to attach; FT4—facultatively periphytic taxa with raphe with fibulae, actively moving; and FT—epipelic taxa with raphe, actively moving, mainly symmetrical biraphid pennate diatoms (see Table S5 for more details). The affiliation of diatom species to functional types is given in Tables S4 and S5. Planktonic and epipelic taxa were present in the diatom samples even though the collected macrophytes were washed in order to remove non-epiphytes. It was decided to keep and classify the whole diatom assemblage in order to get a realistic view of the community.

The patterns in species composition of both diatoms and macrophytes were investigated by non-metric Multidimensional scaling (NMDS). The data in each dataset were standardized by dividing by margin total, and the *metaMDS* function from the *vegan* package [93]) was used. Correlations between the ordination axes and selected environmental and management factors were calculated using the *envfit* function with 999 permutations. Additionally, the impact of the factors "year of the management cycle" and "organic detritus" for vascular plants and "year of sampling" for diatoms was tested by PERMANOVA [94] with 999 permutations. The ordinations of both datasets were plotted, with these factors passively projected to illustrate the associations with gradients in species compositions. For diatom assemblages, selected environmental factors were also plotted as a smooth surface by using the *ordisurf* function. All the analyses were conducted in the R program [95].

Constrained ordination (RDA) was used to evaluate the affinity of diatoms to macrophytes. Here, individual macrophyte species and life history trait- and leaf morphology-based functional groups of macrophytes were used as factors to explain patterns in the ordination, with the year of sampling used as a covariable to filter out interannual variability. To evaluate the affinity of specific diatom functional groups, ordinations were also done for each of the FT2 and FT5 species subsets. p values were calculated based on 999 permutations.

3. Results

3.1. Environmental Parameters

Our data confirmed a high trophy of the aquatic environment in the Dehtář fishpond during the study period (Table S2). The values of total phosphorus (TP), chlorophyll-a concentrations, and water transparency (WT) correspond to hypertrophy (the thresholds between hypertrophy and eutrophy according to the OECD 1982 classification [65] are: TP 0.75 mg/L, chl-a 100 µg/L, and WT 0.5 m; the thresholds adjusted for fishpond conditions [66] are: TP 0.2 mg/L, chl-a 30 µg/L, and WT 0.5 m). Associated criteria such as the values of total nitrogen (TN) are also consistent with this classification. There were only small differences in the analysed parameters between the three sites. BSB showed higher average concentrations of several nutrients (e.g., TP and total Fe) and usually lower water transparency than DSB and FD; however, the highest measured chlorophyll-a concentrations were detected at FD, probably due to the cyanobacterial mats concentrated at this site because of wave action. Most of the trophic parameters show a gradual increase during the study period (Table S2).

The photosynthetically active radiation (PhAR), oxygen concentration, and water temperature measured continually showed large fluctuations (besides the diurnal ones, which are typical of these parameters) in some periods (Figures S16–S20). For instance, in 2016 there was a period of about a week in July with negligible values of PhAR in 0.3 m water depth (Figure S16c). In contrast, in the summer of 2015, the values measured were up to twice as high as in 2016. The 5-day course of PhAR at the time of the diatom sampling in 2015 and 2016 (Figure S17) shows that in 2015 PhAR values were generally higher than in 2016 but also more important fluctuations occurred.

The course of oxygen concentration followed the same trends at all three sites; however, the maxima were the lowest at FD and the highest at DSB (Figure S18). Low oxygen concentrations were also common, particularly at BSB and FD; however, at the beginning of August 2016, serious oxygen depletion was detected at all three sites (Figure S18). Shortly before and during the diatom sampling in June 2015 and 2016, the oxygen concentrations fell into the range of (3.5)5–10(15) mg/L, with a marked difference between FD and BSB in 2016 (Figure S19d,e). Water temperature was rather high during the summers, usually ranging around (15)20–25(30) °C (Figure S20). Although the temperature courses followed the same trends at all three sites, some differences were identified. BSB exhibited the highest temperature fluctuations and most frequent maxima at around 30 °C or more. On the other hand, the DSB temperature curve showed the smallest fluctuations and a particularly low occurrence of high temperatures. There also was a difference between individual years, with 2015 and 2019 having the warmest and 2016 the coldest water in summer (Figure S20).

The data from the meteorological sensors point out high precipitation amounts in the summer of 2014, extremely low precipitation during the whole growing season in 2015 and in the spring of 2016, and a subsequent precipitation extreme in July 2016 (Table A1). Similarly, the mean monthly temperature and temperature course during the growing seasons exhibit extremes, particularly in the summer of 2015, when day temperatures often reached values around 35 °C (Figure S21). In 2014 and 2016, summer temperatures only rarely reached values 30 °C or more. In an open fishpond space, fully exposed to the sun, air temperature was, however, probably much higher in all years (data not available). The large day/night summer temperature fluctuations are remarkable, whereas the night minima in some periods fell below 10 °C (Figure S21). The data on wind speed, available for the whole study period, show high frequency of speed above 5.5 m/s (=the winds raising small waves) in 2014 and 2015 and its lower frequency in 2016, but the year 2015 had a higher proportion of situations without any winds. Nevertheless, all these years may be considered windy, with the year 2015 also having a higher number of strong winds (categories > 8.0 and > 13.9 m/s) with the potential to raise large waves (Table 1, Figure 2, Figure S22). In contrast, the year 2019 had a high proportion of situations without any wind, and winds with speed higher than 5.5 m/s were rather rare.

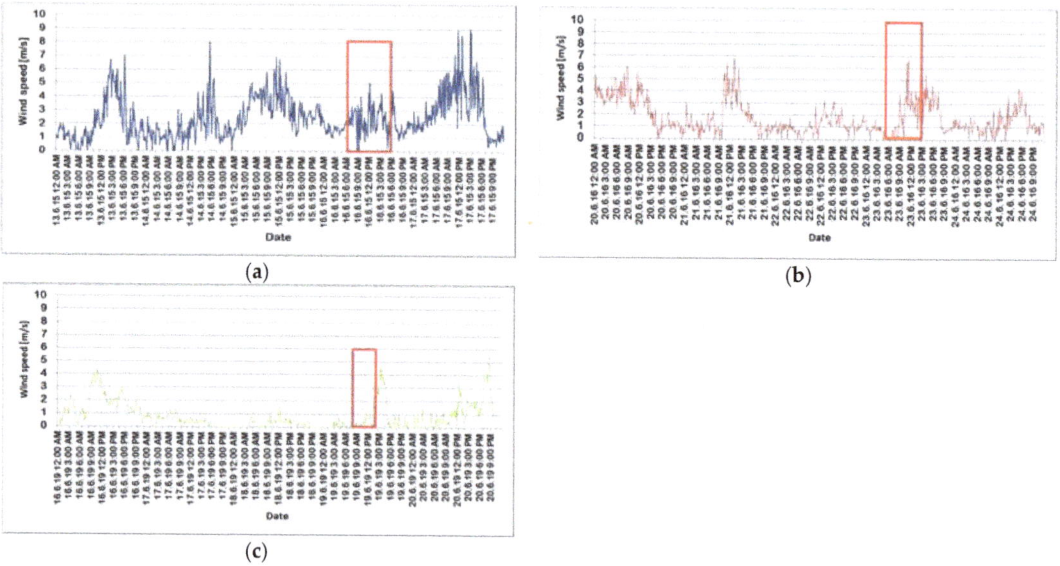

Figure 2. Wind speed course during the period of diatom sampling in the Dehtář fishpond: (**a**) 2015, (**b**) 2016, and (**c**) 2019. The particular date and time of sampling are marked by red rectangles.

Our evaluation of factors with a potentially important impact on vascular plants, particularly the emergent species, showed that the extent of exposed bottom zone was substantially higher in 2015 than in the other years, particularly in the summer of 2016.

This fact is directly reflected in the size of individual mapping segments (=vegetated zone; Table S1), which was several times larger in 2015 than in the other years, particularly in 2014 and 2016. These fluctuations are related to changes in the basic parameters of the first and the second year of the fishpond management cycle, particularly to the water depth and extent of the flooded area (Table S1). However, there were also differences between particular years of the same phase of the management cycle, e.g., 2015 and 2019 (both first years of the management cycle). In 2015, exposed muddy substrata including the relatively deep muds, important for wetland annuals, were more often available. However, the Dehtář fishpond is principally poor in very deep muddy sediments in the vegetated zone, and most of the muddy zones do not reach more than 10 cm in depth. Organic detritus regularly occurred in the bays of both streams, but in 2016 it was also frequent elsewhere (remnants of vegetation from 2015). Some segments were under the direct influence of manuring, liming, or supplemental fish feeding with cereals (Table S1).

3.2. *Vascular Plant and Charophyte Species and Functional Diversity*

3.2.1. Species Richness, Representation of Threatened Species

Altogether, 280 vascular plant taxa and a single charophyte species (*Chara braunii*) were found in the Dehtář fishpond between 2014 and 2016, and in 2019 (Table S4). Total species numbers in individual years varied highly, with the highest species number in 2015 (225 species), followed by 2014 (214 species). In 2016, the last year of intensive research in all 30 segments, only 165 species were recorded. In 2019, however, only 19 segments were investigated, with 92 taxa in total. The median species number per segment was 60 taxa in 2015, about 45 taxa in 2014, but only about 32 and 30 taxa in 2016 and 2019, respectively. However, it varied greatly among the segments (Figure 3).

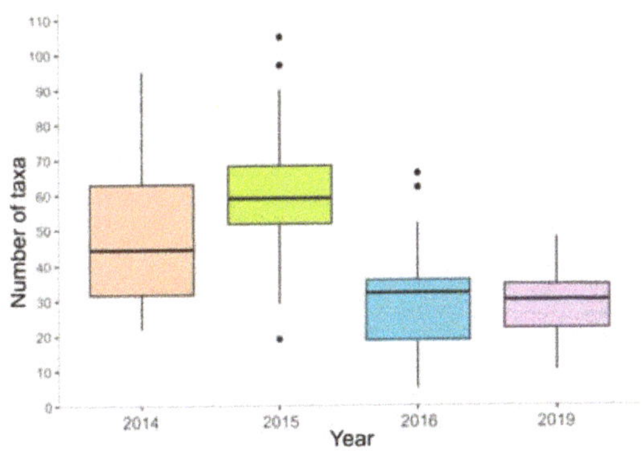

Figure 3. Comparison of the species richness of vascular plants and charophytes between individual years. Median values (=horizontal lines), lower and upper quartiles (=boxes, i.e., 50% of the observed values), 5% and 95% percentiles (=whiskers, i.e., 90% of the observed values), and outliers (=points, i.e., the values outside the 90% range) are displayed.

In total, 28 threatened species have been identified in the Dehtář fishpond habitats during the whole study period. The categories C3 (vulnerable) and C4 (near threatened) were the most numerous, but also the C1 (critically threatened) and C2 (endangered) categories were represented by a few species (Figure 4). Wetland annuals (e.g., *Carex bohemica, Coleanthus subtilis, Lindernia procumbens*, and *Lythrum hyssopifolia*) were the most numerous group of the threatened species, followed by wetland perennials (e.g., *Bolboschoenus laticarpus, B. yagara*, and *Leersia oryzoides*). Among the wetland annuals, *Montia arvensis* (Table S4, Figure S9) is one of the rarest plant species of the Czech Republic. Red-listed aquatics (*Chara braunii*) and amphibious species (*Elatine hydropiper*) were represented as well. Some terrestrial plants also occurred among the threatened taxa, but they mostly appeared very rarely (e.g., *Aphanes arvensis*; Table S4).

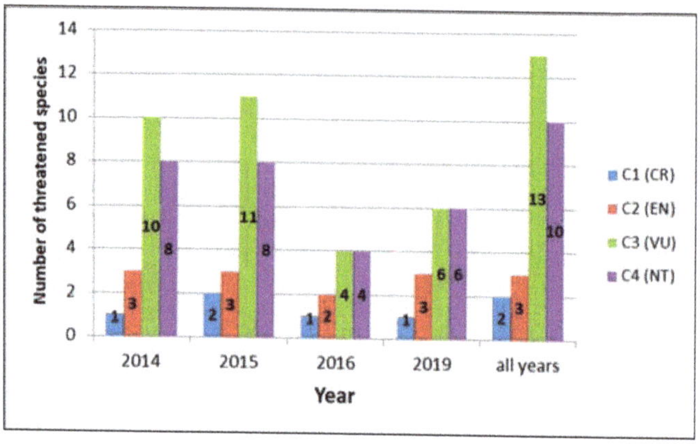

Figure 4. Numbers of threatened species in individual years and within the whole period of the Dehtář research. Four basic threat categories according to the Czech Red Data List, along with their international equivalents, were considered.

3.2.2. Vegetation Structure and Species Composition of Vascular Plant and Charophyte Assemblages

Plant assemblages of the Dehtář fishpond follow a moisture gradient of the freshwater littoral. Three basic zones may be distinguished: 1) Zone of permanent reed and tall sedge beds and of willow shrubs, 2) zone of disturbed grasslands (usually wet), temporary reed beds and wetland annuals, 3) zone of amphibious plants and submerged aquatics. The first zone is placed on the transition between the high fishpond bank and the upper littoral. It is rather narrow, usually formed by *Phalaris arundinacea*, *Carex acuta*, and *Salix cinerea*, more rarely by *Phragmites australis* or *Salix triandra* (Figures S2, S3, S5 and S6). Depending on the littoral morphology and substrate distribution, the second zone may either include several sub-zones (Figures S4 and S5), predominated usually by *Trifolium* spp. (mainly *T. hybridum*, on dry sites *T. arvense*) and *Alopecurus geniculatus* (grasslands), *Bolboschoenus maritimus* agg. (mainly *B. laticarpus*; temporary reed beds) and a mixture of tall and low-growing wetland annuals (e.g., *Bidens radiatus*, *Rumex maritimus*, and *Veronica anagallis-aquatica* in muddy places and *Juncus bufonius*, *Myosurus minimus*, and *Stellaria alsine* at sandy sites), or it is formed by a mosaic of above plant assemblages. The second zone occupies many hectares of the fishpond bottom during low water levels in the growing season (Figures S4, S11 and S12). The third zone usually includes only a narrow strip of vegetated shallow water, predominated by an amphibious species *Elatine hydropiper* (Figure S8) and an aquatic species *Zannichellia palustris*.

While the extent of the first zone was more or less unchanged during the study period, particularly the second but also the third zone varied greatly in size between the years or even in different periods of a single year (Figures S2, S3, S11–S13). For instance, the second zone was best developed during the extremely dry year 2015 which corresponds to the large extent of bottom exposure (Table S1, Figures S4 and S12). On the other hand, the third zone disappeared at the end of the summer of 2015, as it was drained. In contrast, in 2016 the second zone only occurred under a low water level in early spring (low water level continued from 2015) and later disappeared due to extremely high precipitation (Table A1) and increased water levels. The third zone with aquatics, which has a summer phenology, did not appear at all in 2016.

The species which occurred as dominants in the vegetation usually also showed very high frequency within the 30 segments. For instance, the species *Phalaris arundinacea* with the frequency 100%, followed by, e.g., *Alopecurus aequalis*, *Bidens radiatus*, *Carex acuta*, *Juncus bufonius*, *Myosurus minimus* (Figure S10), *Trifolium hybridum*, *Rumex maritimus*, and *Veronica anagallis-aquatica* with the frequency 90–100% in the mapped segments, were also the most common dominants (Table S4). Still, some of the species were an exception to this rule: *Phragmites australis* occurred in only 40% of segments, nearly always forming stands, although they were not very large. Among the species occurring with high frequency (90–100%) but not forming stands were, e.g., common wetland species *Lythrum salicaria*, *Oenanthe aquatica*, and *Persicaria hydropiper*, or *Tripleurospemum inodorum* as a ruderal. Only 50 plant species appeared as dominants in at least one segment. Within the segments the frequency of most of them was at least 50%, although there were some exceptions, e.g., terrestrial species dominating only locally at the dam, such as *Geranium pusillum* and *Lamium purpureum*.

The frequency of many species fluctuated highly between the years (Table S4). While in 2019 many species were not recorded due to lower research intensity, the data from 2014–2016 reflect the presence or absence of some of the vegetation zones in which the particular species usually grow.

3.2.3. Ecological and Functional Groups of Vascular Plants and Charophytes

Only about one-third of the 281 vascular plant and charophyte taxa were classified within some of the groups of wetland and aquatic species (Figure 5a,b). The rest of the species pool was represented by terrestrial plants with an optimum in various types of habitats, mainly in grasslands and in ruderal and arable weed vegetation (Figure 5a,b).

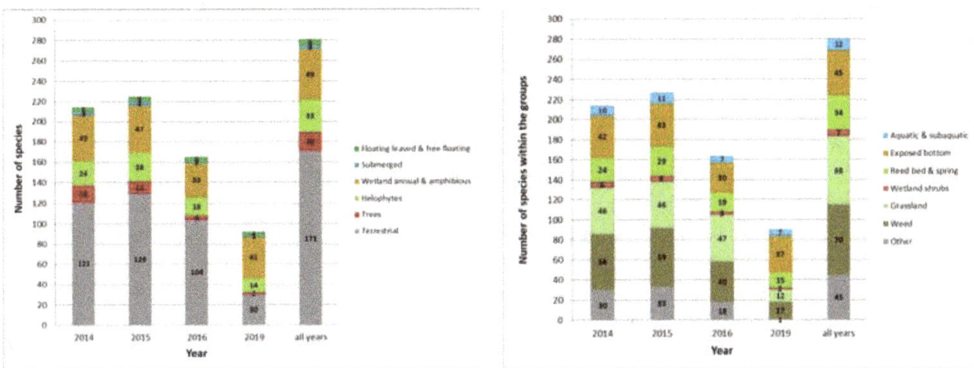

(a) (b)

Figure 5. Numbers of plant species according to their ecology and life-history traits: (**a**) detailed classification of aquatic macrophytes into functional groups based on their position on the moisture gradient and life-history traits; all terrestrial species are included into a single group; (**b**) classification of all vascular plant and charophyte species into the groups according to their habitat and/or vegetation preferences. For a detailed description of each species group and classification of particular species, see Table S4.

Among the wetland plants, annual and shortly perennial species of exposed pond bottoms formed the most numerous group, followed by reed bed species (i.e., helophytes, Figure 5a,b). On the other hand, the number of aquatic species (floating-leaved, free-floating, and submerged plants) was very low. The representation of particular groups varied between the years. This holds true particularly for small wetland annuals, amphibious, and submerged aquatic plants, which were most frequent in 2014 and 2015, some of them also in 2019 (the years with exposed bottom and shallow water available throughout the growing season, see Tables S1 and S4) but rare or even missing in 2016, i.e., the year with high water levels in summer (Table S4, Figure 5a,b). The wetland perennials growing in lower littoral parts (e.g., *Bolboschoenus maritimus* agg.), ruderal and arable weeds, and trees and shrubs (species occurring only as seedlings) also exhibit a strong decline in species numbers in 2016 (Table S4). On the other hand, grassland species, e.g., *Holcus lanatus*, *Lychnis flos-cuculi*, and *Potentilla anserina* profited from a long dry phase in the spring of

2016 and partly in 2014. The year 2019 shows apparently lower proportion of terrestrial species due to the fact that the research focused on the shallow water and exposed bottom zones (Figure 5a,b).

3.2.4. Main Drivers of Species Composition in Vascular Plant and Charophyte Assemblages

Most of the factors tested have been significantly correlated with the NMDS ordination, where the first ordination axis corresponds to the gradient of flood and exposure duration, while the second axis exhibits the relation to the depth of muddy sediment and the distance from large inflows (Figure 6). The extent of summer exposure and the distance from large inflows have been identified as the factors with the highest impact on the species composition of vascular plants and charophytes in the plant assemblages of the Dehtář fishpond. A less important relationship was detected for mud depth and, with a weak significance, for a number of small inflows. Research intensity has a negative relationship with the second ordination axis, but the significance ($p = 0.067$) was below the commonly used threshold. The two remaining factors, i.e., the surrounding landscape with prevailing arable land and shrub share in the buffer zone did not show any relationship to the species composition of the plant assemblages.

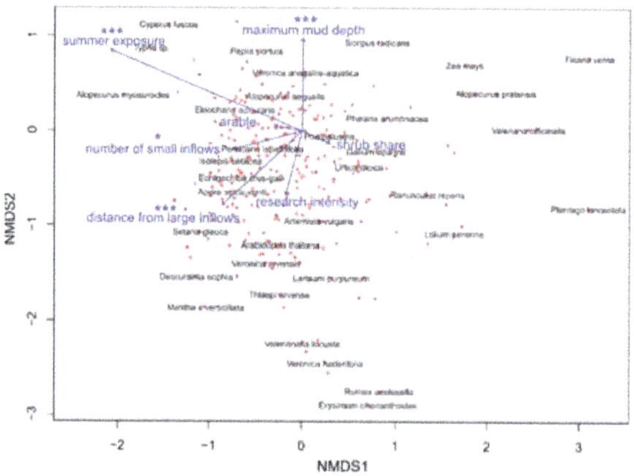

Figure 6. NMDS ordination diagram of species composition of vascular plant and charophyte assemblages of the Dehtář fishpond with passively projected selected environmental and management factors. Significant relationship of the factors to species composition is indicated based on 999 permutations. ***—highly significant ($p \leq 0.001$), *—weakly significant ($p \leq 0.05$). The correlations were calculated using the *envfit* function from the *vegan* package (Oksanen [93]). Explanations of abbreviated names of the selected factors: summer exposure (=extent of bottom exposure in summer), research intensity (=number of visits in the given segment and year), shrub share (% share of shrubs in the buffer zone between open water and the surrounding landscape), arable (=surrounding landscape with prevailing arable land). For more detail, see Table S1.

In the upper left part of the NMDS ordination diagram (Figure 6) the species of exposed substrates, particularly wetland annuals (e.g., *Cyperus fuscus, Isolepis setacea, Persicaria lapathifolia*) and amphibious species (e.g., *Eleocharis acicularis*) are concentrated, with some of them showing relationship to a higher depth of mud (e.g., *Veronica anagallis-aquatica* and *Peplis portula*). On the opposite side of the gradient, mainly grassland species (e.g., *Ranunculus repens, Plantago lanceolata*) and other terrestrial plants (e.g., *Galium aparine*) are displayed. The gradient of the distance from large inflows shows species composition of plant assemblages near the fishpond dam that occurred on dry sandy substrates. These sites were occupied mainly by weeds of arable land (e.g., *Apera spica-venti, Thlaspi arvense*), ruderal weeds (e.g., *Descurainia sophia, Artemisia vulgaris*), and annual ephemeral herbs of dry disturbed habitats (e.g., *Arabidopsis thaliana, Valerianella locusta*). All these terrestrial species have been recorded, particularly in the years with high research intensity. The opposite part of this gradient shows the vicinity of large inflows, with *Scirpus radicans* growing directly in one of them.

The "spider" diagram of the NMDS ordination showing the influence of the year of the management cycle indicates that the samples (species lists collected separately for each segment and year) are well separated between year 1 and 2 (Figure 7a), and the difference in their species composition is significant (PERMANOVA, F = 8.1855, $p = 0.005$).

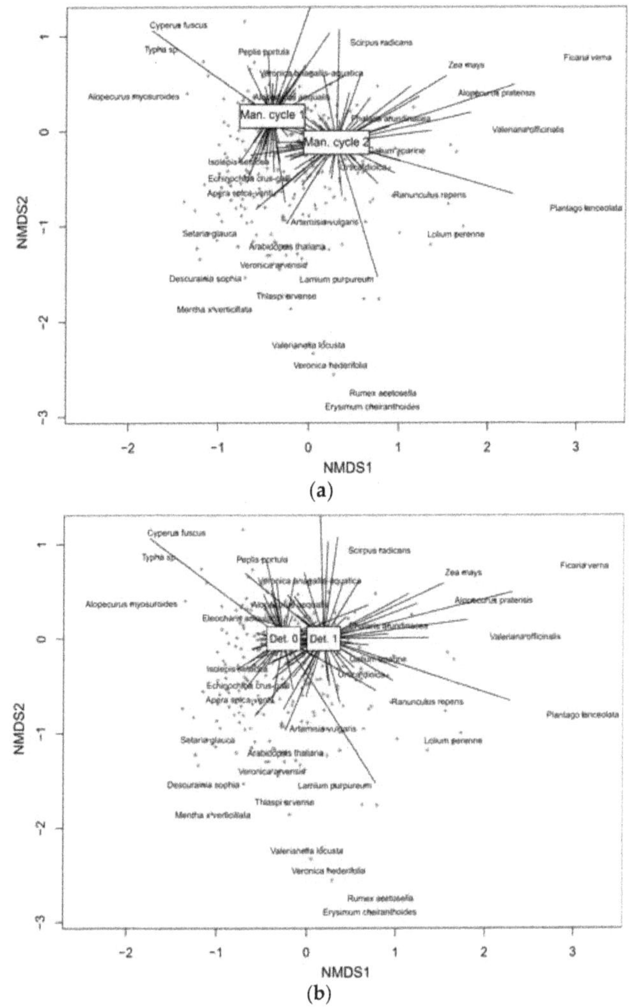

Figure 7. NMDS ordination diagrams of selected environmental factors and their relationship with the species composition of vascular plant and charophyte assemblages of the Dehtář fishpond: (**a**) Factor "year of the management cycle"; Man. cycle 1 = 1st year of the management cycle, Man. cycle 2 = 2nd year of the management cycle; (**b**) Factor "organic detritus"; Det. 1 = organic detritus present in more than 20% of the segment area, man. 0 = organic detritus present in less than 20% of the segment area. For a more detailed explanation of all individual variables, see Table S1.

The figure also clearly demonstrates the predominance of annual species in the species spectrum of year 1, and, by contrast, of perennial species in the samples of year 2, including wetland species of the upper littoral zone. Figure 7b shows the "spider" NMDS ordination diagram with the relationship between organic detritus and species composition. Visually, there are rather large overlaps in the species composition of the segments with and without organic detritus. Nevertheless, the difference in species composition between the contrast sites was significant (PERMANOVA, $F = 6.2197$, $p = 0.012$). Wetland annuals such as *Cyperus fuscus*, *Isolepis setacea*, and seedlings of *Typha* sp. are clearly related to the segments without substantial amounts of organic detritus. On the other hand, nitrophytes such as *Galium aparine* and *Urtica dioica* show the preference for segments with organic detritus deposits.

3.3. Diatoms

3.3.1. Diatom Species and Functional Group Richness

Diatom species richness varied greatly among the samples, particularly in 2015 (min. 14, max. 46 species) and 2016 (min. 13, max. 48 species); in 2019, the differences were smaller (min. 23, max. 35) (Figure 8a–c). The same applied even to samples from the same site (e.g., FD 2015 or BSB 2016) and, in some cases, for the same macrophyte species. Nevertheless, the tendency to support higher or lower diatom species richness was visible in several macrophyte species. Macrophytes with high diatom species richness, reaching the maxima above 40 and the minima higher than 25 diatom species per sample, were, e.g., *Elatine hydropiper* (n = 5, min. 28 and max. 46 diatom species), *Veronica anagallis-aquatica* (n = 4, min. 28, max. 44), and *Zannichellia palustris* (n = 5, min. 29, max. 39). The highest diatom species richness, i.e., 48 taxa, was identified in the single sample of *Persicaria hydropiper*. On the other hand, the samples of some macrophytes reached the maximum richness of 25 diatom species or less. These were, e.g., two species from the *Lemnaceae* family, *Spirodela polyrhiza* (n = 2, min. 14, max. 20) and *Lemna gibba* (n = 2, min. 16, max. 17), *Chara braunii* (n = 2, min. 20, max. 24), and *Persicaria amphibia* (n = 2, min. 22, max. 25). The absolutely lowest diatom species number, i.e., 13, was identified in the single sample of *Alopecurus geniculatus*.

Altogether 160 diatom taxa have been identified in 58 samples from the Dehtář fishpond during three sampling events, with 92 species in 2015, 123 species in 2016, and 62 species in 2019. The median value for species richness per sample in individual years was 34, 24, and 30 taxa, respectively (Figure 9a).

Considering the three sampling sites, the overall species richness in absolute numbers was the highest in BSB 2016 (107 diatom species) and the lowest in FD 2019, site 1 (44 species) (Figure 10a–c). However, the highest median value for species richness was achieved at BSB 2015 (35 species), and only slightly lower values were identified at FD 2015 and DSB 2015 (Figure 9b). The lowest median value for species richness was detected at FD 2016 (21 species), followed by BSB 2016.

All five functional groups of diatoms (i.e., FT1–FT5) were represented in most of the samples (Figure 8a–c). The proportion of particular groups in the species composition was rather stable across the samples, with low species numbers at FT1 and FT4, middle high at FT2 and FT3, and the highest at FT5 (Figure 8a–c). However, FT5 exhibited the largest fluctuations in species numbers among the samples. The extraordinary species-rich samples had all five functional groups, with especially high numbers at FT5, and in 2016 also FT2 and FT3 (Figure 8a–c, see also Figure 10a–c for the summary data of the sites and years). FT5 and sometimes also FT2 was poorly represented in the macrophyte samples with very low diatom species richness (Figure 8a–c).

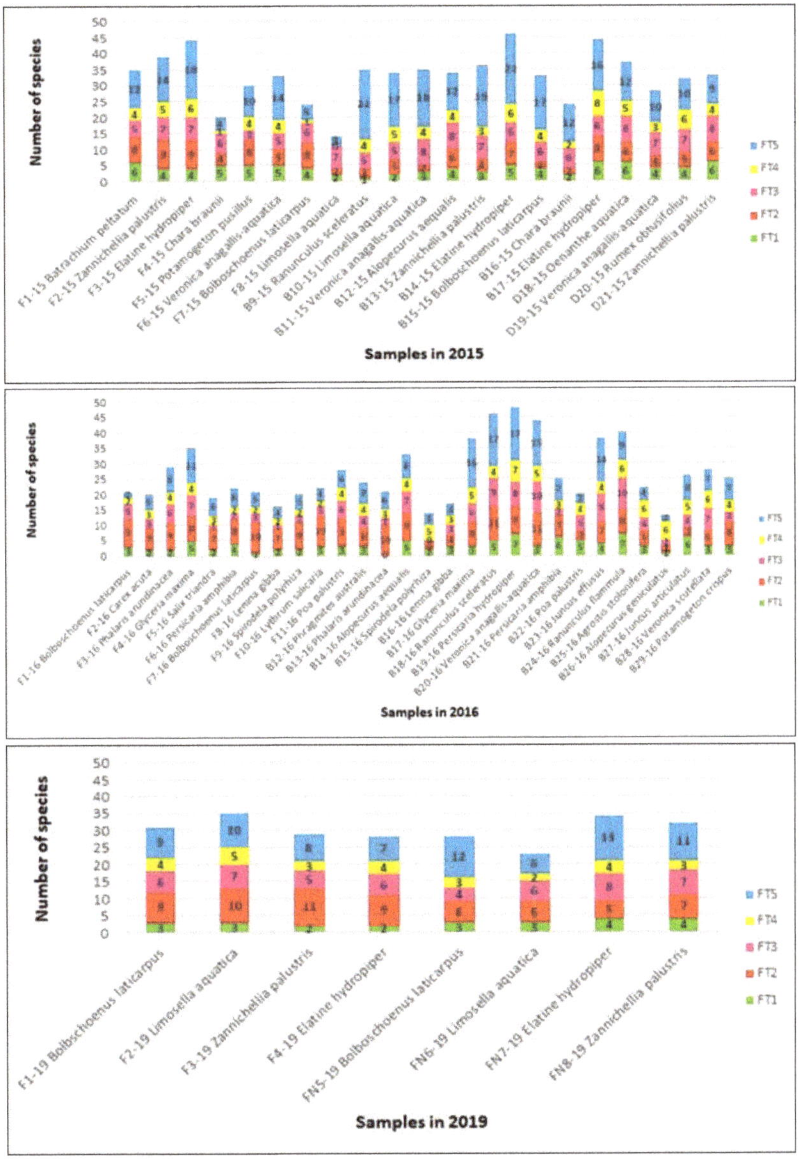

(a) (b) (c)

Figure 8. Diatom species numbers and representation of five diatom functional groups (FT1–FT5) among the species in individual samples (see Table S6) collected in the Dehtář fishpond. (**a**) Year 2015; (**b**) Year 2016; (**c**) Year 2019—only FD was sampled but two different microsites were included. Codes of the samples are according to Table S3. Host plant species is listed at each of the samples.

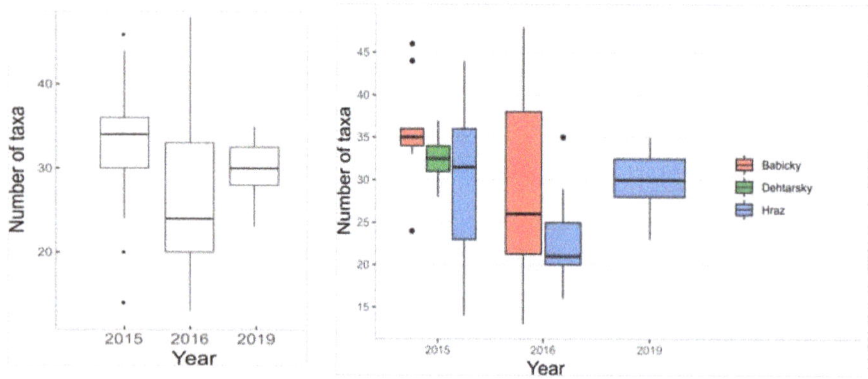

(a) (b)

Figure 9. Comparison of diatom species richness in the Dehtář fishpond. (**a**) Differences between years; (**b**) differences between years and sites. Median values (horizontal lines), lower and upper quartiles (boxes, i.e., 50% of the observed values), 5% and 95% percentiles (whiskers, i.e., 90% of the observed values), and outliers (points, i.e., the values outside the 90% range) are displayed.

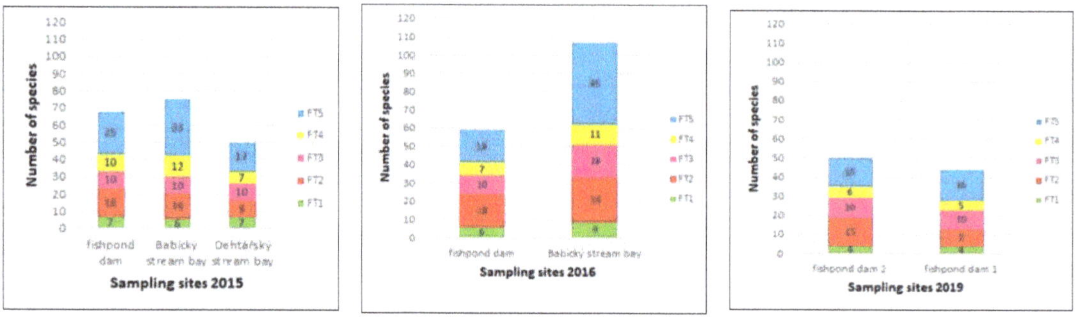

(a) (b) (c)

Figure 10. Diatom species numbers and representation of five diatom functional groups (FT1–FT5) in the species pool of sampling sites in the Dehtář fishpond. (**a**) Year 2015; (**b**) Year 2016; (**c**) Year 2019—only FD was sampled but two different microsites were included (fishpond dam 1 = site 1, fishpond dam 2 = site 2; see Table S3 for more detail).

3.3.2. Quantitative Representation of Diatom Species and Their Functional Groups

Nitzschia palea (FT4 functional group) was the most abundant diatom species in our samples in terms of its frequency and quantity. It occurred in 56 of the total of 58 samples (frequency 96.6%) with 2,986 individuals altogether (Table S5). Its highest concentrations were detected in 2016 when it was found in all the samples and strongly dominated in some of them (Table S6). *Puncticulata balatonis*, a planktonic diatom (FT1), occurred in 54 samples altogether (frequency 93%) but in low quantities (low numbers of individuals of FT1 species are typical for epiphytic diatom samples). Nine additional taxa from various functional groups had a frequency of between 75% and 90% within the whole dataset (=highly frequent taxa). Among them, e.g., *Navicula capitatoradiata* and *Nitzschia archibaldii* also occurred in high quantities, strongly predominating the diatom assemblages of some of the samples. Most of the highly frequent and frequent species (occurring with the frequency between 50% and 75% within the dataset) have shown a strong affinity to a particular year and/or site (Table S6; see also Section 3.3.3). *Pseudostaurosira elliptica*, *P. brevistriata*, and *Staurosirella leptostauron* var. *dubia* in 2015; *Achnanthidium saprophilum* (probably on ESEM photo–Figure 14c), *Cocconeis placentula*, and *Navicula reichardtiana* in 2016, and *Gomphonema parvulum* (Figure 14d) in 2016 and 2019 occurred with a high frequency and, at the same time, in a high number of individuals. The frequent species also included those found in a small number of individuals only, e.g., *Staurosira binodis*, *Navicula gregaria*, and *Placoneis clementiodes*, concentrated into the year 2015, or *Craticula subminuscula* and *Ulnaria acus*, typical for 2016. The species that had, despite low frequencies, relatively high numbers of individuals were exceptional: *Lemnicola hungarica* and *Eunotia bilunaris* (both FT2) occurred only in eight and ten samples, respectively, but especially the former sometimes produced very dense populations (Tables S5 and S6).

The proportion of the functional groups of diatoms strongly varied between the years and sites. In 2015, the groups FT3 and FT5 predominated in the samples, with the former being more frequent at BSB and DSB and the latter in the samples from FD (Figure 11a). In 2016, only FD and BSB were sampled. While typically periphytic diatoms (FT2 group) predominated the samples from FD, the FT4 taxa occurred with the highest quantities in most samples from BSB (Figure 11b). In 2019, when two microsites at FD were sampled, the share of the functional groups among the samples was rather balanced. While the samples from the more muddy site were predominated by the FT2 and FT5 groups (Figure 11c, samples F1–19 to F4–19), the highest share was achieved by FT2 and FT3 in the samples from the sandy site exposed to wave action (Figure 11c, FN5-19 to FN8-19).

The predominance of a single or a few diatom taxa is strongly reflected in the share of their functional species groups within the samples (Figure 11a–c). It is remarkable particularly in species poor samples, e.g., in *Limosella aquatica* (F8-15; Figures 8a and 11a) with an extremely high share of FT3 group (*Pseudostaurosira brevistriata*, *Staurosirella leptostauron*) or *Lemna gibba* and *Spirodela polyrhiza* from FD 2016 (F8-16 and F9-16; Figure 8b, Figure 11b) with a strong predominance of FT2 group (*Lemnicola hungarica*).

3.3.3. Impact of Environmental Factors on Structure and Species Composition of Diatom Assemblages

All the tested factors have been significantly correlated with the NMDS ordination. The first ordination axis corresponds to the gradients of flood and exposure duration, mud depth, and, in the opposite direction, chlorophyll-a concentrations during the growing season. The second axis corresponds to the gradient of the distance from large inflows where the largest distance corresponds to the proximity of the fishpond dam (FD). The other factors correlated with this gradient, i.e., shrub share in the buffer zone, surrounding landscape formed by arable fields, and number of small inflows, are clearly associated with the FD sampling site (Figure 12).

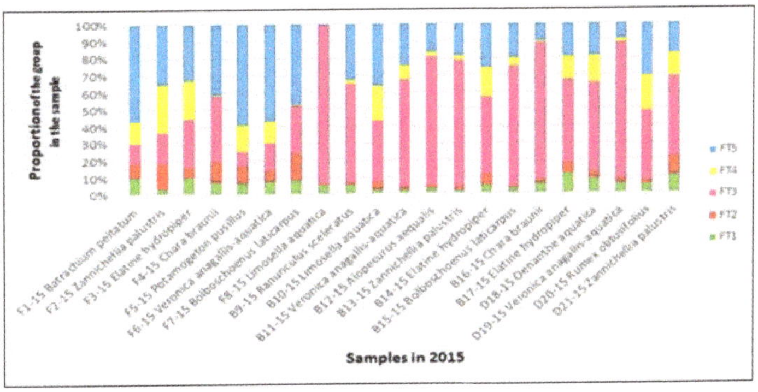

(a)

Figure 11. *Cont.*

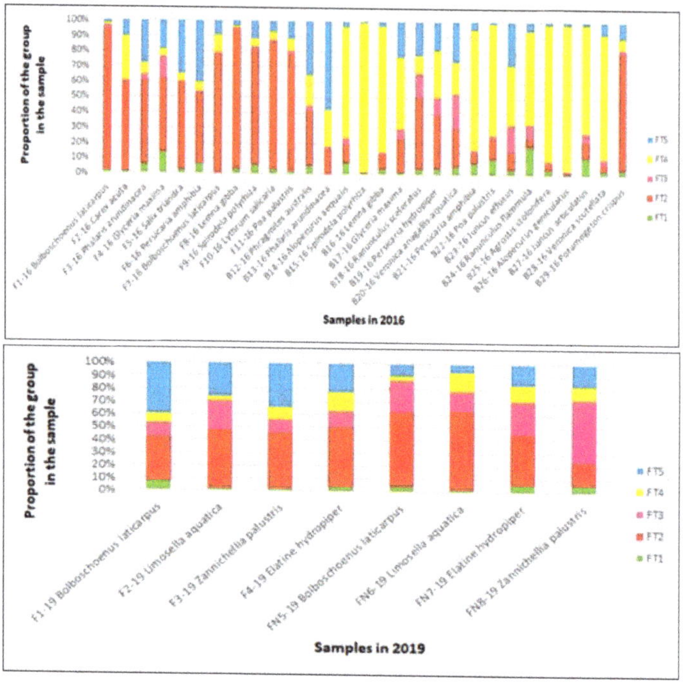

Figure 11. Relative representation of five diatom functional groups (FT1–FT5) in particular samples (based on quantities of diatom individuals within each of the groups). (**a**) Year 2015; (**b**) Year 2016; (**c**) Year 2019—only FD was sampled, but two different microsites were included.

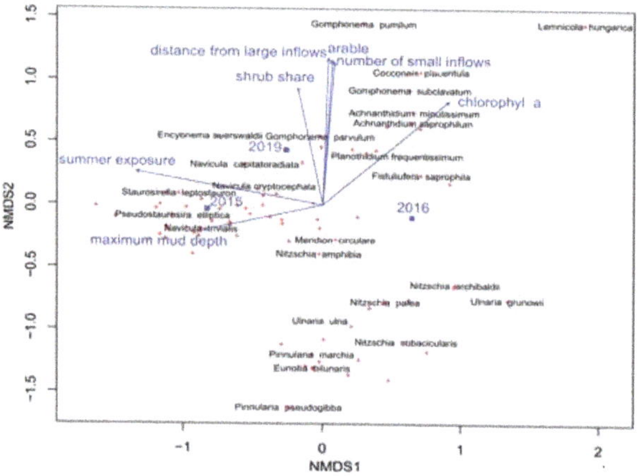

Figure 12. NMDS ordination diagram of species composition of diatom assemblages of the Dehtář fishpond with passively projected selected environmental and management factors. All continuous factors are significantly correlated with species composition ($p \leq 0.01$; 999 permutations). The *envfit* function from the *vegan* package (Oksanen [93]) was used for the calculations. Explanations of abbreviated names of selected factors: summer exposure = extent of bottom exposure in summer, shrub share = % share of shrubs in the buffer zone between open water and the surrounding landscape, arable = surrounding landscape with prevailing arable land, chlorophyll-a = chlorophyll-a concentrations (mean for the relevant site and growing season), 2015, 2016, 2019 = years of the diatom sampling. For a more detailed explanation of individual variables, see Table S1.

The species composition was significantly different between the years of sampling (here a calendar year) (PERMANOVA, F = 34.638, $p = 0.001$). The differences are also clearly demonstrated in the NMDS ordination diagram (Figure 12). For the year 2015, remarkable for its large area of exposed pond bottom and high depth of mud (vegetation zones 2 and 3, see Section 3.2.2), FT3 diatoms such as *Staurosirella leptostauron* and *Pseudostaurosira elliptica* were the most typical, followed by some *Navicula* species (FT5 group), e.g., *N. cryptocephala* and *N. trivialis* (Figure 12, Table S6).

In contrast, the year 2016 was characterized by high water levels, and therefore only the vegetation zones 1 and 2 (deeply flooded) could be sampled. FT4 diatoms of the genus *Nitzschia*, particularly *N. archibaldii* and *N. palea*, as well as the FT5 species such as *Pinnularia marchia* and *Ulnaria ulna* differentiated the sites distant from the fishpond dam at BSB from the sites distant from the large inflows at FD. Samples from FD 2016 were predominated by periphytic diatoms (FT2), e.g., *Achnanthidium minutissimum*, *A. saprophilum*, *Cocconeis placentula*, and *Lemnicola hungarica*. Increased chlorophyll-a concentrations probably modulated the species composition at FD in 2016. The FT2 species *Gomphonema parvulum* and FT5 species *Navicula capitatoradiata* were present in the samples from each year and site, but they were particularly frequent at FD in 2015 (higher share of *N. capitatoradiata*) and 2019 (higher share of *G. parvulum*).

The relationship of diatom assemblages to three important factors, not considered above, is shown in Figure 13: distance from the fishpond bank (a), water depth (b), and exposure to disturbances caused by wave action or streaming (c). These NMDS ordination diagrams show not only the species distribution patterns between the three sites and sampling years, but also the differentiation or overlaps between the samples.

The samples from the three sites collected in 2015 exhibit a high level of similarity in species composition, all being situated a long distance from the bank and in very shallow, disturbed parts of the lower littoral (see Table S3). The two sites sampled in 2016, FD and BSB, exhibit a high level of differentiation between the samples: while the disturbance intensity and distance from the bank were similar for both sites, water depth was markedly higher at FD. At the same time, the samples from both sampling sites in 2016 were clearly differentiated from those from 2015 by a substantially higher water depth, and shorter distance from the bank and disturbance intensity. Finally, the samples from 2019 (only FD) have an intermediate position between FD 2015 and FD 2016, yet showing more similarities to 2015 (Figure 13a–c, Tables S3 and S6).

On a small scale, a particular spatial position of the sampled macrophyte individuals and the distances between them probably also play a role. For instance, diatom assemblages of *Glyceria maxima*, *Persicaria hydropiper*, *Veronica anagallis-aquatica*, and *Ranunculus sceleratus* (samples B17-16 to B20-16) growing at BSB in the distance of up to about 1 m from one another, were highly similar in their species composition, species richness, and proportions of functional groups of diatoms within the species pools (Figure 8b, Table S6). The quantitative share of particular diatom species and functional groups varied, however, even at these spatially close samples. Similarly, the samples of *Phragmites australis* and *Phalaris arundinacea* (B12-16 and B13-16) collected at the same site in a sandy littoral at BSB (segment 6–see Figure 1 and Table S3) strongly differed in the share of functional groups of diatoms (Figure 11b). These differences were also well visible on the microphotographs of plant surfaces with diatom assemblages acquired using ESEM (Figure 14a,b). The overall density of diatom assemblages also strongly varied, as shown by a set of ESEM photographs from the sites with contrasting conditions (e.g., exposure to wave-mediated disturbances) or the different macrophytes from one and the same site and year (Figure 14a–e).

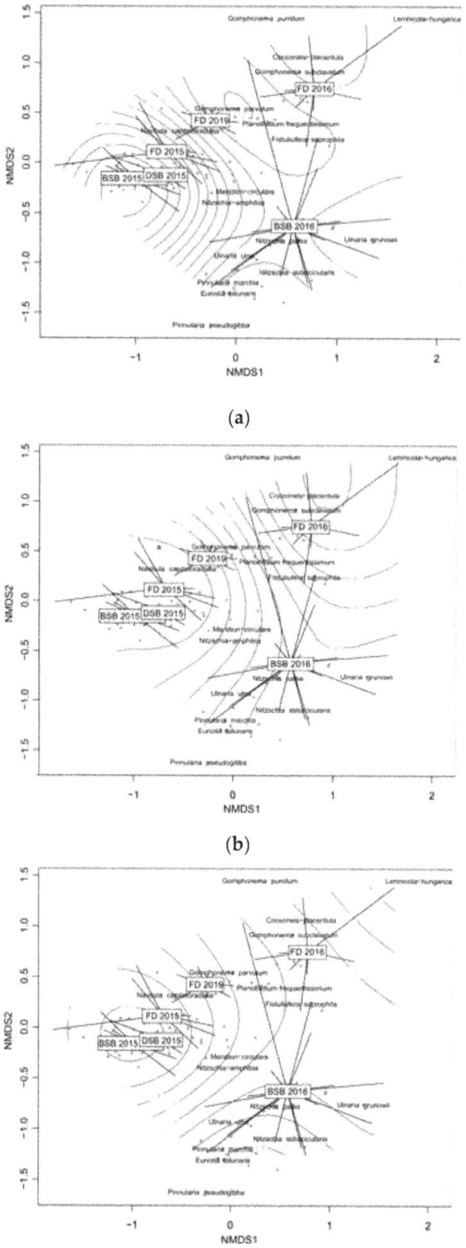

Figure 13. NMDS ordination diagrams of selected environmental factors and their relationship to species composition of diatom assemblages of the Dehtář fishpond. (**a**) Distance from the fishpond bank (m); (**b**) water depth (cm); (**c**) exposure to wave- or streaming-mediated disturbances (ordinal scale from 0—no disturbance, to 3—severe disturbances). The passively projected isolines in the figures demonstrate the scaling of each factor. For more information on environmental factors, see Table S3.

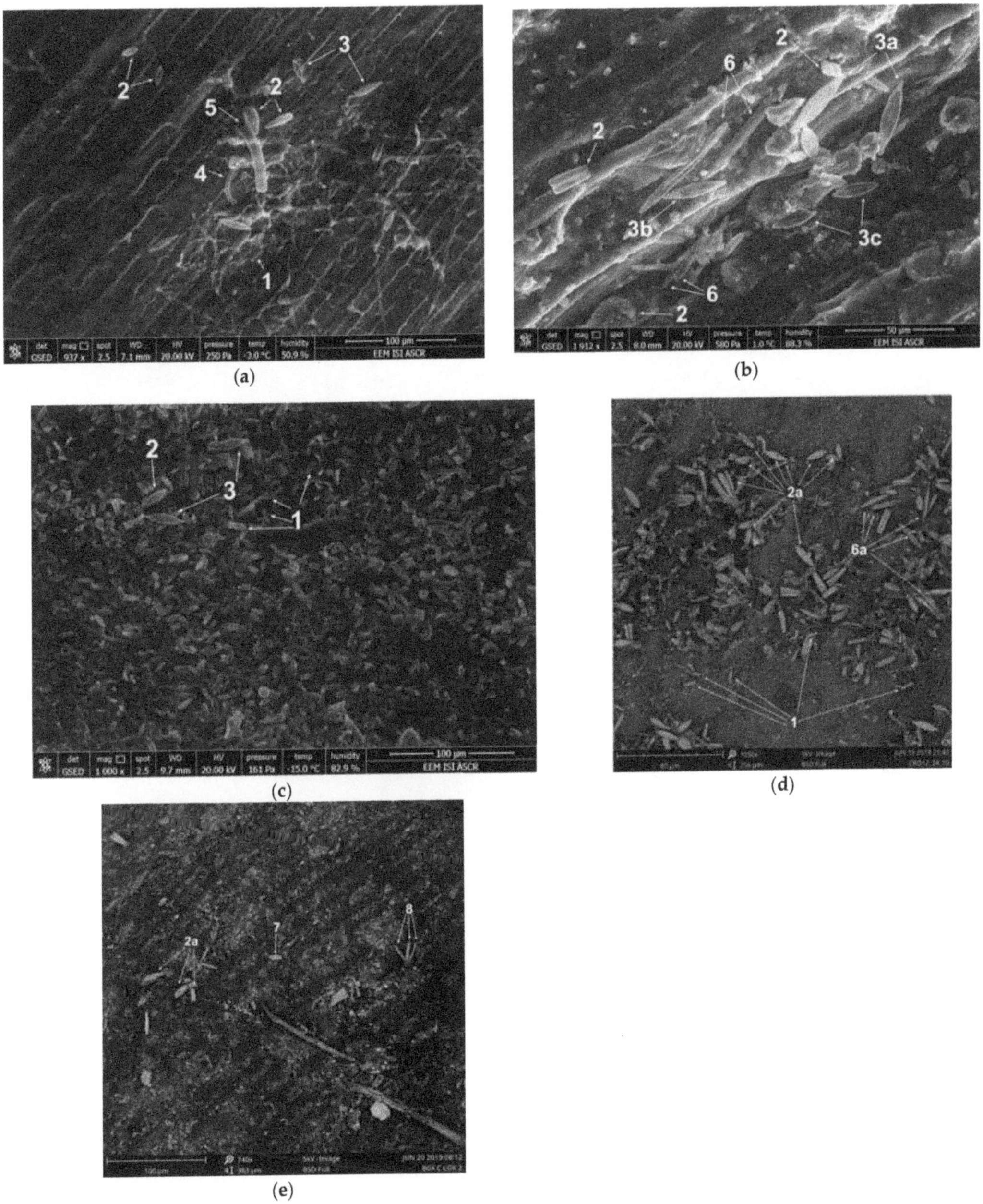

Figure 14. Environmental Scanning Electron Microscopy (ESEM) photographs of the surfaces of selected macrophytes with epiphytic diatom assemblages. (**a**) *Phragmites australis* leaf collected at BSB in 2016 (sample B12-16, 2 m from the bank) with a sparse epiphytic diatom assemblage consisting mainly of the FT2 (*Achnanthidium* sp.—1, *Gomphonema* spp.—2) and FT5 taxa (*Navicula* spp.—3), less frequently of the taxa from FT1 (*Aulacoseira* sp.—4) and FT3 functional groups (*Melosira* sp.–5), (**b**) *Phalaris arundinacea* leaf sheath collected at BSB in 2016 (sample B13-16, 1 m from the bank) near the above sample of *Phragmites australis*. The photo shows the predominance of FT5 diatoms of *Navicula* spp. (*Navicula capitatoradiata*—3a, *N. cryptocephata*—3b, and *N. reichardtiana*—3c) and representatives of FT2 (*Gomphonema* spp.—2) and FT3 (*Nitzschia* spp.—6); (**c**) *Phragmites australis* leaf sheath collected at FD in 2016 is densely covered by *Achnanthidium* spp.—1 (FT2; sample not

analysed by LM; conditions of lower disturbance impact than at BSB). Representatives of FT2 (*Gomphonema* sp.–2) and FT5 (*Navicula* sp.–3) are present in minority; (**d**) *Limosella aquatica* leaf stalk collected at FD in 2019 (sample FN6-19, site 1 with higher influence of wave action) with a dominance of *Gomphonema parvulum*—2a (FT2) and presence of *Achnanthidium minutissimum*—1 (FT2) and *Nitzschia archibaldii*—6a (FT4); (**e**) *Limosella aquatica* leaf stalk collected at FD in 2019 (sample F2-19, sampling site 2 with lower influence of wave action). The photo shows individuals of *Gomphonema parvulum*—2a and *Planothidium* sp.—7 from FT2 and *Pseudostaurosira brevistriata*—8 from FT3 functional groups. For more detail on species composition and the share of individual functional groups in the samples, see Figures 8b,c, 11b,c and 12, and Table S6.

3.3.4. Relationships between Macrophytes and Epiphytic Diatom Assemblages

We did not find any significant relationship between diatom species composition and particular macrophyte species either in the dataset of all macrophyte samples and all diatom functional groups (Permutation test for RDA; $p = 0.72$) or in the test of all macrophyte samples, and FT2 ($p = 0.565$) or FT5 ($p = 0.825$) diatom groups, respectively.

When we used the ecology and life history traits-based functional groups of macrophytes instead of vascular plant and charophyte species, we detected a highly significant response of all diatom species ($p = 0.001$) as well as of diatom species selected only from the functional groups FT2 ($p = 0.002$) and FT5 ($p = 0.001$). The use of leaf morphology-based functional groups of macrophytes did not reveal a significant relationship to species composition within the two functional groups of diatoms ($p = 0.116$ for FT2, $p = 0.127$ for FT5). The relationship between all diatom species and morphology-based macrophyte groups was only weakly significant ($p = 0.048$). The effect of the macrophyte species groups on diatom species richness is shown in Figure 15. In the graph of diatom species richness in the groups of samples sorted according to ecology and trait-based classification (Figure 15a), a trend of higher species richness in amphibious and submerged plants is visible. On the other hand, floating-leaved species exhibit low species richness; however, they were present only in 2016. When the morphological classification of macrophytes was used, no clear relationship between the species richness and a particular macrophyte group was found (Figure 15b).

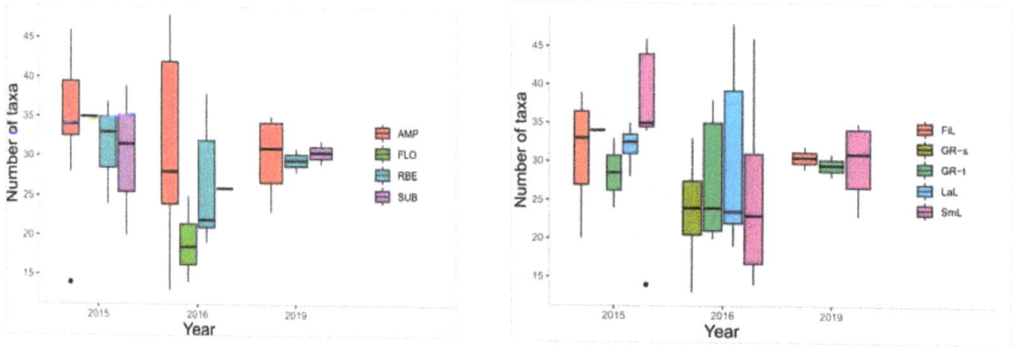

(**a**) (**b**)

Figure 15. Comparison of species richness of diatoms in the macrophyte samples from the Dehtář fishpond classified into species groups. Years 2015, 2016, and 2019: (**a**) Functional groups based on the ecology and life-history traits of macrophytes; (**b**) functional groups based on leaf morphology. Median values (horizontal lines), lower and upper quartiles (boxes, i.e., 50% of the observed values), 5% and 95% percentiles (whiskers, i.e., 90% of the observed values), and outliers (points, i.e., the values outside the 90% range) are displayed. Abbreviations: AMP—amphibious species, FLO—floating-leaved aquatics (both free-floating and rooted); RBE—reed bed species; SUB—submerged aquatics; FiL—filamentous leaves; Gr-s—grasslike small; Gr-t—grasslike tall; LaL—large leaves; SmL—small leaves. For the classification of particular macrophyte species into functional groups, see Table S3.

4. Discussion

4.1. Joint Characteristics of Plant and Epiphytic Diatom Assemblages

Vascular plants (including a single charophyte species) and epiphytic diatoms in the Dehtář fishpond exhibit similarities in several general trends. These are, in particular: (1) the species composition, as well as the proportion of functional and ecological groups, fluctuates significantly between individual years and sites. (2) A substantially higher species richness in 2015 (first year of the fishpond management cycle with extreme summer drought and the largest extent of fishpond bottom exposure) as compared to 2016 (second year of the management cycle with a very dry spring but a very wet summer, causing extremely high water levels): We suppose that these differences are mainly related to the differences in the extent of fishpond bottom exposure, which is typically much larger in the first than in the second year of the management cycle; however, irregularities may also occur due to climatic extremes. Our incomplete data from 2014 (second year of the management cycle but with water levels lower than typical for this period) and 2019 (first year of the management cycle) at least partly support this presumption. For macrophytes, similar results have been obtained also by Francová et al. ([19,20]). (3) The proportion of taxa with high tolerance or preference of water level fluctuations and/or drought tolerance: In particular, there is a high share of non-wetland species among vascular plants, a high frequency of wetland annual species preferring sandy substrata and the occurrence of aerotolerant diatoms, and, in contrast, there is the absence of species preferring a stable water column. The overall species composition and share of functional and ecological species groups of vascular plants and diatoms correspond to a temporary water body (e.g., [18,96–99]. (4) The occurrence of taxa with various nutrient demands at one site or even within a single stand/sample: Particularly in the first year of the management cycle, the joint occurrence of species preferring oligotrophic to mesotrophic habitats with species typical of highly eutrophic habitats was rather common, although the former usually did not exhibit high abundances in case of diatoms. The final trend was (5) an exceptionally high overall species richness in comparison to the studies from similar temperate aquatic habitats (lakes, fishponds, water reservoirs (see [52–54,60] for diatoms and [18–20,26,36,100] for macrophytes).

4.2. Do Macrophytes and Diatoms Follow the Same Environmental Gradients?

4.2.1. Nutritional Gradient

The plant and diatom species and functional diversity of the Dehtář fishpond is very high and complex. We analysed a number of environment- and management-related factors in order to disentangle their influence on the studied groups of autotrophs. Basic nutrients, pH, and conductivity are considered to be important characteristics driving patterns in the distribution of aquatic plants and algae [23,36,49,57,101–105]. This, however, holds true for the level of separated localities, particularly on a larger spatial scale [3,21,23,50,52], rather than within a single water body. Although Dehtář is a large fishpond and its bays are somewhat separated from the main part, which is connected to the dam, spatial differences in water chemistry are, in our opinion, too small to be interpreted as an important source of variability in autotrophic communities occurring at the different sites. As the system is hypertrophic, we do not suppose any limitations by nutrients. Diatoms could be limited by silicon (Si) availability; however, we have no data regarding this nutrient. Yet, as waters in the study region do not suffer from its lack [106], we suppose it is not a limiting element here. Moreover, silicon is usually positively correlated with iron concentrations [107], which were high in some years. Aquatic plants found in the Dehtář fishpond are able to grow in waters with various trophy levels [108]. For macrophytes rooted in the bottom, the sediment is the main source of nutrients; muddy fishpond sediments are generally nutrient-rich [16].

4.2.2. Light Limitation

Light limitation might be crucial for autotrophs in hypertrophic water bodies [47]. Therefore, we involved monthly averages of chlorophyll-a concentrations from each site into the analyses of epiphytic diatom assemblages; this factor corresponds well with nutrient amounts and is negatively correlated with water transparency. Indeed, we found a clear relationship between chlorophyll-a concentrations and the diatom species composition. This result is in agreement with the findings of Poulíčková et al. [105], where, e.g., *Lemnicola hungarica* was shown as an important indicator of high trophy and chlorophyll-a concentrations. These characteristics also depict the high nutrient requirements of the host plants of *Lemnicola hungarica*, in our case *Lemna gibba* and *Spirodela polyrhiza* [109].

A generally low light availability is also illustrated by water transparency (Table S2), which often did not exceed 30 cm at our study sites in the summer months, and PhAR, reaching only negligible values in the depth of 1.5 m; this situation is common in fishponds with adult carp [19]. Although in the depth of 0.3 m, PhAR reached rather high values in some periods, these values did not occur continually throughout the daytime. Thus, light could be an important factor limiting species richness and biomass of submerged aquatic and amphibious plants in the Dehtář fishpond, particularly during the second year of the management cycle, when the water level was high. Although in the first year of the management cycle the areas of shallow water, potentially suitable for submerged vegetation, are fairly large in various parts of the fishpond, the instability of daylight in the water column may be of high importance. Nevertheless, most submerged aquatics and amphiphytes, regularly growing in carp ponds, are rather tolerant of low water transparency and are also known to be able to survive periods of long-term low light availability and high fish stock pressure (i.e., the second year of the management cycle) in the sediment seed bank [19,20,30]. Our data show that these species nearly disappeared from the Dehtář fishpond in 2016 (Table S4, Figure 5) but were present again in 2019. Therefore, the very low species richness of aquatic plants (particularly the submerged ones) and amphiphytes in the Dehtář fishpond is surprising. Only several species of the above functional groups occurred in Dehtář, and only two of them, *Elatine hydropiper* and *Zannichellia palustris*, were rather frequent, forming species-poor stands in shallow waters along the shoreline. Species common in the fishponds in the study region such as *Myriophyllum spicatum* and *Stuckenia pectinata*, which are tolerant of turbid water and high nutrient amounts, did not occur in the Dehtář fishpond at all, and others such as *Potamogeton crispus* and *P. pusillus* were extremely rare here [19,110]. We conclude that light availability is responsible for the interannual population fluctuations of submerged aquatics and amphiphytes, but it is not the decisive factor limiting their overall species richness. It is likely that the propagules of many aquatics are dispersed into the Dehtář fishpond from the surrounding water bodies via waterfowl [111], but site conditions do not enable their establishment. We suppose that the reason for the low species richness of aquatic species could be similar to that identified by Jupp and Spence [42]. They described macrophyte limitation by wave action in a lake shallow water zone, while phytoplankton dominance and associated light shortage restricted macrophyte growth in a deeper water zone protected against the impact of waves.

4.2.3. Moisture and Temperature Fluctuations

Another important factor in wetlands is drought. The strong water level decrease in the Dehtář fishpond in 2015 led to the whole vegetated zone finally being out of the water in the late summer; under such conditions, only amphiphytes survived, while the aquatics died. However, the ability of fishpond macrophytes to survive as seeds or spores in the soil seed bank is usually not eliminated by substrate desiccation [30], as mechanisms of ecological filtering in regularly drained fishponds probably selected species with relevant traits decades or even centuries ago. Nevertheless, in very dry years, the fast desiccation of muddy substrates may be an important factor eliminating not only aquatics and amphiphytes, but also some wetland annuals with high moisture demands (e.g., *Coleanthus subtilis, Eleocharis ovata,* and *Peplis portula;* [18]). In contrast, drought-tolerant wetland annuals such as *Cerastium dubium, Montia arvensis, Myosurus minimus,* and *Spergularia kurkae,* colonising large areas of exposed sandy shores, seem to profit from dry years, although a prolonged dry period may increase the competition by perennials, particularly of clonal grassland species. However, the former three wetland annuals with very early development (March and April) usually make use of a narrow time window between the early beginning of the growing season and the progressive development of grasslands in May. *S. kurkae* is capable of long-term persistence in a soil seed bank [33].

Our analyses revealed that the fishpond summer drainage is one of the most important factors shaping species composition and increasing species richness of wetland and non-wetland vascular plants (see also [19,20]), but also of epiphytic diatoms. We expected that epiphytic diatom communities would be impoverished due to the environmental stress caused by highly unstable moisture during summers and winters (see also [112,113]). However, permanent and temporary wetland habitats usually differ in their diatom species composition, with the taxa dependent on a stable water column or moisture occurring in the former and missing in the latter wetland types [62,114]. Some of the diatom species found in our samples with very low frequency and/or abundance are considered by Evans [114] as species preferring permanent water bodies (e.g., *Melosira varians, Nitzschia acicularis,* and *Gyrosigma acuminatum*) and could be introduced from elsewhere by stream water (see also [52]). In this sense, we did not find analogies between diatoms and aquatic plants: the hydrophytes sensitive to even very short-term water drawdown apparently do not occur in fishponds due to regular drainage. Even the aquatics such as some *Potamogeton* species or *Zannichellia palustris* are supported by a water level drawdown in some of their development phases [115].

While there is a lot of literature on the ecology of vascular plants and charophytes (e.g., [28–30,99,108,115–117]), this is only partly true for diatoms; unresolved taxonomy is one of the reasons why details on the ecology of some diatom species may highly differ between papers [52,105]. An FT2 diatom species *Achnanthidium minutissimum,* recently regarded as species aggregate, is a typical example: while many authors consider it as an indicator of good water quality and, at the same time, tolerant of stress caused by water level fluctuations (e.g., [62,112,118]), we found its highest density in 2016, when the nutrient concentrations were particularly high and the water level was, after the bottom re-flooding, more stable than in 2015 and 2019. Recent studies prove the existence of several ecologically distinct microspecies [105,119], which makes *A. minutissimum* agg. an apparently euryvalent species. We propose that the mechanisms that allow the survival of epiphytic diatoms under the conditions of a temporary water unavailability could be one of the crucial attributes of distinct microspecies within the species aggregates. At the same time, these mechanisms should be seen as important determinants shaping the species composition in temporary wetlands [114,120]. Our data suggest that the diatom taxa such as *Achnanthidium minutissimum* agg., *A. saprophillum,* and *Gomphonema parvulum* agg. are able to tolerate high water level fluctuations but prefer a long phase of flooding (see also [98,121]). Their highest densities were identified at FD, and the only sample with abundant *G. parvulum* agg. from BSB came directly from the Babický brook mouth, which holds water even in the driest years. All these taxa belong to the most frequent dominants of epiphytic diatom assemblages [52,53,105,122,123] where they probably are highly competitive, as we have also shown on ESEM photographs (Figure 14c,d). According to some authors, high water temperatures and nutrient amounts are important for *G. parvulum* agg. [52,124,125], which corresponds to our water chemistry and temperature data (Table S2, Figure S20j).

At the FD site, diatoms were better indicators of habitat conditions than macrophytes, because sites with quickly increasing water levels are only suitable for a limited number of macrophytic species. Nevertheless, the narrow bay at FD is one of the few sites in the littoral zone of the Dehtář fishpond with deep sapropelic sediments and high substrate moisture, predominated by nutrient- and moisture-demanding wetland annuals and perennials (e.g., *Bidens* spp., *Bolboschoenus laticarpus, Persicaria* spp., *Oenanthe aquatica;* Figure S3a) in late summer. This site is also exceptional for its lack of drought-tolerant small wetland annuals and aerotolerant diatoms. The BSB sampling site had the highest share of species with a more frequent occurrence in temporary than in permanent waters, e.g., *Eunotia bilunaris, Nitzschia archibaldii, N. palea* (present there in the highest abundances), and *Navicula reichardtiana*, and species classified as aerotolerant, i.e., highly tolerant of a long phase without flooding, e.g., *Hantzschia abundans, Pinnularia borealis, P. marchia,* and *P. obscura* [58,96,114,120,122]. Particularly the species of the latter group occurred at sites close to the bank. Considering vascular plants, a mosaic of moisture demanding and moisture tolerant species occurred at BSB, as the sites with rather deep mud (segments 7, 8, and 9) and sandy and stony fishpond margins (segments 6, 10, and 11) are situated side-by-side there. This part of the fishpond is very shallow and is the last to be overflooded. Therefore, it is an important refuge for small wetland annuals such as *Coleanthus subtilis* (Figure S7) even in wet years when suitable sites at FD and DSB are flooded early.

The temperature course reasonably modulated the overall species richness of vascular plants of the Dehtář fishpond during the study period. While diurnal temperature fluctuations are crucial for the germination and early development of a summer annual *Coleanthus subtilis*, winter annuals such as *Cerastium dubium, Montia arvensis*, and *Myosurus minimus* have profited from mild winters in the last decade [116] (see also [86] and Table A1 for climatic data). The frequency of *Cyperus fuscus* and *Lythrum hyssopifolia* markedly increased in 2019, the year with extraordinary high summer temperatures but also sufficient precipitation (Table A1 and Table S4). Both of these species are relatively thermophilous and late-germinating [116], but the former has recently been spreading in colder regions such as southern Bohemia [99]. *Lindernia procumbens* (Table S4) appeared only in the summer of 2015, probably as a consequence of extremely hot and dry weather. This highly thermophilous species otherwise mostly survives in the soil seed bank at many of its sites in Central Europe [110,117].

4.2.4. Omitted Role of Wave Action in Fishponds

Wave action is an important factor in large water bodies including fishponds [126]. Among all the large fishponds in the Czech Republic, there are probably only few in which winds causing wave action occur regularly [9]. The size as well as overall volume, water body shape, and position in the landscape are all important predictors of wave action [43,48]. Indirectly, the waves act through the transport of fine sediment particles into deeper and/or calm parts of the water body, thus forming pure sandy or stony beaches in the upper littoral zone [126]. This effect is particularly important for plant species with specific edaphic requirements [48]. The direct impact of wave action on macrophytes proceeds through the uprooting and fragmenting of submerged species, disturbance of the root system of helophytes, burial of plant parts with sand or organic detritus, and decrease in water transparency [42,127]. Macrophytes able to form creeping morphotypes are at an advantage. In the Dehtář fishpond, *Zannichellia palustris, Eleocharis acicularis*, and *Elatine hydropiper* possess the necessary morphological traits and occupy places under a strong wave action [116,128]. Nevertheless, at some sites or in some years, wave-mediated disturbances are probably too strong for any macrophyte species. The substantially lower incidence of strong winds in 2019 (as compared to 2015) probably contributed to the colonisation of the sites that were previously without any submerged vegetation by *Zannichellia palustris*.

For diatoms, the published information on wave mediated disturbance is scarce, but the FT2 species attached to the substrate with mucilaginous stalks, e.g., *Achnanthidium* spp. and *Gompohonema* spp., are reported as more disturbance-sensitive than the adnate FT2 species or the species from other functional groups [98]. This would be in accordance with our results, although it seems to be difficult to separate the influence of wave action and water level fluctuations/water depth in the above species. The year 2016, with the highest and most stable water level, and high *Achnanthidium* spp. and medium high *Gomphonema parvulum* agg. abundances, was rather windy (Table 1, Figure 2 and Figure S22). Our ESEM photographs (Figure 14a,c) of diatom assemblages in reed (*Phragmites australis*) samples from two sites at which the level of exposure to wave actions was different suggest, however, that strong wave-mediated disturbances may have a rather devastating effect on the FT2 stalked taxa. It is also likely that the influence of wave action on the littoral plant and algal communities is different under different water levels [126].

Among vascular plants, even *Phragmites australis*, which is one of the most common helophytes in fishponds [20], forms stand only at those sites of the Dehtář fishpond that are at least partly protected against wave action. Nevertheless, its overall cover is negligible in comparison to other large fishponds without intensive wave-mediated disturbances (see, e.g., [12,13]). Similarly, *Typha latifolia*, very common even in fishponds with high nutrient levels [20,129], occurred in the Dehtář fishpond only as young plants during the exposure of muddy sediments. Although Gaberščik et al. [130] have demonstrated the negative impact of water level fluctuations on *Phragmites australis* in an intermittent lake, studies from fishponds show the positive effect of regular summer drainage on the reed bed regeneration [12], which is why we do not consider this factor to be responsible for the very limited extent of reed beds in the Dehtář fishpond. *Lemnaceae*, known for being very sensitive to streaming and wave action [131], occurred in small quantities only, usually within reed bed stands or in small inflows.

To a certain extent, wave action probably decreases the impact of high nutrient amounts and enables side-by-side coexistence of vascular plant and diatom species with contrasting nutrient demands, increasing the overall plant and diatom species diversity. According to Morris et al. [48], disturbances caused by wave action might even support rare and threatened plants. Also in our case, species that are the rarest in the Czech Republic grow preferentially on shores under strong wave action. In the case of *Montia arensis*, the Dehtář fishpond is one of the four sites in the Czech Republic and a single fishpond locality where this species occurs. *Spergularia kurkae* is more widespread [110], but the population from Dehtář is probably one of the richest in the country. *Cerastium dubium* is considered to be secondary in Southern Bohemia, colonising exclusively the sandy shores of large fishponds there [132].

Among diatoms, Staurosirella leptostauron var. dubia, Pseudostaurosira brevistriata, and P. elliptica are considered as species preferring nutrient poor habitats [55,122], while, e.g., Nitzschia palea, Fistulifera saprophila, and Planothidium frequentissimum are classified as the species of highly eutrophic to hypertrophic water bodies [58,122,133]. Most species on our list, however, occur between the two extremes, e.g., Cocconeis placentula, Navicula capitatoradiata, and Nitzschia archibaldii [122]. While the species with lower nutrient demands seem to be also highly tolerant of wave action and low water level (Figure 13b,c), those preferring highly eutrophic habitats occurred more frequently in places with fine sediment and organic detritus sedimentation and in the absence of wave-mediated disturbances. It is clearly visible not only when comparing findings from individual years and sites with varying disturbance intensities, but also when comparing two sites at FD (2019), differing in disturbance levels and also in quantities of, e.g., Pseudostaurosira spp. on the one hand and Planothidium frequentissimum on the other. Surprisingly, stalked FT2 diatom Gomphonema parvulum agg., predominating most of the 2019 samples, occurred in higher quantities on the more disturbed shore; however, its abundance varied between samples. Species from all the above groups occurred frequently in a single sample, but the samples dominated by species with low nutrient demand/higher disturbance tolerance were generally richer in species and had more balanced species abundances than the samples dominated by species with high nutrient demands/lower disturbance tolerance (Table S6, Figure 8). Analogically, the non-disturbed reed beds, e.g., stands of Phalaris arundinacea (zone 1), a strong competitor preferring nutrient-rich soils [134], were extremely species-poor (data not displayed). The phenomenon of mechanical disturbances and their positive influence on vascular plant species diversity in eutrophic wetlands has been already discussed (e.g., [18,20,68]).

4.3. A Real Biennial Fluctuation or a Trend of Hypertrophy Progression?

Potužák et al. [16] published substantially lower values of TN, TP, and chlorophyll-a concentrations from the Dehtář fishpond for the growing seasons of 2011 and 2012 than was detected during our research. The years 2016 and 2019 in particular show a great increase of trophy levels, potentially associated with the mineralisation of nutrients in the fishpond sediment during bottom exposure in dry summers. There is a high nutrient input related to the fish farming management, the surrounding agricultural landscape, and an outflow from a small sewage plant treating waste water from the surrounding settlements, but these factors have been constant over at least two decades. We suggest that sediment desiccation accompanied by nutrient mineralisation and the subsequent nutrient release into a water column may accelerate the influence of the above factors. Such a situation occurred at the Dehtář fishpond after the long dry phase in 2015 and the spring of 2016, followed by a water level increase in the summer of 2016. However, also a continual release of phosphorus from the submerged sediment due to high water temperatures is likely, as summer water temperatures measured by automatic sensor stations often reached 30 °C (see, e.g., [135,136] for the description of these processes).

There are no data available for an earlier period (2011–2012) on epiphytic diatom and macrophyte assemblages of the Dehtář fishpond. Therefore, we cannot be sure if the supposed continual increase in trophic levels in the Dehtář fishpond already caused sub-recent associated change in species composition of biotic communities or not. Only scarce data on selected vascular plant species of exposed pond bottoms were published several decades ago [69,70]. Thanks to these papers we know that, e.g., *Juncus tenageia* and *Pseudognaphalium luteoalbum*, the wetland annuals of nutrient-poor habitats [46,116] not found during our research, grew in the Dehtář fishpond in the past, but we have no information on what exactly preceded the disappearance of these species. On the one hand, potential hypertrophy-related changes may be very fast, particularly in algal assemblages [85]. On the other hand, frequent wind-mediated disturbances might have reduced the relative importance of nutrients in the hypertrophic ecosystem. Moreover, a parametes that could seem to be a consequence of recent global changes, namely a rather frequent prolonged summer drainage due to summer drought, has, according to our information, been occurring in the Dehtář fishpond for at least several decades [67]. Nevertheless, manuring and fish feeding based on the real needs of this aquacultural system would not only reduce the nutrient load but would also save money for the fish farm, as a large part of these inputs are not involved in the food chain and thus cannot be reflected in the fish production [37,68].

4.4. Diatoms and Macrophytes—Do Direct Relationships Exist?

Most published papers confirm the so-called "neutral substrate hypothesis", i.e., the absence of diatom assemblages strictly specific to a particular macrophyte species (plant substrate; [137]); our own early study from the Dehtář fishpond brought similar results [55]. On the other hand, we cannot say that there are no relationships between epiphytic diatom assemblages and their host plants. We proved that the ecological and life-history traits of the host plant species are an important factor influencing species richness, structure, and species composition of associated diatom assemblages. In some cases, the whole vegetation zone probably functions as a refuge for a certain epiphytic diatom species pool. This fact is reflected in the high similarity in diatom assemblages of the samples from different macrophyte species collected within a few square meters in a single zone. The importance of vegetation zonation for diatom assemblages is also mentioned by, e.g., Letáková et al. [60].

On the other hand, sometimes even the samples collected close to each other within a single zone showed very important differences in epiphytic diatom assemblages. For instance, our analyses revealed substantial differences between the samples from *Phragmites australis* and *Phalaris arundinacea* collected during the same sampling event at BSB. Similarly, diatom assemblages recorded in the samples from *Spirodela polyrhiza* and *Lemna gibba* differed markedly from those from reed bed species growing in the same stand: while *Lemnaceae* were densely colonised by a specialised diatom *Lemnicola hungarica*, only rarely occurring elsewhere [138], reed bed species samples were mainly predominated by *Achnanthidium* spp. and *Cocconeis placentula*. It is supposed that plant macro-morphology (i.e., stem branching, leave dissection) and surface microstructure (e.g., trichomes, spikes) have a selective effect on diatom assemblages, where e.g., species of a certain cell size, shape, or type of attachment may better cope with conditions offered by particular macrophytes [53,60].

The filtering of particular diatom size categories by macrophytes occurred also in some of our samples: while small (up to 12.5 μm) and middle-sized (up to 37 μm) diatoms such as *Achnanthidium* spp. predominated at FD in 2016, large species (>40–45 μm) such as *Pinnularia* spp. and *Ulnaria* spp. were common at BSB in 2016. In 2015, there was a mixture of medium-sized (e.g., *Staurosira leptostauron*) and large diatoms (e.g., *Craticula cuspidata*) at all the study sites. While the large and medium-sized diatoms seemed to be related particularly to small growing wetland annuals, shortly perennial species, and submerged plants, the small-sized species were more common on large wetland perennials (with the exception of *Lemnicola hungarica* on *Lemnaceae*). As among the small-sized species, the attached periphytic taxa dominated, one of the possible explanations might be that the periphytes need to attach to a stable structure in order to be able to persist in conditions of strong wave action. Remarkably, in 2019 when hydrological conditions at the time of diatom sampling were similar to those in 2015, but the year was markedly less windy than 2015 (Table 1, Figure 2 and Figure S22b,d), small-sized species such as *Gomphonema parvulum* occurred in high densities in FD samples from various types of macrophytes, mainly the non-reed bed species (Figure 14d, Table S6). *G. parvulum* is considered to be a species highly sensitive to mechanical disturbances due to its long stalks [139], and thus the course of the weather in 2019 with less frequent wave action could be the decisive factor in colonising the potentially unstable substrate such as *Zannichellia palustris*. On the other hand, epipelic species of the *Navicula* genus (FT5) were among the large-sized diatoms. Their higher species richness and/or quantities in the samples of low-growing macrophytes may be related to their occurrence on the surface of muddy sediments and to the active movement between the sediment and the plants [140]. This strategy would enable the diatoms to profit from their host plants by, e.g., gaining better access to light (the diatom may move upwards on the plant).

4.5. Importance of Research Intensity

After the end of the research grant in 2016, our data sampling in the Dehtář fishpond was limited. Therefore, the data on vascular plants after 2017 are rather incomplete, as we preferentially recorded taxa in the second and third zones, because many of them are of high conservation value. Despite this incompleteness, we included the data from 2019 when we also collected a limited number of diatom samples. These data serve partly for comparison with data from the research period of 2014–2016, but they particularly highlight the necessity of more intensive research in large water bodies. Although due to higher substrate moisture, the conditions in 2019 were even more suitable to plant development than in the very dry year 2015, during three short visits in 2019, we recorded less than 1/2 of the vascular plant and charophyte species listed for 2015 (Table S4). We excluded some difficult-to-access segments at the Dehtářský stream inflow from our 2019 research; however, this part of the Dehtář fishpond hosts only a few species that only grow there. Thus, it seems that a substantially higher time investment would be necessary to obtain data fully comparable with the research grant period.

For diatoms, the number of samples is crucial. There was a similarly significant decrease in the total species number in 2019 as compared to 2015 and especially to 2016 (the highest amounts of samples). However, the most important reduction of information was related to the large-scale distribution patterns of diatoms, because all 2019 samples came only from FD. Another possible source of bias in diatom data is related to differences in the zones sampled in particular years: while in 2015 and 2019, all the samples were collected on the transition between the second and the third zone, in 2016, the first and the second zones were sampled. Therefore, somewhat different sets of diatom substrates were sampled each year. The sampling of the same plant species at each site and each year would allow us to formulate more general conclusions; however, it was not realistic due to the differences in macrophytic species composition between the sites and years and because not all of the plant species occurring at particular sites were always available growing in water. Thus, some rare macrophyte species could be collected only once. In some cases, our sample was probably the first one of the given macrophyte species (e.g., *Veronica scutellata*, *Alopecurus geniculatus*) that was ever analysed for epiphytic diatoms.

5. Conclusions

Macrophytes and diatoms exhibit similar responses to some of the basic, usually locally acting environmental factors, such as water depth, water level fluctuations, nutrient amounts, water transparency, type of bottom sediment, occurrence of organic detritus, morphology of littoral zone, and mechanical disturbance intensity. Locally as well as regionally, important environmental and management-related factors such as the yearly course of weather conditions, type of management (e.g., fish farming), and landscape structure usually act in close synergy.

Our study of the Dehtář fishpond offers a complex view of multiple factors and their interactions within a large hypertrophic water body. For instance, wind-mediated wave action is related to specific weather conditions, landscape structure, and fishpond morphology, but the controlled manipulation of the water level may largely modify its influence on macrophyte and diatom species in particular littoral habitats. Similarly, the timing and intensity of water level fluctuations in a water body are basically determined by weather conditions, the hydrology of a local catchment basin, and the type of surrounding landscape. In fishponds specifically, however, the overall hydrological conditions may also be greatly modified by an intended drainage before fish harvesting and the continual filling of a fishpond after restocking. Due to the synergy between hydrological, climatic, and management factors, the functioning of the littoral zones of these fishponds resembles temporary wetlands, as we have also shown with our complex data. Our results also reveal other stress factors, such as the eutrophication and mechanical disturbances with a joint effect on water transparency and, consecutively, on macrophyte and microalgal assemblages. Yet, the Dehtář fishpond is an important hotspot of wetland biodiversity and a refuge of threatened taxa.

The study period was characteristic of various extremes in climate and related local factors such as water level and substrate moisture fluctuations, with associated responses in plant and diatom assemblages. As such extremes are supposed to be more frequent in the future due to the progressing global climate change, our study from the Dehtář fishpond may be considered as the basis for predictive model constructions in similar hypertrophic water bodies, including for proposals of mitigation measures preventing accelerated habitat deterioration.

Supplementary Materials: The following are available online at https://www.mdpi.com/article/10.3390/w13111569/s1: Figures S1–S15: Photographs documenting the environment, selected vascular plant species and their assemblages, and fish farming practices in the Dehtář fishpond, Figures S16–S22: Line graphs documenting a course of selected environmental factors in the Dehtář fishpond, Table S1: Summary of the factors used in NMDS ordinations of macrophyte and diatom species assemblages; Table S2: Basic physico-chemical parameters of water; Table S3: List of epiphytic diatom samples; Table S4: List of vascular plant species and charophytes along with their frequencies, quantities and classification to species groups; Table S5: Frequency of occurrence and cumulative numbers of diatom taxa in the samples from the Dehtář fishpond; Table S6: List of diatom taxa found in particular samples of epiphyton of the Dehtář fishpond.

Author Contributions: Conceptualisation, K.Š., M.F. (Markéta Fránková), O.V. and J.P.; methodology, K.Š., M.F. (Markéta Fránková), O.V. and J.P.; formal analysis, O.V., K.Š. and M.F. (Markéta Fránková); investigation, M.D., K.Š., M.F. (Martina Fabšičová), M.F. (Markéta Fránková) and J.P.; resources, K.Š.,

M.D., J.P., M.F. (Martina Fabšičová) and M.F. (Markéta Fránková); data curation, K.Š., M.D., M.F. (Markéta Fránková) and J.P.; writing—original draft preparation, K.Š, M.F. (Markéta Fránková), and O.V.; writing—review and editing, M.F. (Martina Fabšičová), J.P. and M.D.; visualisation, O.V., M.F. (Markéta Fránková), K.Š., M.D. and J.P.; supervision, K.Š.; project administration, K.Š.; funding acquisition, K.Š., M.F. (Markéta Fránková) and M.D. All authors have read and agreed to the published version of the manuscript.

Funding: This study is a result of the research grant nr. LD 14045 provided by the Ministry of Education, Youth, and Sports of the Czech Republic in the frame of the research program COST CZ, to K. Šumberová and financing the involvement of M. Fabšičová, M. Fránková, M. Ducháček, and J. Potužák. The research was related to the goals of overarching COST Action NETLAKE (ES 1201). M. Fabšičová, M. Fránková, K. Šumberová, and O. Vild were further supported by the Institute of Botany of the Czech Academy of Sciences under the long-term research development project no. RVO 67985939. M. Ducháček was supported by the National Museum under a long-term research development project no. 00023272 (National Museum, DKRVO 2019–2023/4.I.c) of the Ministry of Culture of the Czech Republic.

Institutional Review Board Statement: Not applicable.

Informed Consent Statement: Not applicable.

Data Availability Statement: Data are stored within the documentation of LD 14045 project. Data are available upon a reasonable request.

Acknowledgments: We would like to thank the colleagues who revised the herbarium specimens of taxonomically complex or difficult-to-identify taxa: Jiří Danihelka (*Epilobium*), Zdeněk Kaplan (*Lemna, Potamogeton, Zannichellia*), Jan Prančl (*Batrachium, Callitriche*), and Jitka Štěpánková (*Bidens*). Our great thanks go to the staff of the fish farm Rybářství Hluboká nad Vltavou Cz, s.r.o., for the logistical support of our research and information about management measures applied in the Dehtář fishpond, and to Karel Císař who kindly controlled and cleaned the sensors of our automatic measuring stations. We are very grateful to the Vltava River Authority, state enterprise, for providing us with the data from their own monitoring of the Dehtář fishpond, and to Markéta Letáková for sharing her detailed unpublished data on epiphytic diatom assemblages from other Czech fishponds for comparative purposes. We are thankful to Vilém Neděla and Eva Tihlaříková from the Environmental Electron Microscopy Group, Institute of Scientific Instruments of the Czech Academy of Sciences, Brno, for taking ESEM photographs of epiphyton in 2016 (Figure 14a–c). Our thanks for linguistic correction go to Mai Havrdová Fathi. We are also thankful to three anonymous reviewers and the Academic Editor for valuable comments to an earlier version of the manuscript.

Conflicts of Interest: The authors declare no conflict of interest. The funders had no role in the design of the study; in the collection, analyses, or interpretation of data; in the writing of the manuscript; or in the decision to publish the results.

Appendix A

Table A1. Monthly sums of precipitation and temperature means at Dehtář, based on the data from automatic sensor stations, or the closest meteorological station (values with *) in České Budějovice-Rožnov (CHMI 2021). Precipitation sum and mean temperature averages for the whole South Bohemian region in the study period and long-term normals (1981–2010) are given for a better overview of meteorological situation in individual months.

Month	Precipitation—Dehtář (mm)	Precipitation—S Bohemia (mm; Average)	Precipitation Normal—S Bohemia (mm; Average)	Temperatures—Dehtář (°C)	Temperature—S Bohemia (°C; Average)	Temperature Normal—S Bohemia
April 2014	21.9 *	35	41	10.6 *	9.0	7.2
May 2014	116.2 *	126	71	12.8 *	11.3	12.5
June 2014	41.4	33	85	17.6 *	15.7	15.3
July 2014	128.8	120	92	19.7 *	18.3	17.3
August 2014	139.2	104	85	16.3	15.0	16.7
April 2015	17.4	28	41	8.1	7.2	7.2
May 2015	59.1 *	64	71	13.0	12.0	12.5
June 2015	54	68	85	16.8	15.7	15.3
July 2015	41.4	30	92	21.1	20.0	17.3
August 2015	31.6	42	85	21.0	20.5	16.7
April 2016	33.6 *	35	41	8.8 *	7.1	7.2
May 2016	108.8 *	95	71	13.9 *	12.5	12.5
June 2016	88.8	94	85	17.6	16.3	15.3
July 2016	177.7	143	92	19.5	18.1	17.3
August 2016	90.7	35	85	17.9	16.4	16.7
April 2019	10.4 *	16	41	10.3 *	8.6	7.2
May 2019	74.5 *	85	71	11.5 *	9.9	12.5
June 2019	85.5 *	69	85	21.9 *	20.0	15.3
July 2019	104.6 *	69	92	20.5 *	18.6	17.3
August 2019	76.1 *	70	85	20.1 *	18.3	16.7

References

1. Sand-Jensen, K.; Riis, T.; Vestergaard, O.; Larsen, S.E. Macrophyte decline in Danish lakes and streams over the past 100 years. *J. Ecol.* **2000**, *88*, 1030–1040. [CrossRef]
2. Declerck, S.; Vandekerkhove, J.; Johansson, L.; Muylaert, K.; Conde-Porcuna, J.M.; Van der Gucht, K.; Pérez-Martínez, C.; Lauridsen, T.; Schwenk, K.; Zwart, G.; et al. Multi-group biodiversity in shallow lakes along gradients of phosphorus and water plant cover. *Ecology* **2005**, *86*, 1905–1915. [CrossRef]
3. Alahuhta, J.; Rosbakh, S.; Chepinoga, V.; Heino, J. Environmental determinants of lake macrophyte communities in Baikal Siberia. *Aquat. Sci.* **2020**, *82*, 39. [CrossRef]
4. Eriksson, F.; Hornstrom, E.; Mossberg, P.; Nyberg, P. Ecological effects of lime treatment of acidified lakes and rivers in Sweden. *Hydrobiologia* **1983**, *101*, 145–164. [CrossRef]
5. Murphy, K.J. Plant communities and plant diversity in softwater lakes of northern Europe. *Aquat. Bot.* **2002**, *73*, 287–324. [CrossRef]
6. Bakker, E.S.; Van Donk, E.; Immers, A.K. Lake restoration by in-lake iron addition: A synopsis of iron impact on aquatic organisms and shallow lake ecosystems. *Aquat. Ecol.* **2016**, *50*, 121–135. [CrossRef]
7. Hilt, S.; Alirangues Nuñez, M.M.; Bakker, E.S.; Blindow, I.; Davidson, T.A.; Gillefalk, M.; Hansson, L.-A.; Janse, J.H.; Janssen, A.B.G.; Jeppesen, E.; et al. Response of submerged macrophyte communities to external and internal restoration measures in north temperate shallow lakes. *Front. Plant Sci.* **2018**, *9*, 194. [CrossRef]
8. Jeppesen, E.; Kronvang, B.; Meerhoff, M.; Søndergaard, M.; Hansen, K.M.; Andersen, H.E.; Lauridsen, T.L.; Liboriussen, L.; Beklioglu, M.; Özen, A.; et al. Climate change effects on runoff, catchment phosphorus loading and lake ecological state, and potential adaptations. *J. Environ. Qual.* **2009**, *38*, 1930–1941. [CrossRef]
9. Liebscher, P.; Rendek, J. *Rybníky České Republiky*; Academia: Prague, Czech Republic, 2014; pp. 1–584.
10. Hoffmann, R.C. Economic development and aquatic ecosystems in medieval Europe. *Am. Hist. Rev.* **1996**, *101*, 631–669. [CrossRef]
11. Turoňová, D. (Ed.) Národní přírodní rezervace Novozámecký rybník: Přírodovědecké průzkumy a péče o chráněné území. *Příroda* **2002**, *20*, 1–176.
12. Zákravský, P.; Hroudová, Z. Vliv řízeného rybničního managementu na obnovu rákosin v NPR Velký a Malý Tisý. *Zprávy Čes. Bot. Společ.* **2007**, *42 Mater. 22*, 167–196.
13. Miklín, J.; Macháček, P. Birds of Lednické rybníky Fishponds (Czech Republic). *J. Maps* **2016**, *12* (Suppl. 1), 239–248. [CrossRef]
14. Meerhoff, M.; Jeppesen, E. Shallow lakes and ponds. In *Encyclopedia of Inland Waters*; Likens, G.E., Ed.; Elsevier: Oxford, UK, 2009; Volume 2, pp. 645–655.
15. Søndergaard, M.; Johansson, L.S.; Levi, E.E.; Lauridsen, T.L.; Jeppesen, E. Lake types and their definition: A case study from Denmark. *Inland Waters* **2020**, *10*, 227–240. [CrossRef]
16. Potužák, J.; Duras, J.; Drozd, B. Mass balance of fishponds: Are they sources or sinks of phosphorus? *Aquac. Int.* **2016**, *24*, 1725–1745. [CrossRef]
17. Rutegwa, M.; Potužák, J.; Hejzlar, J.; Drozd, B. Carbon metabolism and nutrient balance in a hypereutrophic semiintensive fishpond. *Knowl. Manag. Aquat. Ecosyst.* **2019**, *420*, 49. [CrossRef]
18. Šumberová, K.; Lososová, Z.; Fabšicová, M.; Horáková, V. Variability of vegetation of exposed pond bottoms in relation to management and environmental factors. *Preslia* **2006**, *78*, 235–252.
19. Francová, K.; Šumberová, K.; Kučerová, A.; Čtvrtlíková, M.; Šorf, M.; Borovec, J.; Drozd, B.; Janauer, G.A.; Vrba, J. Macrophyte assemblages in fishponds under different fish farming management. *Aquat. Bot.* **2019**, *159*, 103131. [CrossRef]
20. Francová, K.; Šumberová, K.; Kučerová, A.; Šorf, M.; Grill, S.; Exler, N.; Vrba, J. Drivers of plant species composition of ecotonal vegetation in two fishpond management types. *Wetl. Ecol. Manag.* **2021**, *29*, 93–100. [CrossRef]
21. Schäfer-Guignier, O. Weiher in der Franche-Comté: Eine floristisch-ökologische und vegetationskundliche Untersuchung. *Diss. Bot.* **1994**, *213*, 1–239.
22. Wezel, A.; Oertli, B.; Rosset, V.; Arthaud, F.; Leroy, B.; Smith, R.; Angélibert, S.; Bornette, G.; Vallod, D.; Robin, J. Biodiversity patterns of nutrient-rich fish ponds and implications for conservation. *Limnology* **2014**, *15*, 213–223. [CrossRef]
23. Bubíková, K.; Hrivnák, R. Artificial ponds in Central Europe do not fall behind the natural ponds in terms of macrophyte diversity. *Knowl. Manag. Aquat. Ecosyst.* **2018**, *419*, 8. [CrossRef]
24. Waldon, B. The conservation of small water reservoirs in the Krajeńskie Lakeland (North-West Poland). *Limnologica* **2012**, *42*, 320–327. [CrossRef]
25. Scheffer, M.; van Nes, E.H. Shallow lakes theory revisited: Various alternative regimes driven by climate, nutrients, depth and lake size. *Hydrobiologia* **2007**, *584*, 455–466. [CrossRef]
26. Waldon-Rudzionek, B. Is the flora of oxbow lakes different from that of fishponds? A comparison of two types of water reservoirs in the Noteć River valley and Bydgoszcz Canal valley (NW Poland). *Ecol. Quest.* **2017**, *25*, 27–40. [CrossRef]
27. Broyer, J.; Curtet, L. Biodiversity and fish farming intensification in French fishpond systems. *Hydrobiologia* **2012**, *694*, 205–218. [CrossRef]
28. Müller-Stoll, W.R.; Pietsch, W. Ökologische Untersuchungen über die Gesellschaft des *Eleocharito-Caricetum bohemicae* auf wasserfrei gewordenen Teichböden in Zentraleuropa. *Verh. Zool. Bot. Ges. Osterr.* **1985**, *123*, 51–70.

29. Richert, E.; Achtziger, R.; Dajdok, Z.; Günther, A.; Heilmeier, H.; Hübner, A.; John, H.; Šumberová, K. Rare wetland grass *Coleanthus subtilis* in Central and Western Europe—Current distribution, habitat types, and threats. *Acta Soc. Bot. Pol.* **2016**, *85*, 1–16. [CrossRef]
30. Poschlod, P.; Rosbakh, S. Mudflat species: Threatened or hidden? An extensive seed bank survey of 108 fish ponds in Southern Germany. *Biol. Conserv.* **2018**, *225*, 154–163. [CrossRef]
31. Zając, M.; Zając, A. Zbiorowiska z klasy *Isoëto-Nanojuncetea* na dnach wysychających stawów w południowej części Kotliny Oświęcimskiej. *Zesz. Nauk. Uniw. Jagiell. Pr. Bot.* **1988**, *17*, 155–160. (In Polish)
32. Bernhardt, K.-G.; Koch, M.; Kropf, M.; Ulbel, E.; Webhofer, J. Comparison of two methods characterising the seed bank of amphibious plants in submerged sediments. *Aquat. Bot.* **2008**, *88*, 171–177. [CrossRef]
33. Kúr, P.; Pachschwöll, C.; Štech, M. Notes on the distribution of *Spergularia echinosperma* and the newly recognized species *S. kurkae* in Austria. *Neilreichia* **2018**, *9*, 269–282.
34. Lemmens, P.; Mergeay, J.; De Bie, T.; Van Wichelen, J.; De Meester, L.; Declerck, S.A. How to maximally support local and regional biodiversity in applied conservation? Insights from pond management. *PLoS ONE* **2013**, *8*, e72538. [CrossRef] [PubMed]
35. Hroudová, Z.; Zákravský, P. Vegetation dynamics in a fishpond littoral related to human impact. *Hydrobiologia* **1999**, *415*, 139–145. [CrossRef]
36. Grzywna, A.; Sender, J.; Bronowicka-Mielniczuk, U. Physical and chemical variables, species composition and coverage of macrophytes in ponds (case study in Eastern Poland). *Appl. Ecol. Environ. Res.* **2018**, *16*, 2129–2139. [CrossRef]
37. Potužák, J.; Hůda, J.; Pechar, L. Changes in fish production effectivity in eutrophic fishponds—Impact of zooplankton structure. *Aquac. Int.* **2007**, *15*, 201–210. [CrossRef]
38. Duras, J.; Potužák, J.; Marcel, M.; Pechar, L. Rybníky a jakost vody. *Vodní Hospodářství* **2015**, *7*, 16–24.
39. Potužák, J.; Šumberová, K.; Fránková, M.; Fabšičová, M.; Ducháček, M.; Císař, K.; Duras, J. Tříleté dobrodružství výzkumu rybníka Dehtář aneb s rybníkem mezi "jezerní" badatele projektu NETLAKE/Three years long investigation of fishpond Dehtář—First fishpond in the NETLAKE COST ACTION project. In *Rybníky 2017, Sborník Příspěvků Odborné Konference, 15–16 June 2017, Česká Zemědělská Univerzita, Praha, Czech Republic*; David, V., Davidová, T., Eds.; Česká Společnost Krajinných Inženýrů (ČSKI): Prague, Czech Republic, 2017; pp. 73–81.
40. Kubelka, V.; Havlíček, J.; Mikeš, V. (Eds.) Zajímavá a vzácná ornitologická pozorování v jižních Čechách IV. *Sbor. Jihočes. Muz. Čes. Budějovicích, Přír. Vědy* **2017**, *57*, 99–125.
41. Hroudová, Z.; Kloubec, B.; Zákravský, P. Dynamika litorální vegetace a avifauny Opatovického rybníka u Třeboně. *Příroda* **1999**, *14*, 73–98.
42. Jupp, B.P.; Spence, D.H.N. Limitations on macrophytes in a eutrophic lake, Loch Leven: II. Wave action, sediments and waterfowl grazing. *J. Ecol.* **1977**, *65*, 431–446. [CrossRef]
43. Azza, N.; van de Koppel, J.; Denny, P.; Kansiime, F. Shoreline vegetation distribution in relation to wave exposure and bay characteristics in a tropical great lake, Lake Victoria. *J. Trop. Ecol.* **2007**, *23*, 353–360. [CrossRef]
44. Keddy, P.A. Quantifying within-lake gradients of wave energy: Interrelation- ships of wave energy, substrate particle size and shoreline plants in Axe Lake, Ontario. *Aquat. Bot.* **1982**, *14*, 41–55. [CrossRef]
45. Kolář, J.; Kučerová, A.; Jakubec, P.; Vymazal, J. Seed bank of *Littorella uniflora* (L.) Asch. in the Czech Republic, Central Europe: Does burial depth and sediment type influence seed germination? *Hydrobiologia* **2017**, *794*, 347–358. [CrossRef]
46. Lepší, M.; Boublík, K.; Štech, M.; Hans, V. (Eds.) *Červená kniha květeny jižní části Čech*; Jihočeské muzeum v Českých Budějovicích: České Budějovice, Czech Republic, 2013; pp. 235–236, 333. (In Czech)
47. Jupp, B.P.; Spence, D.H.N. Limitations on macrophytes in a eutrophic lake, Loch Leven: I. Effects of phytoplankton. *J. Ecol.* **1977**, *65*, 175–186. [CrossRef]
48. Morris, P.A.; Hill, N.M.; Reekie, E.G.; Hewlin, H.L. Lakeshore diversity and rarity relationships along interacting disturbance gradients: Catchment area, wave action and depth. *Biol. Conserv.* **2002**, *106*, 79–90. [CrossRef]
49. Poulíčková, A.; Hájková, P.; Křenková, P.; Hájek, M. Distribution of diatoms and bryophytes on linear transects through spring fens. *Nova Hedwigia* **2004**, *78*, 411–424. [CrossRef]
50. Fránková, M.; Bojková, J.; Poulíčková, A.; Hájek, M. The structure and species richness of the diatom assemblages of the Western Carpathian spring fens along the gradient of mineral richness. *Fottea* **2009**, *9*, 355–368. [CrossRef]
51. Marvan, P.; Komárek, J.; Ettl, H.; Komárková, J. Structure and functioning of algal communities in fishponds. 4.1. Structural Elements. Principal Populations of Algae. Spatial Distribution. In *Pond Littoral Ecosystems. Structure and Functioning. Methods and Results of Quantitative Ecosystem Research in the Czechoslovakian IBP Wetland Project*; Dykyjová, D., Květ, J., Eds.; Springer: Berlin/Heidelberg, Germany; New York, NY, USA, 1978; pp. 295–313.
52. Kollár, J.; Fránková, M.; Hašler, P.; Letáková, M.; Poulíčková, A. Epiphytic diatoms in lotic and lentic waters—Diversity and representation of species complexes. *Fottea* **2015**, *15*, 259–271. [CrossRef]
53. Mutinová, P.T.; Neustupa, J.; Bevilacqua, S.; Terlizzi, A. Host specificity of epiphytic diatom (Bacillariophyceae) and desmid (Desmidiales) communities. *Aquat. Ecol.* **2016**, *50*, 697–709. [CrossRef]
54. Letáková, M.; Poulíčková, A. Epifytické rozsivky vybraných rybníků v České republice. *Zpr. Vlastiv. Muz. Olomouci* **2017**, *313*, 44–58.
55. Fránková, M.; Šumberová, K.; Potužák, J.; Vild, O. The role of plant substrate type in shaping the composition and diversity of epiphytic diatom assemblages in a eutrophic reservoir. *Fundam. Appl. Limnol.* **2017**, *189*, 117–135. [CrossRef]

56. Caput, K.; Plenković-Moraj, A. Epiphytic diatoms on sawgrass (*Cladium mariscus*) in the karstic Plitvice lakes (Croatia). *Biologia* **2000**, *55*, 343–350.
57. Cejudo-Figueiras, C.; Álvarez-Blanco, I.; Bécares, E.; Blanco, S. Epiphytic diatoms and water quality in shallow lakes: The neutral substrate hypothesis revisited. *Mar. Freshwater Res.* **2010**, *61*, 1457–1467. [CrossRef]
58. Kapetanović, T.; Jahn, R.; Redžić, S.; Carić, M. Diatoms in a poor fen of Bijambare protected landscape, Bosnia & Herzegovina. *Nova Hedwigia* **2011**, *93*, 125–151.
59. Blanco, S.; Cejudo-Figueiras, C.; Álvarez-Blanco, I.; van Donk, E.; Gross, E.M.; Hansson, L.-A.; Irvine, K.; Jeppesen, E.; Kairessalo, T.; Moss, B.; et al. Epiphytic diatoms along environmental gradients in western European shallow lakes. *CLEAN Soil Air Water* **2014**, *42*, 229–235. [CrossRef]
60. Letáková, M.; Cantonati, M.; Hašler, P.; Nicola, A.; Poulíčková, A. Substrate specificity and fine-scale distribution of epiphytic diatoms in a shallow tarn in the Brenta Dolomites (south-eastern Alps). *Plant Ecol. Evol.* **2016**, *149*, 144–156. [CrossRef]
61. Letáková, M.; Fránková, M.; Poulíčková, A. Ecology and applications of freshwater epiphytic diatoms—Review. *Cryptogam. Algol.* **2018**, *39*, 3–22. [CrossRef]
62. Evtimova, V.V.; Donohue, I. Quantifying ecological responses to amplified water level fluctuations in standing waters: An experimental approach. *J. Appl. Ecol.* **2014**, *51*, 1282–1291. [CrossRef]
63. Česká Geologická Služba. Geovědní Mapy 1:25 000. Available online: https://mapy.geology.cz/geocr25 (accessed on 25 March 2021).
64. Geoportal Inspire. Půdní Mapa ČR 1:250 000—Klasifikace dle TKSP a WRB. Available online: https://geoportal.gov.cz/web/guest/map (accessed on 25 March 2021).
65. Organization for Economic Cooperation and Development (OECD). *Eutrophication of Waters. Monitoring, Assessment and Control*; Final Report; OECD Cooperative Programme on Monitoring of Inland Waters (Eutrophication Control); OECD: Paris, France, 1982; pp. 1–154.
66. Hartman, P.; Přikryl, I.; Štědronský, E. *Hydrobiologie*; Informatorium: Prague, Czech Republic, 2005; pp. 1–359.
67. Klepal, V.; Hluboká nad Vltavou Fish Farm, Hluboká nad Vltavou, Czech Republic. Personal communication, 2016.
68. Francová, K.; Šumberová, K.; Janauer, G.A.; Adámek, Z. Effects of fish farming on macrophytes in temperate carp ponds. *Aquac. Int.* **2019**, *27*, 413–436. [CrossRef]
69. Jílek, B. K fytocenologii rybničních společenstev. *Preslia* **1956**, *28*, 66–77.
70. Gazda, J. Příspěvek k fytocenologii jihočeských rybníků. *Sborn. Kraj. Vlastiv. Muz. České Budějovice Přír. Vědy* **1958**, *1*, 64–76.
71. Knižátková, E.; Lacina, D. Souhrny doporučených opatření pro ptačí oblasti. *Ochr. Přír.* **2010**, *10*, 12–14.
72. Thiers, B.M. *Index Herbariorum*; The New York Botanical Garden: Bronx, NY, USA; Available online: http://sweetgum.nybg.org/science/ih/ (accessed on 20 March 2021).
73. Taylor, J.C.; Harding, W.R.; Archibald, C.G.M. *A Methods Manual for Collection, Preparation and Analysis of Diatom Samples*; Version 1.0; WRC Report TT 281/07; Water Research Commission: Gezina, South Africa, 2007; pp. 1–49. Available online: http://docs.niwa.co.nz/library/public/1770054839.pdf (accessed on 14 March 2016).
74. Krammer, K.; Lange-Bertalot, H. Bacillariophyceae: Naviculaceae. In *Süßwasserflora von Mitteleuropa*; Ettl, H., Gerloff, J., Heynig, H., Mollenhauer, D., Eds.; Gustav Fischer Verlag: Jena, Germany, 1986; 1. Teil; pp. 1–876.
75. Krammer, K.; Lange-Bertalot, H. Bacillariophyceae: Epithemiaceae, Surirellaceae. In *Süßwasserflora von Mitteleuropa*; Ettl, H., Gerloff, J., Heynig, H., Mollenhauer, D., Eds.; Gustav Fischer Verlag: Jena, Germany, 1988; 2. Teil; pp. 1–876.
76. Krammer, K.; Lange-Bertalot, H. Bacillariophyceae: Centrales, Fragilariaceae, Eunotiaceae. In *Süßwasserflora von Mitteleuropa*; Ettl, H., Gerloff, J., Heynig, H., Mollenhauer, D., Eds.; Gustav Fischer Verlag: Jena, Germany, 1991; 3. Teil; pp. 1–576.
77. Krammer, K.; Lange-Bertalot, H. Bacillariophyceae: Achnanthaceae, Kritische Ergänzungen zu Navicula (Lineolatae), Gomphonema. In *Süßwasserflora von Mitteleuropa*; Ettl, H., Gerloff, J., Heynig, H., Mollenhauer, D., Eds.; Gustav Fischer Verlag: Jena, Germany, 1991; 4. Teil; pp. 1–437.
78. Krammer, K. The Genus Pinnularia. In *Diatoms of Europe*; Lange-Bertalot, H., Ed.; A. R. G. Gantner Verlag K. G.: Rugell, Switzerland, 2000; Volume 1, pp. 1–703.
79. Krammer, K. Cymbella. In *Diatoms of Europe*; Lange-Bertalot, H., Ed.; A. R. G. Gantner Verlag K. G.: Rugell, Switzerland, 2002; Volume 3, pp. 1–584.
80. Krammer, K. Cymbopleura, Delicata, Navicymbula, Gomphocymbellopsis, Afrocymbella Supplements to cymbelloid taxa. In *Diatoms of Europe*; Lange-Bertalot, H., Ed.; A. R. G. Gantner Verlag K. G.: Rugell, Switzerland, 2003; Volume 4, pp. 1–530.
81. Lange-Bertalot, H. Navicula sensu stricto, 10 genera separated from Navicula sensu stricto, Frustulia. In *Diatoms of Europe*; Lange-Bertalot, H., Ed.; A. R. G. Gantner Verlag K. G.: Rugell, Switzerland, 2001; Volume 2, pp. 1–526.
82. Hofmann, G.; Werum, M.; Lange-Bertalot, H. *Diatomeen im Süßwasserbenthos von Mitteleuropa. Bestimmungsflora Kieselalgen für Die Ökologische Praxis. Über 700 der Häufigsten Arten und ihre Ökologie*; Koeltz Scientific Books: Königstein, Germany, 2013; pp. 1–908.
83. Lange-Bertalot, H.; Hofmann, G.; Werum, M.; Cantonati, M. *Freshwater Benthic Diatoms of Central Europe: Over 800 Common Species Used in Ecological Assessment*; English Edition with Updated Taxonomy and Added Species; Koeltz Botanical Books: Schmitten-Oberreifenberg, Germany, 2017; pp. 1–942.
84. Fránková, M.; Poulíčková, A.; Neděla, V.; Tihlaříková, E.; Šumberová, K.; Letáková, M. The low temperature method for environmental scanning electron microscopy—A new method for observation of diatom assemblages in vivo. *Diatom Res.* **2019**, *33*, 397–403. [CrossRef]

85. Kairesalo, T. Dynamics of epiphytic communities on *Equisetum fluviatile* L. In *Periphyton of Freshwater Ecosystems*; Wetzel, R.G., Ed.; Developments in Hydrobiology; Springer: Dordrecht, Germany, 1983; Volume 17, pp. 153–160.
86. CHMI (Czech Hydrometeorological Institute). Historical Data—Meteorology and Climatology. Available online: https://www.chmi.cz/historicka-data/pocasi/uzemni-teploty (accessed on 10 March 2021).
87. Pyšek, P.; Danihelka, J.; Sádlo, J.; Chrtek, J., Jr.; Chytrý, M.; Jarošík, V.; Kaplan, Z.; Krahulec, F.; Moravcová, L.; Pergl, J.; et al. Catalogue of alien plants of the Czech Republic (2nd edition): Checklist update, taxonomic diversity and invasion patterns. *Preslia* **2012**, *84*, 155–255.
88. Grulich, V. Red List of vascular plants of the Czech Republic: 3rd edition. *Preslia* **2012**, *84*, 631–645.
89. Grulich, V.; Chobot, K. (Eds.) Červený seznam cévnatých rostlin ČR [The Red List of vascular plants of the Czech Republic]. *Příroda* **2017**, *35*, 75–132.
90. Wild, J.; Kaplan, Z.; Danihelka, J.; Petřík, P.; Chytrý, M.; Novotný, P.; Rohn, M.; Šulc, V.; Brůna, J.; Chobot, K.; et al. Plant distribution data for the Czech Republic integrated in the Pladias database. *Preslia* **2019**, *91*, 1–24. [CrossRef]
91. Danihelka, J.; Chrtek, J., Jr.; Kaplan, Z. Checklist of vascular plants of the Czech Republic. *Preslia* **2012**, *84*, 647–811.
92. Caisová, L.; Gąbka, M. Charophytes (*Characeae*, Charophyta) in the Czech Republic: Taxonomy, autecology and distribution. *Fottea* **2009**, *9*, 1–43. [CrossRef]
93. Oksanen, J.; Blanchet, F.; Friendly, M.; Kindt, R.; Legendre, P.; McGlinn, D.; Minchin, P.; O'Hara, R.; Simpson, G.L.; Solymos, P.; et al. Vegan: Community Ecology Package, R Package Version 2.5.-6. 2019. Available online: https://cran.r-project.org/web/packages/vegan/index.html (accessed on 30 March 2021).
94. Anderson, M.J. A new method for non-parametric multi-variate analysis of variance. *Austral Ecol.* **2001**, *26*, 32–46.
95. R Core Team. *R: A Language and Environment for Statistical Computing*; R Foundation for Statistical Computing: Vienna, Austria, 2020. Available online: https://www.R-project.org/ (accessed on 30 March 2021).
96. Riato, L.; Van Ginkel, C.; Taylor, J.C. Zooplankton and diatoms of temporary and permanent freshwater pans in the Mpumalanga Highveld region, South Africa. *Afr. Zool.* **2014**, *49*, 113–127. [CrossRef]
97. Popiela, A.; Łysko, A.; Molnár, V.A.; Kącki, Z.; Lukács, B.A. Distribution, morphology and habitats of *Elatine triandra* (*Elatinaceae*) in Europe, with particular reference to the central part of the continent. *Acta Bot. Gallica* **2015**, *162*, 325–337. [CrossRef]
98. Riato, L.; Della Bella, V.; Leira, M.; Taylor, J.C.; Oberholster, P.J. A diatom functional-based approach to assess changing environmental conditions in temporary depressional wetlands. *Ecol. Indic.* **2017**, *78*, 205–213. [CrossRef]
99. Kúr, P.; Píšová, S.; Tremetsberger, K.; Dřevojan, P.; Kącki, Z.; Hroudová, Z.; Böckelmann, J.; Bernhardt, K.G.; Šumberová, K. Ecology and genetics of *Cyperus fuscus* in Central Europe—A model for ephemeral wetland plant research and conservation. *Water* **2021**, *13*, 1277. [CrossRef]
100. Juříček, M. Flora and vegetation of fishponds in the Bohemian-Moravian Highlands. *Acta Rer. Natur.* **2012**, *13*, 33–51.
101. Cantonati, M. Diatom communities of springs in the southern Alps. *Diatom Res.* **1998**, *13*, 201–220. [CrossRef]
102. Cantonati, M.; Corradini, G.; Jüttner, I.; Cox, E.J. Diatom assemblages in high mountain streams of the Alps and the Himalaya. *Nova Hedwigia* **2001**, *123*, 37–61.
103. Cantonati, M.; Gerecke, R.; Bertuzzi, E. Springs of the Alps, sensitive ecosystems to environmental change: From biodiversity assessment to long-term studies. *Hydrobiologia* **2006**, *562*, 59–96. [CrossRef]
104. Pätzig, M.; Kalettka, T.; Glemnitz, M.; Berger, G. What governs macrophyte species richness in kettle hole types? A case study from Northeast Germany. *Limnologica* **2012**, *42*, 340–354. [CrossRef]
105. Poulíčková, A.; Letáková, M.; Hašler, P.; Coxb, E.; Duchoslav, M. Species complexes within epiphytic diatoms and their relevance for the bioindication of trophic status. *Sci. Total Environ.* **2017**, *599–600*, 820–833. [CrossRef] [PubMed]
106. Kopp, R. *Hydrochemie Nejen Pro Rybáře*; Mendelova Univerzita v Brně: Brno, Czech Republic, 2015.
107. Björnerås, C.; Weyhenmeyer, G.A.; Evans, C.D.; Gessner, M.O.; Grossart, H.-P.; Kangur, K.; Kokorite, I.; Kortelainen, P.; Laudon, H.; Lehtoranta, J.; et al. Widespread increases in iron concentration in European and North American freshwaters. *Global Biogeochem. Cycles* **2017**, *31*, 1488–1500. [CrossRef]
108. Lacoul, P.; Freedman, B. Environmental influences on aquatic plants in freshwater ecosystems. *Environ. Rev.* **2006**, *14*, 89–136. [CrossRef]
109. Emson, D.; Sayer, C.D.; Bennion, H.; Patmore, I.R.; Rioual, P. Mission possible: Diatoms can be used to infer past duckweed (lemnoid *Araceae*) dominance in ponds. *J. Paleolimnol.* **2018**, *60*, 209–221. [CrossRef]
110. Kaplan, Z.; Danihelka, J.; Štěpánková, J.; Ekrt, L.; Chrtek, J., Jr.; Zázvorka, J.; Grulich, V.; Řepka, R.; Prančl, J.; Ducháček, M.; et al. Distributions of vascular plants in the Czech Republic. Part 2. *Preslia* **2016**, *88*, 229–322.
111. Lovas-Kiss, Á.; Vizi, B.; Vincze, O.; Molnár, V.A.; Green, A.J. Endozoochory of aquatic ferns and angiosperms by mallards in Central Europe. *J. Ecol.* **2018**, *106*, 1714–1723. [CrossRef]
112. Cantonati, M.; Scola, S.; Angeli, N.; Guella, G.; Frassanito, R. Environmental controls of epilithic diatom depth-distribution in an oligotrophic lake characterized by marked water-level fluctuations. *Eur. J. Phycol.* **2009**, *44*, 15–29. [CrossRef]
113. Carmignani, J.R.; Roy, A.H. Ecological impacts of winter water level drawdowns on lake littoral zones: A review. *Aquat. Sci.* **2017**, *79*, 803–824. [CrossRef]
114. Evans, J.H. The survival of freshwater algae during dry periods: Part I. An investigation of the algae of five small ponds. *J. Ecol.* **1958**, *46*, 149–167. [CrossRef]

115. Hejný, S. *Ökologische Charakteristik der Wasser und Sumpfpflanzen in den Slowakischen Tiefebenen*; Verlag der Slowakischen Akademie der Wissenschaften: Bratislava, Czechoslovakia, 1960.
116. Von Lampe, M. Wuchsform, Wuchsrhythmus und Verbreitung der Arten der Zwergbinsengesellschaften. *Diss. Bot.* **1996**, *266*, 1–353.
117. Šumberová, K.; Lososová, Z.; Ducháček, M.; Horáková, V.; Fabšičová, M. Distribution, habitat ecology, soil seed bank and seed dispersal of threatened *Lindernia procumbens* and alien *Lindernia dubia* (*Antirrhinaceae*) in the Czech Republic. *Phyton* **2012**, *52*, 39–72.
118. Leira, M.; Meijide-Failde, R.; Torres, E. Diatom communities in thermo-mineral springs of Galicia (NW Spain). *Diatom Res.* **2017**, *32*, 29–42. [CrossRef]
119. Ponader, K.C.; Potapova, M.G. Diatoms from the genus *Achnanthidium* in flowing waters of the Appalachian Mountains (North America): Ecology, distribution and taxonomic notes. *Limnologica* **2007**, *37*, 227–241. [CrossRef]
120. Souffreau, C.; Vanormelingen, P.; Verleyen, E.; Sabbe, K.; Vyverman, W. Tolerance of benthic diatoms from temperate aquatic and terrestrial habitats to experimental desiccation and temperature stress. *Phycologia* **2010**, *49*, 309–324. [CrossRef]
121. Vasiljević, B.; Simić, S.B.; Paunović, M.; Zuliani, T.; Krizmanić, J.; Marković, V.; Tomović, J. Contribution to the improvement of diatom-based assessments of the ecological status of large rivers–The Sava River Case Study. *Sci. Total Environ.* **2017**, *605*, 874–883. [CrossRef] [PubMed]
122. Falasco, E.; Piano, E.; Bona, F. Diatom flora in Mediterranean streams: Flow intermittency threatens endangered species. *Biodivers. Conserv.* **2016**, *25*, 2965–2986. [CrossRef]
123. Ács, É.; Bíró, T.; Berta, C.; Duleba, M.; Földi, A.; Grigorszky, I.; Hidas, A.; Knisz, J.; Korponai, J.L.; Trábert, Z.; et al. Long-term changes of species composition and functional traits of epiphytic diatoms in the Szigetköz region (Hungary) of the Danube River. *Water* **2020**, *12*, 776. [CrossRef]
124. Patrick, R. The effects of increasing light and temperature on the structure of diatom communities. *Limnol. Oceanogr.* **1971**, *16*, 405–421. [CrossRef]
125. Carrick, H.J.; Lowe, R.L. Benthic algal response to N and P enrichment along a pH gradient. *Hydrobiologia* **1989**, *179*, 119–127. [CrossRef]
126. Kaczmarek, H.; Mazaeva, O.A.; Kozyreva, E.A.; Babicheva, V.A.; Tyszkowski, S.; Rybchenko, A.A.; Brykała, D.; Bartczak, A.; Słowiński, M. Impact of large water level fluctuations on geomorphological processes and their interactions in the shore zone of a dam reservoir. *J. Great Lakes Res.* **2016**, *42*, 926–941. [CrossRef]
127. Van Zuidam, B.G.; Peeters, E.T.H.M. Wave forces limit the establishment of submerged macrophytes in large shallow lakes. *Limnol. Oceanogr.* **2015**, *60*, 1536–1549. [CrossRef]
128. Bertrin, V.; Boutry, S.; Alard, D.; Haury, J.; Jan, G.; Moreira, S.; Ribaudo, C. Prediction of macrophyte distribution: The role of natural versus anthropogenic physical disturbances. *Appl. Veg. Sci.* **2018**, *21*, 395–410. [CrossRef]
129. Kaplan, Z.; Danihelka, J.; Chrtek, J., Jr.; Zázvorka, J.; Koutecký, P.; Ekrt, L.; Řepka, R.; Štěpánková, J.; Jelínek, B.; Grulich, V.; et al. Distributions of vascular plants in the Czech Republic. Part 8. *Preslia* **2019**, *91*, 257–368. [CrossRef]
130. Gaberščik, A.; Grašič, M.; Abram, D.; Zelnik, I. Water level fluctuations and air temperatures affect common reed habitus and productivity in an intermittent wetland ecosystem. *Water* **2020**, *12*, 2806. [CrossRef]
131. Hillman, W.S. The Lemnaceae, or Duckweeds: A Review of the Descriptive and Experimental Literature. *Bot. Rev.* **1961**, *27*, 221–287. [CrossRef]
132. Kolář, F.; Šumberová, K.; Ducháček, M. Cerastium dubium—Rožec pochybný. In *Nálezy Zajímavých a Nových Druhů v Květeně Jižní Části Čech XXI*; Lepší, M., Lepší, P., Eds.; Jihočeské Muzeum v Českých Budějovicích: České Budějovice, Czech Republic, 2015; Volume 55, pp. 16–17.
133. Poulíčková, A.; Neustupa, J.; Špačková, J.; Škaloud, P. Distribution of epipelic diatoms in artificial fishponds along environmental and spatial gradients. *Hydrobiologia* **2009**, *624*, 81–90. [CrossRef]
134. Kaplan, Z.; Danihelka, J.; Šumberová, K.; Chrtek, J., Jr.; Rotreklová, O.; Ekrt, L.; Štěpánková, J.; Taraška, V.; Trávníček, B.; Prančl, J.; et al. Distributions of vascular plants in the Czech Republic. Part 5. *Preslia* **2017**, *89*, 333–439. [CrossRef]
135. Feuchtmayr, H.; Moran, R.; Hatton, K.; Connor, L.; Heyes, T.; Moss, B.; Harvey, I.; Atkinson, D. Global warming and eutrophication: Effects on water chemistry and autotrophic communities in experimental hypertrophic shallow lake mesocosms. *J. Appl. Ecol.* **2009**, *46*, 713–723. [CrossRef]
136. Lu, J.; Bunn, S.E.; Burford, M.A. Nutrient release and uptake by littoral macrophytes during water level fluctuations. *Sci. Total Environ.* **2018**, *622*, 29–40. [CrossRef]
137. Blindow, I. The composition and density of epiphyton on several species of submerged macrophytes—The neutral hypothesis tested. *Aquat. Bot.* **1987**, *29*, 157–168. [CrossRef]
138. Buczkó, K. The occurrence of the epiphytic diatom *Lemnicola hungarica* on different European *Lemnaceae* species. *Fottea* **2007**, *7*, 77–84. [CrossRef]
139. Riato, L.; Leira, M. Heterogeneity of epiphytic diatoms in shallow lakes: Implications for lake monitoring. *Ecol. Indic.* **2020**, *111*, 105988. [CrossRef]
140. Totti, C.; Poulin, M.; Romagnoli, T.; Perrone, C.; Pennesi, C.; De Stefano, M. Epiphytic diatom communities on intertidal seaweeds from Iceland. *Polar Biol.* **2009**, *32*, 1681–1691. [CrossRef]

water

MDPI

Article

Ecology and Genetics of *Cyperus fuscus* in Central Europe—A Model for Ephemeral Wetland Plant Research and Conservation

Pavel Kúr [1], Soňa Píšová [2,3], Karin Tremetsberger [4,*], Pavel Dřevojan [5], Zygmunt Kącki [6], Jörg Böckelmann [4,7], Karl-Georg Bernhardt [4], Zdenka Hroudová [2], Attila Mesterházy [8] and Kateřina Šumberová [9]

1 Department of Botany and Biodiversity Research, University of Vienna, Rennweg 14, 1030 Vienna, Austria; kur.pavel@gmail.com
2 Institute of Botany, The Czech Academy of Sciences, Zámek 1, 252 43 Průhonice, Czech Republic; sonka.krl@gmail.com (S.P.); zdenka.hroudova@ibot.cas.cz (Z.H.)
3 Department of Forest Biodiversity and Nature Conservation, Austrian Research Centre for Forests, Seckendorff-Gudent-Weg 8, 1131 Vienna, Austria
4 Institute of Botany, University of Natural Resources and Life Sciences, Vienna, Gregor-Mendel-Straße 33, 1180 Vienna, Austria; joerg.boeckelmann@univie.ac.at (J.B.); karl-georg.bernhardt@boku.ac.at (K.-G.B.)
5 Department of Botany and Zoology, Faculty of Science, Masaryk University, Kotlářská 2, 611 37 Brno, Czech Republic; pavel.drevojan@seznam.cz
6 Botanical Garden, University of Wrocław, Sienkiewicza 23, 50-335 Wrocław, Poland; zygmunt.kacki@uwr.edu.pl
7 Division of Tropical Ecology and Animal Biodiversity, Department of Botany and Biodiversity Research, University of Vienna, Rennweg 14, 1030 Vienna, Austria
8 Hunyadi utca 55, 9500 Celldömölk, Hungary; amesterhazy@gmail.com
9 Institute of Botany, The Czech Academy of Sciences, Lidická 25/27, 602 00 Brno, Czech Republic; katerina.sumberova@ibot.cas.cz
* Correspondence: karin.tremetsberger@boku.ac.at; Tel.: +43-1-47654-83113

Citation: Kúr, P.; Píšová, S.; Tremetsberger, K.; Dřevojan, P.; Kącki, Z.; Böckelmann, J.; Bernhardt, K.-G.; Hroudová, Z.; Mesterházy, A.; Šumberová, K. Ecology and Genetics of *Cyperus fuscus* in Central Europe—A Model for Ephemeral Wetland Plant Research and Conservation. *Water* **2021**, *13*, 1277. https://doi.org/10.3390/w13091277

Academic Editor: Alenka Gaberščik

Received: 30 March 2021
Accepted: 27 April 2021
Published: 30 April 2021

Abstract: The ecology and species diversity of ephemeral wetland vegetation have been fairly well studied, but the biology of its characteristic species has rarely been investigated holistically. Here we combine previous results on the genetic diversity of a suitable model species (the diploid *Cyperus fuscus*) with new data on its historical and recent occurrence, its ecological and climatic niche, and the associated vegetation. Analysis of phytosociological relevés from Central Europe revealed a broad ecological niche of *C. fuscus* with an optimum in the Isoëto-Nanojuncetea class, extending to several other vegetation types. Overall species composition in the relevés highlight *C. fuscus* as a potential indicator of habitat conditions suitable for a range of other threatened taxa. Analysis of historical records of *C. fuscus* from the Czech Republic showed an increasing trend in the number of localities since the 1990s. It seems that recent climate warming allows the thermophilous *C. fuscus* to expand its range into colder regions. Isoëto-Nanojuncetea and Bidentetea species are well represented in the soil seed bank in both riverine and anthropogenic habitats of *C. fuscus*. Vegetation diversity has a weak negative effect and anthropogenic (compared to riverine) habitats have a strong negative effect on genetic diversity in this species.

Keywords: chromosome number; climatic niche; distribution trends; ecological niche; fishpond habitats; genome size; genetic diversity; Isoëto-Nanojuncetea; soil seed bank; species diversity

1. Introduction

Wetlands and their biota currently suffer under multiple threats including global eutrophication, climate change, direct habitat destruction, and land use change [1–4]. Frequent episodes of droughts, typical of the last two decades, may lead not only to release of nutrients from drying subhydric soils and increase of trophy levels but also to severe water level fluctuations and/or substrate desiccation, and, consecutively, to damage of populations of some plant and animal species with high moisture demands [5–7]. Ephemeral wetland plant species, sometimes called mudflat plants and usually classified within the syntaxonomic class Isoëto-Nanojuncetea [8], are a group of organisms with population frequency and/or density strongly fluctuating according to the conditions of the given growing season, particularly temperature and precipitation [9]. On the one hand, mudflat plants possessing persistent soil seed banks may easily balance the loss of a part of their populations in the years with unfavorable conditions. On the other hand, any slight but permanent shift in, e.g., the course of temperatures and precipitations, particularly during such a sensitive development phase like seed germination and/or seedling recruitment, may quickly lead to soil seed bank depletion and a species' disappearance [9,10]. Many mudflat species are listed in national Red Data Lists and are considered as threatened and worth protecting, some of them even on a continental scale [11,12]. As it is not realistic to study all of these species in detail in a short period of time, we decided to select a model that could be used to explain processes in vegetation and soil seed banks, including genetic aspects, in ecologically similar species. The methodology we adapted to *Cyperus fuscus* could be used as a standardized protocol in studies of other short-lived wetland species.

Although *Cyperus fuscus* is rather widespread in some parts of Europe [13], it is considered as rare and threatened in some countries (critically endangered—Denmark [14]; vulnerable—Austria [15], Great Britain [16], Switzerland [17]; threatened and declining—Germany [18], near threatened—Czech Republic [19], Belgium [20], Estonia [21]; extremely rare—Luxembourg [22]). However, our recent investigations suggest that the species might, at least in some regions, profit from global environmental changes, particularly climate change [23]. Frequent availability of suitable exposed muddy bottoms because of summer drought may support more frequent occurrence of the species on already known sites. As high temperatures enhance faster growth and reproduction of this thermophilous species [24], higher soil seed bank and population density of *C. fuscus* can be expected. For most plant species, the higher their frequency and the propagule pressure in the environment, the higher the probability of their dispersal within the landscape and vice versa [25,26]. Hence, also for *C. fuscus* we could expect higher probability of its further spread to not yet colonized sites. Therefore, we decided to test the hypothesis that *C. fuscus* is recently (last ~20 years) spreading in central Europe, compared to the situation in the 20th century, particularly in its first half. Independent of the validity of this hypothesis, habitat preferences of the species could also change, particularly the vegetation types in which the species usually occurs. A similar trend has been observed e.g., in *Crassula (Tillaea) aquatica*, another ecologically specialized wetland annual herb [10]. We suggest that the vegetation types with recent occurrence of *C. fuscus* include a higher amount of alien species and ruderals than the analogical plant communities documented in the past.

Böckelmann et al. [27,28] studied phenotypic and genotypic characteristics of populations of *Cyperus fuscus* in Central Europe. Three habitat types with various intensities of anthropogenic influence inhabited by *C. fuscus* were in the focus of the studies: the near-natural riverbanks, the fishponds with intermediate levels of human impact, and the fish storage ponds with a strongly anthropogenic disturbance regime. The phenotypes and the genetic structure in the soil seed bank of *C. fuscus* were also investigated. No phenotypic and only weak genotypic differentiation (in the same order of magnitude as between spatially separated sampling plots within one and the same above-ground population) was found between the soil seed bank and the above-ground population [27,28]. This was equally true for the three habitat types: rivers, fishponds, and fish storage ponds. On the level of the species composition of the plant community, the highly dynamic habitats are usually colonized by annuals, which are also present in the soil seed bank; the perennial vegetation on the more stable sites, which are not deeply flooded, however, differs more from its soil seed bank, which consists of annuals of earlier successional stages [29,30]. We used the similarity in species composition between the soil seed bank and the above-ground vegetation as an indicator of site dynamics in the three habitat types studied [31]. To explain any differences in species composition, we investigated which phytosociological species groups were missing from the soil seed bank or vegetation.

Two different chromosome numbers were so far reported for *Cyperus fuscus* in the Central European region. Plants with $2n = 36$ chromosomes have been reported from the Czech Republic (fishpond Řemínek in the district Náchod in East Bohemia) [32]; plants with $2n = 72$ chromosomes have been reported from Slovakia (gravel pit near Lakšárska Nová Ves in the district Senica in West Slovakia) [33]. Further reports are spread randomly across Europe and the Mediterranean. Plants with $2n = 36$ chromosomes have been reported from England (Breamore, South Hampshire) [34] and Israel (Hadera river, Sharon plain) [35]; plants of unknown origin cultivated at the Botanical Garden in Uppsala, Sweden, had $2n = 72$ chromosomes [36]. As reported for other plant species, ploidy level variation might be related to the geographical origin of populations [37]. Cytotypes may also be differentiated ecologically and higher polyploids may be associated with habitats strongly influenced by anthropogenic activities [38,39]. To assess the distribution and habitat preference of the two cytotypes, we analyzed the genome size and chromosome number in a larger number of populations in Central Europe.

We were also interested whether there is an effect of plant community (vegetation) diversity on within-population genetic diversity. It is suggested that parallel processes, as caused by, for instance, size, area, land-use history, or isolation, on the two levels of diversity may result in a positive relationship between them [40,41]. Böckelmann et al. [28] found significantly reduced values of within-population genetic diversity in anthropogenic habitats (fishponds and fish storage ponds) compared to river habitats. We investigated whether the habitat type has same-directional effects on the two levels of diversity and thus establishes a positive relationship between them. For example, founder effects when anthropogenic habitats were created, restricted connections between anthropogenic and river habitats and/or the selection regime could reduce both levels of diversity. However, plant community diversity may also have a direct effect on within-population genetic diversity, which may vary for species with poor versus good competitive abilities (or rare versus common species [40]). Overall, we aimed to integrate the different levels of information to gain a better understanding of the functioning and current status of ephemeral wetland plants.

2. Materials and Methods

2.1. Sampling of Plant Material and Vegetation Data in the Field, Cultivation of Seeds from Soil

The localities with *Cyperus fuscus* belonging to the three habitat types (rivers: R1-R11; fishponds: F1-F10; fish storage ponds: S1-S10) are the same as those used for phenotypic and genetic analyses [27,28]. In addition, a fish rearing pond (RP) and an old quarry (Q) were also included for cultivation of seeds from the soil (Table S1). Large populations of *C. fuscus* were found on all these sites and therefore we selected them for complex study, including the soil seed bank and genetic analyses. Soil samples were taken from 1 m^2 plots from each of these sites in summer 2012 in the depth of up to 15 cm; the soil surface with seeds matured in 2012 was scraped before the sampling. After six months of storage at 6 °C in the dark, the soil samples were spread out as a 5 mm thin layer on sterile sand (seedling emergence method). The temperature in the glasshouse was set to 28 °C to stimulate the germination of *C. fuscus*, and the trays were watered daily until August 2013. All species emerging from the soil samples were recorded and *C. fuscus* plants germinated from the soil seed bank were collected for the phenotypic and genetic analysis [27,28]. Phytosociological relevés were collected using the method of the Zurich-Montpellier school prior to the soil sampling on the same 1 m^2 plots from which the soil samples were taken. Additionally, on four plots the relevés were collected, but these plots were not included in the soil seed bank and genetic analyses (Table S1). For the evaluation of cover of individual species, the nine-grade modified Braun-Blanquet scale [42] was used.

2.2. Vegetation Data Processing and Analyses

2.2.1. Data Sources

Altogether three basic sources of vegetation data have been used: (1) Our own relevés from plots for complex investigation of *Cyperus fuscus* (genetics, soil seed bank, vegetation; see Section 2.1). In total, 37 sites in AT, CZ, PL, and SK were sampled (31 of them also served for sampling of plant material for genetic analyses and 32 of them were analyzed for their soil seed bank), with 3 relevés for each of them, i.e., 111 relevés in total (July–October 2012; Table S1); (2) Additional relevés collected by K.G.B., K.B., K.Š., K.T., or P.K. between 2000 and 2019 in AT, CZ, or SK; 37 relevés in total, usually a single relevé per site; (3) Data with the occurrence of *C. fuscus* from the national phytosociological databases [43] of Austria (Austrian Vegetation Database—AVD [44]; 50 relevés), the Czech Republic (Czech National Vegetation Database—CNPD [45]; 516 relevés), Poland (Polish Vegetation Database —PVD [46]; 337 relevés), and Slovakia (Slovak Vegetation Database, also called Central Database of Phytosociological Samples—CDF [47]; 147 relevés), and from the Gravel Bar Vegetation Database (GBVD [48]; 11 relevés). The initial number of relevés from all sources was 1206.

2.2.2. Compilation of the Vegetation Dataset

All the vegetation data were stored using Turboveg 2.0 database software [49], at the beginning as separated databases, and checked for possible inconsistencies. To concentrate our attention on the vegetation types, where *Cyperus fuscus* finds its optimum, we removed all relevés where the species occurred with less than 1% cover. These were mainly the mosaics of wet and mesic meadows, forests with small patches of springs, and other more or less terrestrial habitats. Further, we removed all relevés with too small or too large plot sizes. As the plot sizes were very diverse, it was difficult to comply with the rules suggested by Chytrý & Otýpková [50], without substantial loss of the data. Therefore, we accepted relevés with plot sizes between 0.5 to 50 m^2, as the lower and upper extreme values were still sufficiently frequent, occupying more than 5% of the relevés in some of the national datasets. In the next step, we removed all relevés without sufficiently precise localization, i.e., without coordinates and/or the nearest settlement name, and the relevés from other countries than AT, CZ, PL, and SK. To reduce oversampling of some sites, we randomly selected 3 relevés as maximum for one site defined by coordinates and/or verbal description. Using these selection principles, additional relevés were accepted if they documented different vegetation type(s) of an alliance or higher syntaxonomic rank or if they were related to another period of time, with a minimum interval of 5 years between the samplings (this interval is based on our personal observation of speed of changes in wetland habitats, which may enhance the spread of *C. fuscus*).

The final Turboveg database included altogether 925 relevés, of which 37 (4%) came from Austria (32 from AVD and the rest from the other databases), 555 (60%) from the Czech Republic (446 from CNPD and the rest from the other databases), 227 (24.5%) from Poland (218 from PVD and the rest from the other databases), and 106 (11.5%) from Slovakia (96 from the CDF and the rest from the other databases). Detailed information about the structure of original databases, as well as the relevés used after the selection procedure, including the most important authors represented in the final dataset, are available in the Appendix A.

The final database in Turboveg was unified under the species list of Czechia_Slovakia_2015 (https://www.sci.muni.cz/botany/juice/KUBAT2015.txt, accessed on 10 December 2020). For further processing and analyses, the data were imported into JUICE software, version 7.0.45 [51]. After importing into the JUICE software, we unified the layers for certain species as follows: for all herbal species and charophytes, including their juveniles, the herb layer was selected. All tree and shrub species were included into the juvenile layer and all bryophytes and algae (except for charophytes) into the moss layer. To simplify the dataset, in the next step we merged some of the taxa into a single species, aggregate, or other higher taxon; for their list see Table S2. However, in the case that the relative or morphologically similar species are ecologically distinct, having an important indicator potential, we made efforts to keep the taxonomic treatment as detailed as possible. During the merging of taxa, we also corrected some highly probable identification errors (see the comments in Table S2). Finally, we unified the nomenclature of taxa according to the Euro+Med PlantBase [52], as this online source reflects the most recent taxonomic treatment. This nomenclature is used throughout the whole paper.

2.2.3. Vegetation Classification and Ordinations

For classification of the final vegetation dataset, we used an automatic expert system (ESY) designed for vegetation classification of large data packages and working under JUICE software. From the expert systems available, we selected the expert system recently developed for Poland [53], which, in contrast to, e.g., ESY for the Czech Republic [54], also includes some additional wetland vegetation types relevant for our study and missing in the Czech Republic. Two rounds of the classification procedure were applied: up to class level for basic orientation in the vegetation types and up to association level for the class represented by the highest amount of the relevés. The first classification round should give us a basic overview of vegetation patterns in the communities with *Cyperus fuscus*, while the second round was applied to describe the vegetation optimal for the occurrence of the target species in more detail. Syntaxonomic nomenclature of the classes and alliances corresponds to Mucina et al. [8]; the nomenclature of the associations is based on Kącki et al. [53].

As we wanted to determine what factors influence the vegetation composition of the communities with *Cyperus fuscus*, a canonical correspondence analysis (CCA) was performed. Weighted mean Ellenberg indicator values (EIVs, see Ellenberg et al. [55], i.e., the ordinal estimation scales for moisture, soil reaction, light, nutrients, continentality, temperature) calculated in the JUICE software for each of the relevés, number and cover of archaeophytes and neophytes, cover of *C. fuscus*, and sampling year were used as the explanatory variables. The cover of *C. fuscus* was excluded from the EIV calculations and from the response variables in the ordination. The significance of the model as well as the significance of the variables were tested using the Monte Carlo permutation tests (999 permutations). In addition, the correlation of time, expressed as the sampling year, with vegetation composition was tested using generalized additive models (GAM; tested models with 1–6 degrees of freedom; the best model selected based on the AIC) and then projected into the ordination space as contour lines.

To have an alternate view of the relationships of the independent explanatory variables, whose mutual correlations can be easily distorted in ordination diagrams, a set of pairwise correlation plots for all the variables was produced. The significance of the relationships between any two variables was tested using GAM in the *mgcv* R package [56–60].

Identification of alien species and their classification as either archaeophytes or neophytes was based on two complementary published sources. The checklist of alien species of the Czech Republic [61], suitable also for other "southern" countries (AT, SK), was combined with the alien species checklist for Poland [62]. These checklists overlap by approximately 90–95%; some of the thermophilous species that are considered native or archaeophytes in the southern countries are classified as archaeophytes or neophytes in Poland. In the case of differences in the evaluation of a species between the two checklists, our final classification was based on the country of origin of the majority of relevés with the species in question.

2.3. Species Composition of the Soil Seed Bank

The final dataset based on the cultivation of seeds from the soil and vegetation data collected in the field comprised a species list of the soil seed bank (presence/absence) and the associated above-ground vegetation (relevés) from one sampling plot for each of 32 sites (Tables S1 and S3). To assess the similarity between the soil seed bank and the above-ground vegetation, we used Sørensen's similarity index [63], defined as $S = 2c/(a + b + 2c)$, where a is the number of species present only in the relevé, b is the number of species present only in the soil seed bank and *c* is the number of species present in both the relevé and the soil seed bank. Species were assigned to seven groups according to phytosociological classes (Table S3). A linear model was used to estimate the effect of the belonging of a species to one of the seven groups on its proportion of occurrence in the soil (function *lm* in R [64]).

2.4. Chromosome Counts

Seeds from various localities in Austria, Croatia, the Czech Republic, Hungary, Italy, Poland, and Slovakia (Table S4) were germinated on filter paper in Petri dishes in a Sanyo MLR-352 environmental test chamber ("day": 35 °C/12 h/light with 15 fluorescent lamps FL40SS-W/37 on; "night": 15 °C/12 h/dark) and watered with distilled water. After 7-8 days, the radicle was cut from the seedlings and pre-treated with 0.002 M 8-hydroxyquinoline at 12 °C for 4 h or, alternatively, on ice for 4 h and then at 4 °C overnight. After pre-treatment, the radicles were fixed in freshly made absolute ethanol: glacial acetic acid (3:1) and kept at -20 °C until preparation.

Chromosomes were prepared by enzymatic digestion and squashing [65,66]. Fixed root tips were kept in citric acid-trisodium citrate buffer (pH 4.8) for 20 min. Then the citrate buffer was replaced by enzyme mixture (1% (*w/v*) cellulase Onozuka (Serva, Heidelberg, Germany), 0.4% (*w/v*) pectolyase (Sigma-Aldrich, Vienna, Austria), 0.4% (*w/v*) cytohelicase (Sigma-Aldrich), in citrate buffer, pH 4.8, pre-warmed at 37 °C) and the root tips were incubated at 37 °C for 30 min. After removal of the enzyme mixture, the root tips were washed in citrate buffer. For squashing, root tips were transferred to a drop of 60% acetic acid on a slide, dissected using entomological needles and squashed under a cover slip. The coverslips were removed, and preparations were air-dried and stored at 4 °C. The material was then stained with 2 µg/mL DAPI and mounted in antifade Vectashield mounting medium (Vector Laboratories, Burlingame, CA, USA) and analyzed under a Zeiss (Carl Zeiss, Vienna, Austria) Axio Imager.M2 epifluorescence microscope equipped with an AxioCam HRm camera. Images were acquired with the Zeiss AxioVision SE64 software.

2.5. Flow Cytometry

The relative and absolute DNA content was measured from fresh leaves of individual plants (collected in the field or grown from seeds; Table S4) using two different flow cytometric methods: (1) 4′,6-diamidino-2-phenylindole (DAPI) staining for a rough overview of the relative DNA content of a large number of individuals and (2) propidium iodide (PI) staining for determination of the absolute genome size of a limited number of individuals. For both methods, fresh leaves of *Solanum pseudocapsicum* (1.295 pg/1C [67]) were used as internal size standard.

For the DAPI measurements, the leaves were prepared with the CyStain UV Precise P kit (Partec, Münster, Germany), according to the manufacturer's instructions. Leaves of the sample and internal size standard were chopped together with a razor blade in 400 µL of extraction buffer. After 1 min of incubation, the suspension with cell walls and cell contents including nuclei was filtered through a CellTrics filter (Partec; mesh size 50 µm). The filtered suspension was incubated with 1.6 mL staining buffer for 1 min or longer. Measurements were done in a Partec Ploidy Analyzer equipped with an HBO 100 long life mercury lamp.

For the PI measurements, we used Otto et al.'s [68] buffer for extraction of nuclei following the chopping method of Galbraith et al. [69]. After chopping, the suspension was filtered through a 30 µm nylon mesh (Sefar, Rüschlikon, Switzerland). Then 50 µL RNase A was added. Digestion of RNA took place at 37 °C for 30 min. After digestion, the suspension was supplied with 4 mL propidium iodide solution (pH ~ 9.5) and incubated in the dark at room temperature for at least 20 min or at 4 °C overnight. Measurements were carried out in a CyFlow ML flow cytometer (Partec) equipped with a Samba 532 nm green laser (Cobolt, Stockholm, Sweden).

For both the DAPI and PI methods, the fluorescence of 5000 particles was recorded and the values for mean and CV of peaks were determined with the software FloMax ver. 2.81 (Partec).

2.6. Relationship between Plant Community (Vegetation) and Genetic Diversity

The Pearson correlation was calculated between the Shannon diversity index (H) and the average expected heterozygosity under Hardy–Weinberg equilibrium (H_S; taken from [28] as a measure of within-population genetic diversity) for the same 31 sites belonging to the three habitat types (rivers: R1-R11; fishponds: F1-F10; fish storage ponds: S1-S10) as those used in Böckelmann et al. [27,28]. A structural equation model was used to test the hypothesis that the habitat type simultaneously influences H (which in turn influences H_S) and H_S in Amos 24.0.0 software (Amos Development Corporation, Wexford, PA, USA).

2.7. Species Distribution in the Czech Republic

The distributional data for *Cyperus fuscus* for the territory of the Czech Republic were obtained by excerption from all major Czech public herbaria and some Austrian and Slovak herbaria (BRA, BRNM, BRNU, CB, CBFS, CESK, FMM, GM, HOMP, HR, CHEB, CHOM, LIT, MJ, MMI, MP, MZ, NJM, OL, OLM, OP, OSM, PL, PR, PRA, PRC, ROZ, SAV, SLO, SOB, SUM, VM, VYM, W, WU, ZMT; acronyms see Thiers [70]) as well as some private herbaria and from the Czech National Phytosociological Database [45]. In addition, our personal field observations, as well as the observations of a few other field botanists, were also included. All the data on *C. fuscus* are currently stored in the database PLADIAS (www.pladias.cz, accessed on 21 February 2021), which, among others, summarizes all the digitized records on the species of the Czech flora [71]. Only those records that could be unambiguously georeferenced were retained for further analyses.

Records with the information about the collection year were classified into three groups based on their age reflecting the (in our opinion) most important breakpoints in the management of rivers and water reservoirs in Central Europe. The year 1900 was chosen as the first turning point as the scale of river engineering works in Europe was at its peak and most large European rivers had been channelized by this date [72]. The year 2000 was chosen as the other turning point to test the hypothesis that *Cyperus fuscus* might benefit from climate change.

2.7.1. Temporal Changes of Species Distribution

To assess the changes of the species' distribution over time, the numbers of individual records per year were plotted as simple frequency curves. Only dated records could be used; all the records without at least a year of sampling were removed (these were mainly old herbarium records). In the case of duplicates, i.e., two or more records with the same date, locality, and collector, only one record was retained for the analyses and the rest was deleted. The search of the duplicates was done throughout all the sources used, as e.g., some records from the Czech National Phytosociological Database were documented also by herbarium specimens. For the records from different parts of the same fishpond, river etc., at least 0.5 km distance was requested if the site should be considered separately. To account for random fluctuations in the records, the data were smoothed using 11-year moving averages [73]. Importantly, the distributional data were corrected for botanical activity based on a dataset of 13 wetland species (*Chenopodium ficifolium*, *Coleanthus subtilis*, *Cyperus flavescens*, *C. michelianus*, *Gnaphalium uliginosum*, *Laphangium luteoalbum*, *Leersia oryzoides*, *Lythrum hyssopifolia*, *Oxybasis glauca*, *O. rubra*, *Persicaria dubia*, *Pulicaria dysenterica*, *P. vulgaris*) from 36 Czech public herbaria (BRNL, BRNM, BRNU, CB, CBFS, CESK, FMM, GM, HOMP, HR, CHEB, CHOM, LIM, LIT, MJ, MMI, MP, MZ, NJM, OL, OLM, OMJ, OP, OSM, OVMB, PL, PR, PRA, PRC, ROZ, SOB, SOKO, SUM, VM, VYM, ZMT). The same additional data sources as for *C. fuscus* were used for these species; all the data are stored in the PLADIAS database. Similar procedure of elimination of duplicate records as in the case of *C. fuscus* was applied for each of the 13 species. To account for uncertainty in botanical activity during periods with an overall low number of records, 50% jackknifing was performed 10,000 times on the input data tables, and means and 99% percentiles of the corrected number of records were calculated for each year.

2.7.2. Climatic Conditions

The relationship between climatic variables and the species' distribution in the Czech Republic was analyzed from all georeferenced records. The values of the 19 bioclimatic variables contained in the WorldClim database [74] downloaded in the highest available resolution (30 arc seconds ≈ 1 km^2) were extracted for each locality of *Cyperus fuscus* using the R package *raster* [75]. The same set of bioclimatic variables was also extracted for the centers of all quadrants of the Central European recording grid [76] located in the Czech Republic and without any records of *C. fuscus*. These values served as pseudo-absence data to test for climatic variables correlated with the occurrence of *C. fuscus*. A canonical discriminant analysis (CDA) in Canoco 5 [77] was used to test for the differences in the climatic conditions between locations with and without the presence of *C. fuscus*.

2.7.3. Climatic Niche Modelling

For predicting the occurrence of *Cyperus fuscus* in the Czech Republic and identifying any climate-driven limits of its distributional range, we employed the maximum entropy modelling approach implemented in maxent 3.4.1 [78]. The R package *wallace* was used [79]. We used the same 19 bioclimatic variables previously used in CDA. Duplicate data were removed by spatial thinning at a 1 km distance, and 7000 background points (pseudo-absences) were randomly sampled without replacement from the study area. The occurrences were partitioned into testing and training bins using the checkerboard2 method for fourfold cross-validation (aggregation factor 2). We ran the maxent models with the regularization multiplier (RM) values ranging from 0.5 to 4 (incremented by 0.5) and five alternative settings for feature classes (i.e., L, LQ, H, LQH, LQHP) and used the "area under the receiver operating characteristic curve" (AUC [80]) statistics to select the model with the highest predictive power.

3. Results

3.1. Community Affiliation of Cyperus fuscus and Vegetation Species Composition

Semi-supervised classification of altogether 925 relevés from AT, CZ, PL and SK identified occurrence of *C. fuscus* in plant communities of altogether ten syntaxonomic classes, with the highest share of the vegetation of the classes Isoëto-Nanojuncetea (622 relevés, i.e., ca 67% of the classified dataset), Bidentetea tripartitae (109 relevés, ca 12%) and Phragmito-Magnocaricetea (73 relevés, ca 8%). The nine other classes, i.e., Molinio-Arrhenatheretea, Scheuchzerio palustris-Caricetea nigrae, Crypsietea aculeatae, Potamogetonetea, Festuco-Puccinellietea, Charetea intermediae, Papaveretea rhoeadis, Littorelletea uniflorae, and Epilobietea angustifolii, were represented by altogether 33 relevés, with 1–9 relevés for each class (Table 1). The rest, 88 relevés, remained unclassified, representing mainly the communities transitional between two or more classes (data not displayed).

When we allowed classification of the relevés belonging to the Isoëto-Nanojuncetea class up to the association level, altogether 462 relevés were assigned to one of the seven detected associations and the rest, 160 relevés, was classified into the class or alliance (Eleocharition soloniensis (=Eleocharition ovatae), Verbenion supinae) level only. Among the associations, Cyperetum micheliani and Polygono-Eleocharitetum ovatae were the most frequent, documented by 329 (ca 36% of the classified dataset) and 103 (ca 11%) relevés, respectively (Table S5). All the other associations, i.e., Cerastio-Ranunculetum sardoi, Veronico anagalloidis-Lythretum hyssopifoliae, Cyperetum flavescentis, Pulicario-Menthetum, and Eleocharito-Schoenoplectetum supini, were represented by only a few relevés each.

Table 1. Percentage synoptic table of the vegetation with *Cyperus fuscus* (short version; for full version see Table S5). Classes distinguished: (1) Isoëto-Nanojuncetea; (2) Bidentetea tripartitae; (3) Phragmito-Magnocaricetea; (4) Molinio-Arrhenatheretea; (5) Scheuchzerio-Caricetea nigrae; (6) Crypsietea aculeatae; (7) Potamogetonetea; (8) Festuco-Puccinellietea; (9) Charetea intermediae; (10) Papaveretea rhoeadis; (11) Littorelletea uniflorae; (12) Epilobietea angustifolii. Only the diagnostic species occurring in at least 30 relevés and the other species occurring in at least 50 relevés are displayed. Layers: 6—herb layer, 7—juvenile trees and shrubs, 9—bryophyte and algal layer.

Vegetation Units	Layer	1	2	3	4	5	6	7	8	9	10	11	12
Number of relevés		622	109	73	9	5	5	4	4	2	2	1	1
Isoëto-Nanojuncetea													
Cyperus fuscus	6	100	100	100	100	100	100	100	100	100	100	100	100
Plantago major subsp. *intermedia*	6	49	54	16	11	40	60	25			100		
Gnaphalium uliginosum	6	50	38	16							50		
Juncus bufonius	6	42	28	12	22						50		
Limosella aquatica	6	36	16	12	11			50					
Eleocharis acicularis	6	28	3	8				25					
Potentilla supina	6	21	25	7		20	60		50		50		
Eleocharis ovata	6	23	4	5								100	
Carex bohemica	6	19	7	4				50					
Lythrum portula	6	21	1	3							50		
Riccia cavernosa/crystallina	9	14	1	4									
Lindernia procumbens	6	7		3									
Pulicaria vulgaris	6	5	6		11				25				
Lythrum hyssopifolia	6	4	3	4									
Bidentetea tripartitae													
Persicaria lapathifolia s. l. (excl. *P. lapathifolia* subsp. *brittingeri*)	6	51	61	42	33	40	20	25	25		100	100	100
Rumex maritimus	6	41	67	47	33	60	40	25	50		50		100
Rorippa palustris	6	47	51	29	22	20		50			50		100
Ranunculus sceleratus	6	33	36	22	11	20		50	25			100	100
Persicaria hydropiper	6	26	39	34	22					50	50		100
Bidens tripartitus	6	26	36	23	11				25		50	100	100
Alopecurus aequalis	6	27	21	19	22	60			25		50		
Oxybasis rubra	6	20	48	26		20	40		25				
Bidens frondosus	6	15	40	23	11	20					50	100	
Bidens radiatus	6	13	10	16									
Oxybasis glauca	6	11	23	3			40		25				
Bidens cernuus	6	10	15	7	22						50		
Persicaria maculosa	6	9	14	8			20		25			100	
Persicaria minor	6	11	4	5							50		
Lipandra polysperma	6	6	24	8					25		50		
Atriplex prostrata	6	5	25	5			60	25	25				
Persicaria dubia	6	5	5	1		20							
Xanthium orientale agg.	6	3	9	5									
Chenopodium ficifolium	6	4	7	1			40						
Phragmito-Magnocaricetea													
Veronica anagallis-aquatica	6	31	32	22	11	40	20	25	25		50	100	
Alisma plantago-aquatica	6	27	13	52	22	20		50		50		100	100
Leersia oryzoides	6	26	8	30		20					50		
Oenanthe aquatica	6	20	32	27	11	60		50	50			100	
Rorippa amphibia	6	14	34	26	22				25				
Bolboschoenus maritimus agg.	6	12	12	34	11	40	60		100		50		

Table 1. *Cont.*

Vegetation Units	Layer	1	2	3	4	5	6	7	8	9	10	11	12
Phalaroides arundinacea	6	10	31	21	22	20		25					100
Eleocharis palustris agg.	6	15	8	11					25			100	
Typha latifolia	6	12	15	19	11	20		25					
Phragmites australis	6	9	17	12	22	20	40		25			100	
Typha angustifolia	6	7	7	8	11	20			25	50			
Poa palustris	6	6	11	5	11	20							
Veronica beccabunga	6	5	5	5	11					100		100	
Butomus umbellatus	6	2	12	11				25					
Molinio-Arrhenatheretea													
Lythrum salicaria	6	27	22	25	56		20					100	
Taraxacum sect. *Ruderalia*	6	19	20	8	33	20	20	75	25				
Agrostis stolonifera	6	19	17	12	44	20	40		25				
Juncus effusus	6	10	11	5	22								
Trifolium hybridum	6	12	5	4	11				25				
Myosotis palustris agg.	6	10	8	11	11								
Trifolium repens	6	9	12	1	33						50		
Juncus compressus	6	8	6	7	22	20	40		50				
Argentina anserina	6	7	10	7									
Rorippa sylvestris	6	7	6	8	22	20			25				
Ranunculus repens	6	6	9	4	33	20							
Alopecurus geniculatus	6	7	4	5									
Poa trivialis	6	2	11	3	33	20		25					
Juncus inflexus	6	3	4	1	44	20							
Scheuchzerio-Caricetea nigrae													
Juncus articulatus	6	34	15	29	56	100			50	50	50		
Potamogetonetea													
Callitriche palustris agg.	6	18	6	7									
Papaveretea rhoeadis													
Echinochloa crus-galli	6	41	43	36	22	20	60	25	75		100	100	
Tripleurospermum inodorum	6	16	23	10		20	20		25		50		
Epilobietea angustifolii													
Urtica dioica	6	8	28	5	22	20		25			50		
Myosoton aquaticum	6	9	19	5	11	20		25	25		50		
Epilobium hirsutum	6	6	6	8									100
Others													
Salix sp.	7	34	38	18	22			75	25			100	
Lycopus europaeus	6	16	18	18	22	20		25					
Nostoc sp.	9	14	4	3		20							
Polygonum aviculare agg.	6	9	19	8									
Epilobium ciliatum	6	11	5	5	11								
Plantago major	6	9	6	10	11				25				
Ochlopoa annua	6	8	9	3	22						50		
Epilobium sp.	6	8	4	5	22			25			50		
Sagina procumbens	6	9	5	3	11						50		
Bryum argenteum	9	9	6		11						50		

Altogether 589 taxa of vascular plants, bryophytes, and algae were identified in the 925 relevés with *Cyperus fuscus* (Table S5; only the 549 taxa occurring in the classified relevés are presented). Among the taxa often co-occurring with *C. fuscus*, some diagnostic species of the classes Isoëto-Nanojuncetea (e.g., *Plantago uliginosa* subsp. *intermedia*, *Gnaphalium uliginosum*, *Juncus bufonius*, and *Limosella aquatica*) and Bidentetea (e.g., *Persicaria lapathifolia* s. l., *Rumex maritimus*, *Rorippa palustris*, and *Ranunculus sceleratus*) were the most frequent (frequency >25% in the whole dataset). Very common were also *Echinochloa crus-galli* (frequency >25%) from the class Papaveretea rhoeadis, *Alisma plantago-aquatica*, *Leersia oryzoides*, and *Oenanthe aquatica* (frequency of each species >20%) from the class Phragmito-Magnocaricetea, *Lythrum salicaria* from the class Molinio-Arrhenatheretea (frequency >20%), and several other species (Table 1 and Table S5). Most of the species across all the classes had an overall frequency lower than 10%, whereas large parts of them occurred in only one or a few relevés.

Altogether, 91 alien species (i.e., ca 15% of the total number of species) were identified in the vegetation with *Cyperus fuscus*, with 53 and 38 taxa assigned to archaeophytes and neophytes, respectively. Most of them were rare, occurring in less than 5% of the relevés. Among the aliens with rather high frequency were *Echinochloa crus-galli* (arch), *Bidens frondosus* (neo), *Xanthium orientale* agg. (neo), and *Epilobium ciliatum* (neo).

3.2. Factors Influencing Species Composition of the Vegetation with *Cyperus fuscus*

The relation between the selected explanatory variables and the species composition of the plots was highly significant (test on the first axis: $F = 1.6$, $p < 0.001$; test on all axes: $F = 7.6$, $p < 0.001$). The proportion of variation explained by the explanatory variables was 8.93% (adjusted explained variation). The marginal effects of all the explanatory variables were also highly significant (Table 2) although the proportions of explained variation were relatively low. The general additive model (GAM) testing the relation of year and the vegetation composition was significant ($F = 8.3$, $p < 0.001$ in the ordination space represented by the 1st and 2nd axes; $F = 37.0$, $p < 0.001$ in the ordination space represented by the 1st and 3rd axes). The model (Figure 1) indicates the decline of some rather sensitive wetland species with lower nutrient and temperature demands (*Drepanocladus aduncus*, *Eleocharis acicularis*, *Juncus bulbosus*, *Lythrum portula*) and the increase of some taller, nutrient demanding species from the Bidentetea and Phragmito-Magnocaricetea classes (*Bidens frondosus*, *Persicaria lapathifolia* s. l., *Phalaroides arundinacea*, *Rumex maritimus*, *Rorippa amphibia*) and willow seedlings (*Salix* sp.) in the newer relevés.

Table 2. Explanatory variables used in the canonical correspondence analysis (CCA) and their marginal effects. The *p*-values were adjusted using the false discovery rate. Environmental variables nutrients, moisture, soil reaction, continentality, temperature, and light correspond to relevant Ellenberg indicator values.

Variable	Explained Variation [%]	*F*	*p* (adj.)
Nutrients	1.94	16.0	<0.001
Moisture	1.75	14.5	<0.001
Soil reaction	1.59	13.1	<0.001
Continentality	1.50	12.4	<0.001
Temperature	1.11	9.1	<0.001
Number of archaeophytes	1.08	8.9	<0.001
Number of neophytes	0.92	7.6	<0.001
Light	0.88	7.2	<0.001
Year	0.85	6.9	<0.001
Total cover of archaeophytes	0.63	5.2	<0.001
Total cover of neophytes	0.49	4.0	<0.001
Cyperus fuscus cover	0.27	2.2	<0.001

Correlations among all the explanatory variables are displayed in the correlograms (Figure S1). These results show significant relationships between *Cyperus fuscus* cover and EIVs for temperature, moisture, light, nutrients, and continentality, while the relationships to the other factors including the year, number of species, and cover of archaeophytes and neophytes, and EIVs for soil reaction were not significant (Figure 2 and Figure S1). Optima of *C. fuscus*, according to EIVs derived from the species co-occurring in the relevés, show *C. fuscus* to be a highly light demanding species. Its optima for moisture and temperature correspond to 7.5 and 6, respectively, i.e., slightly higher values than the mean of the relevant scale (= 6.5 on the 12-degree scale for moisture and 5 on the 9-degree scale for temperature). In the case of nutrients, the optimum of *C. fuscus* corresponds to 5, i.e., the mean value, and the value 4 for continentality is slightly lower than the mean (both EIVs have a 9-degree scale). Archaeophytes and neophytes show rather different relationships to the environmental factors and time than *C. fuscus* (Figure S1). Neophyte species numbers significantly increased since the 1930s, while their cover change was not significant. In contrast, the highest species numbers and covers for archaeophytes were recorded in the historical relevés from the 1940s to the 1960s. Both groups of aliens exhibit substantially higher values for nutrients and lower values for moisture than *C. fuscus*. EIVs of the aliens for light show the opposite trend than *C. fuscus* cover. In contrast, similarities between the response curves of aliens and *C. fuscus* are shown at EIVs for soil reaction, but the relationship is not significant for *C. fuscus* cover ($p = 0.092$; Figure S1). Similar positions of optima are also found in the response curves of the EIVs for continentality. Finally, the response curves for EIVs for temperature also exhibit similar trends for number of neophyte species and *C. fuscus* cover; the optimum EIVs for temperature for archaeophytes is substantially higher (Figure S1).

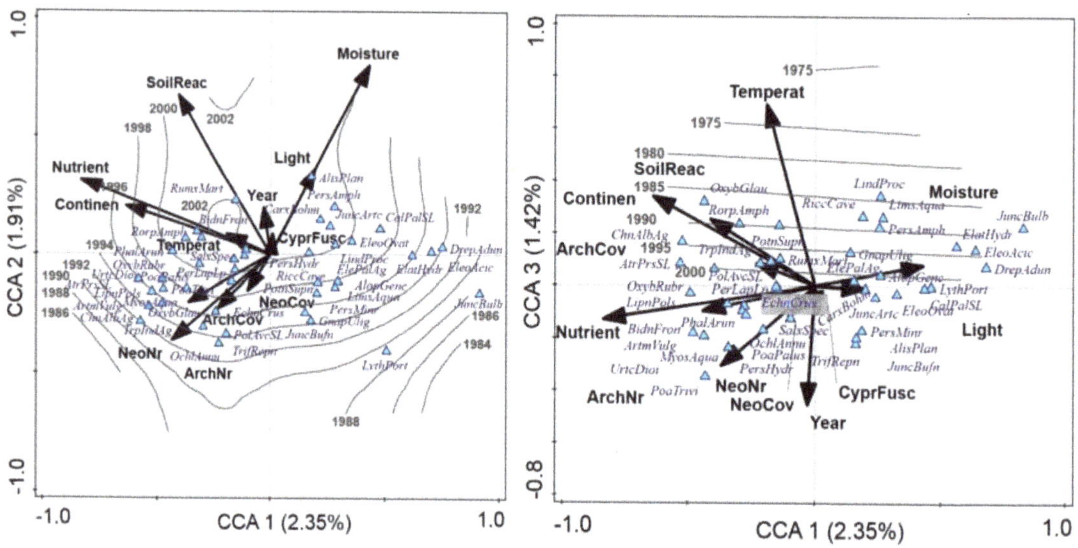

Figure 1. Canonical correspondence analysis (CCA) ordination diagram of species and independent variables supplemented with contour lines representing the year. Only 42 species with the highest contribution on the horizontal axis are shown.

3.3. Similarity between Soil Seed Bank and Above-Ground Vegetation (Relevés)

There were 16.7 species on average in the relevés across the three habitat types (16.2 in river habitats, 17.6 in fishpond habitats, and 17.0 in fish storage pond habitats, $n = 32$). In the soil seed bank, there was an average of 12.5 species per m^2 (14.0 in river habitats, 11.0 in fishpond habitats, and 11.9 in fish storage pond habitats, $n = 32$; Figure S2). Sørensen's similarity index (the similarity between the vegetation and the soil seed bank) for the habitat type river was 0.57 ± 0.12 (mean $\pm$ SD; median = 0.60, $n = 11$); for the habitat type fishpond 0.55 ± 0.15 (median = 0.53, $n = 10$); and for the habitat type fish storage pond 0.51 ± 0.11 (median = 0.49, $n = 9$; Figure 3a). There was no significant difference in Sørensen's similarity index among the three habitat types.

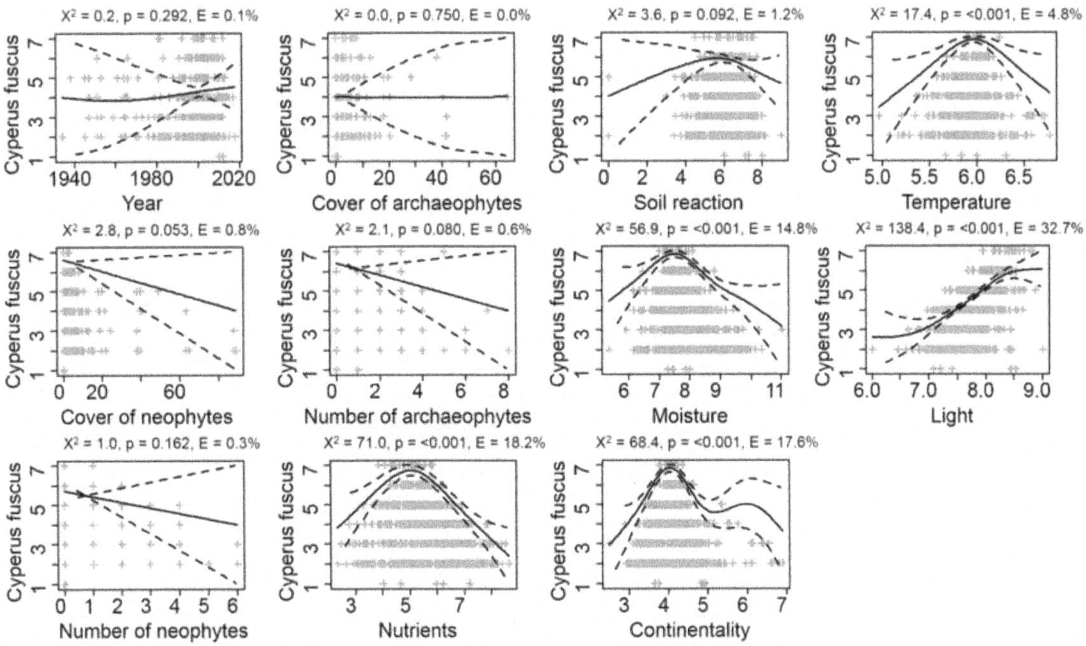

Figure 2. The relationships between the *Cyperus fuscus* cover in the phytosociological relevés and selected independent variables fitted by generalized additive models (GAM). The comma-separated values below each subplot denote: the chi-square test statistics of the model, model significance (p-value), explained deviance (variation).

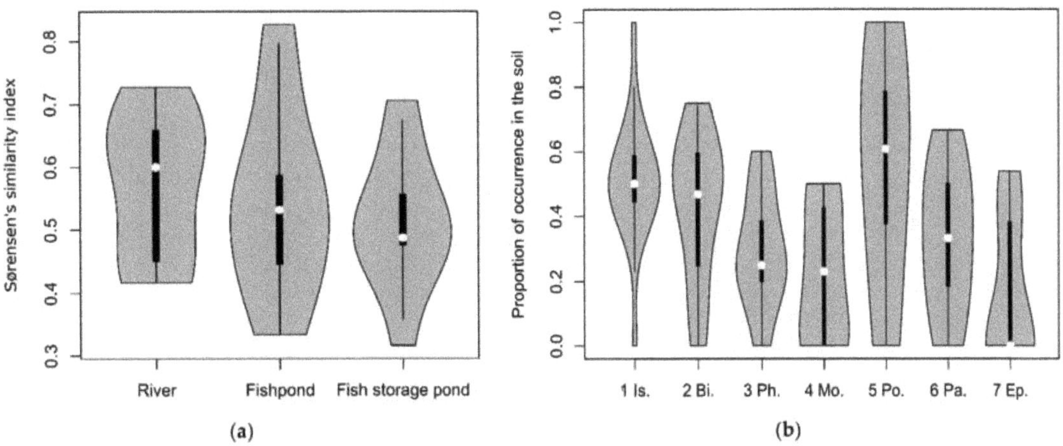

Figure 3. (**a**) Violin plots of Sørensen's similarity index showing the similarity in species composition between the soil seed bank and the above-ground vegetation (relevé) for three habitat types: river (n = 11), fishpond (n = 10), and fish storage pond (n = 9). (**b**) Violin plots showing the proportion of occurrence in the soil of the species of seven groups according to phytosociological classes: (1) Isoëto-Nanojuncetea and Crypsietea (n = 17), (2) Bidentetea (n = 18), (3) Phragmito-Magnocaricetea and Scheuchzerio-Caricetea (n = 17), (4) Molinio-Arrhenatheretea and other grassland classes (n = 15), (5) Potamogetonetea and Lemnetea (n = 4), (6) Papaveretea and other classes of annual ruderal vegetation (n = 11), and (7) Epilobietea and other classes of perennial ruderal vegetation (n = 11).

Across all habitat types, Sørensen's similarity index showed a mean similarity between the soil seed bank and the above-ground vegetation of 0.55 (SD = 0.13; n = 32). The belonging of species to groups according to phytosociological classes may be used to explain the remaining dissimilarity, as not all species occurring in mudflat habitats are equally likely to form a soil seed bank there. The proportion of variance explained by the linear model used to estimate the effect of the belonging of a species to one of seven groups according to phytosociological classes on its proportion of occurrence in the soil (multiple R-squared) is 0.25. Isoëto-Nanojuncetea and Crypsietea species (the reference; mean proportion of occurrence in the soil = 0.509, SD = 0.194, n = 17), as well as Potamogetonetea and Lemnetea species (mean proportion = 0.554, SD = 0.422, n = 4), are equally abundant in the soil and in the above-ground vegetation (Table 3; Figure 3b). Bidentetea species (mean proportion = 0.423, SD = 0.224, n = 18) are somewhat less frequently represented in the soil seed bank. Phragmito-Magnocaricetea and Scheuchzerio-Caricetea species (mean proportion = 0.293, SD = 0.176, n = 17, $p < 0.01$), as well as Molinio-Arrhenatheretea and other grassland species (mean proportion = 0.215, SD = 0.213, n = 15, $p < 0.001$), are significantly less frequent in the soil. Papaveretea and other annual ruderal species (mean proportion = 0.330, SD = 0.228, n = 11, $p < 0.05$) are less common in the soil than the reference. Lastly, Epilobietea, and other perennial ruderal species are the least abundant in the soil (mean proportion = 0.164, SD = 0.232, n = 11, $p < 0.001$).

Table 3. Linear model of the proportion of occurrence in the soil depending on the belonging of species to one of seven groups according to phytosociological classes. Isoëto-Nanojuncetea and Crypsietea species serve as reference. The estimate of the other groups indicates the deviation from the reference. Significance codes: * <0.050; ** <0.010; *** <0.001.

Coefficient	Estimate	t Value	p Value	
Isoëto−Nanojuncetea and Crypsietea (intercept)	0.509	9.53	0.000	***
Bidentetea	−0.086	−1.16	0.251	
Phragmito-Magnocaricetea and Scheuchzerio-Caricetea	−0.216	−2.86	0.005	**
Molinio−Arrhenatheretea and other grassland classes	−0.294	−3.77	0.000	***
Potamogetonetea and Lemnetea	0.045	0.36	0.717	
Papaveretea and other classes of annual ruderal vegetation	−0.179	−2.10	0.038	*
Epilobietea and other classes of perennial ruderal vegetation	−0.345	−4.05	0.000	***

3.4. Chromosome Number, Genome Size, and Ploidy Level

Unequivocal chromosome counts with $2n$ = 36 chromosomes were obtained from one plant each of seven populations of *Cyperus fuscus* (one from Austria, three from the Czech Republic, one from Hungary, and two from Slovakia; Figure 4; Table S4). This chromosome number corresponds to a genome size of 0.231 pg/1C (population SK1) or 0.232 pg/1C (population HU2). The mean and standard deviation calculated across 29 additional population 1C-values is 0.233 ± 0.004 pg/1C (range of population 1C-values = 0.227–0.243 pg; Table S4). The flow cytometric measurements (DAPI) of 223 individuals representing 50 populations from Austria, Croatia, the Czech Republic, Hungary, Italy, Poland, and Slovakia also revealed a uniform relative DNA content (Table S4) suggesting that all plants represent the same ploidy level.

3.5. Effect of Plant Community (Vegetation) Diversity on Genetic Diversity in River and Anthropogenic Habitats

There is no significant correlation between Shannon diversity (H) and average expected heterozygosity (for above-ground H_S: r = −0.111, p = 0.554, n = 31; for soil H_S: −0.233, p = 0.207, n = 31). The structural equation model (standardized solution, n = 31) shows the regression coefficient of Shannon diversity on average expected heterozygosity (R-squared of H_S = 0.70) in consideration of the direct and indirect effects of the third variable habitat type (river habitat = reference; Figure 5). There is a moderately significant regression path from Shannon diversity to H_S with a standardized effect size of −0.291 ($p = 0.013$). The bivariate correlation between the two variables is weaker and not significant, because it is reduced by the effect of the third variable, habitat type, which has a negative effect on both Shannon diversity (weak and not significant effect) and H_S (strong and significant effect). The third variable effect of habitat type, thus, creates a weak positive relationship between Shannon diversity and H_S, which weakens the negative correlation between the two variables.

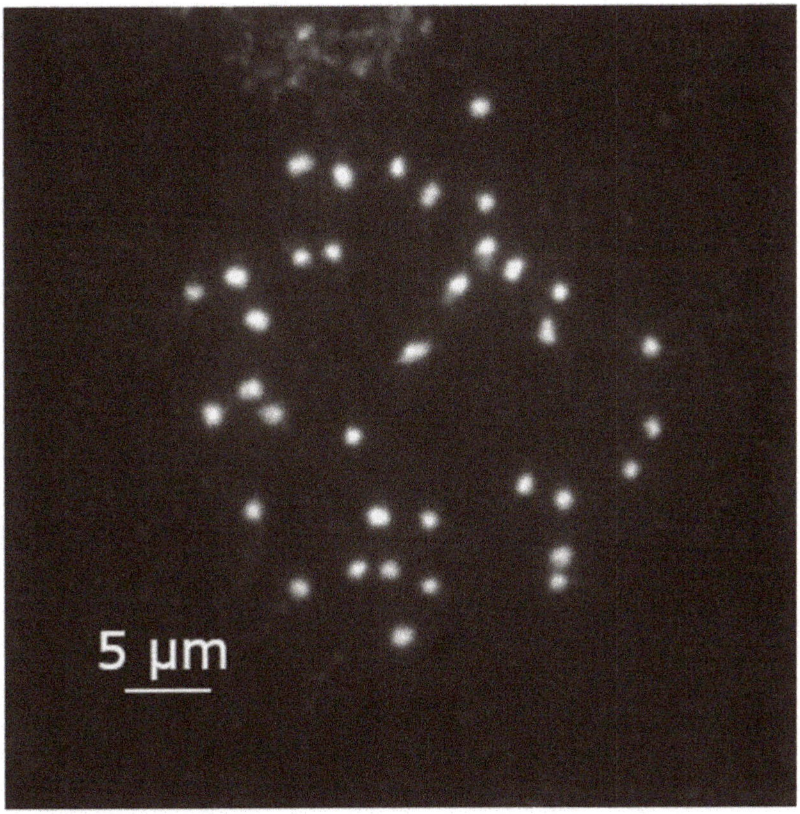

Figure 4. Mitotic metaphase chromosomes of *Cyperus fuscus* from the dam reservoir Vodná nádrž Horná Studená voda NW of Tomky, Slovakia ($2n = 36$; P. Kúr, K. Tremetsberger, 20.9.2013).

The relative proportion (species richness and cover) of Isoëto-Nanojuncetea species is somewhat higher in the anthropogenic (fishpond and fish storage pond) habitats in comparison to river habitats (Figures 6 and 7). The mean overall cover of the herbaceous layer does also not differ much among the three habitat types. Plots with *Cyperus fuscus* in river habitats have a mean cover of 55.2% (SD = 17.9), fishpond habitats of 66.2% (SD = 17.6), and fish storage pond habitats of 55.7% (SD = 18.8).

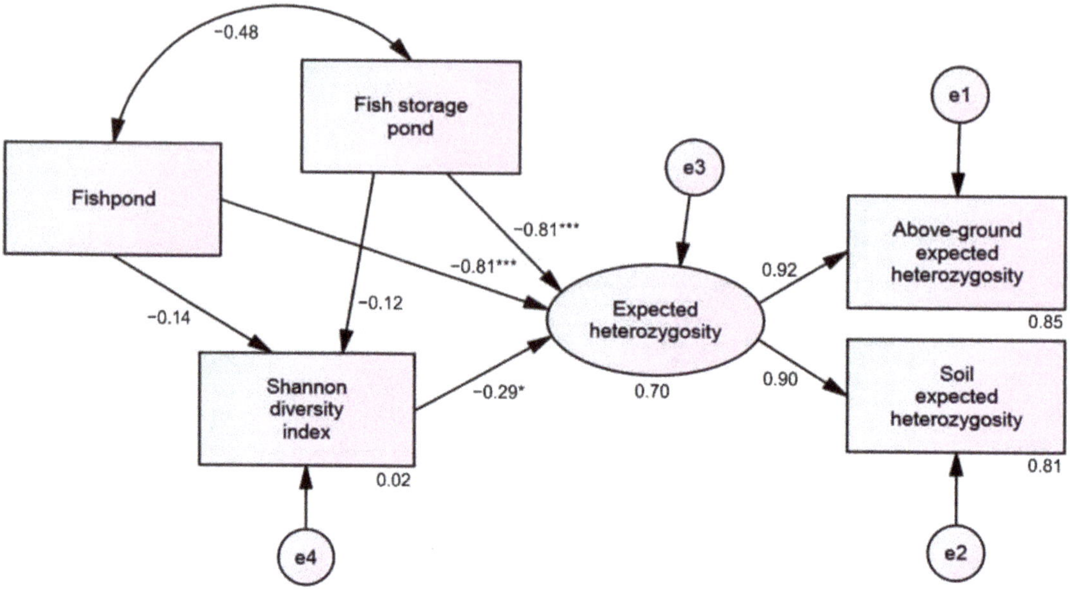

Figure 5. Standardized solution of the structural equation model investigating the effect of Shannon diversity (H) on average expected heterozygosity (H_S) in consideration of the direct and indirect effects of the third variable habitat type (river = reference, n = 31). Measures of model fit: $\chi^2 = 5.057$, $\chi^2/\mathrm{df} = 2.528$, RMSEA = 0.226, TLI = 0.778. Significance codes: * <0.050; *** <0.001.

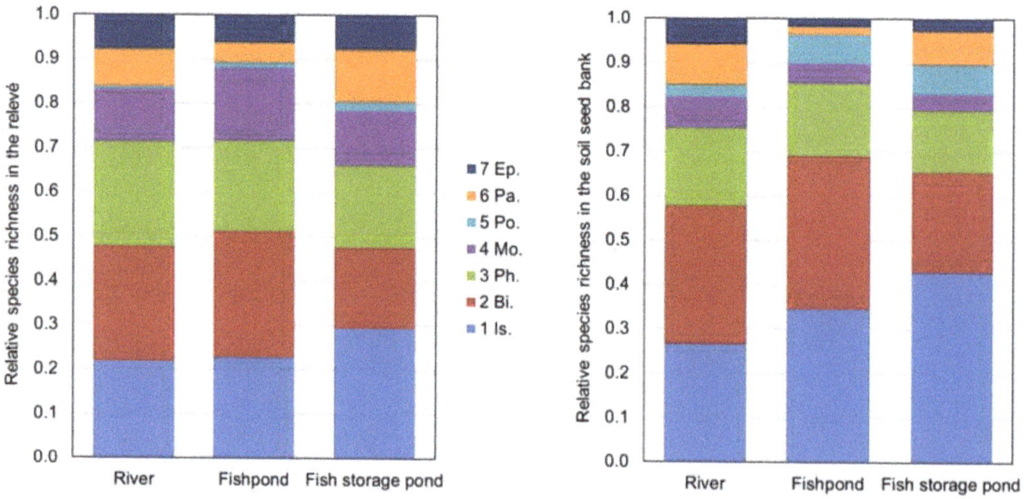

Figure 6. Relative species richness in the relevé and in the corresponding soil seed bank of species groups according to vegetation classes in the three habitat types, rivers (n = 11), fishponds (n = 10), and fish storage ponds (n = 9), based on the species list used for comparison of the relevé with the soil seed bank on the same 1 m^2 plot (Table S3). (1) Isoëto-Nanojuncetea and Crypsietea, (2) Bidentetea, (3) Phragmito-Magnocaricetea and Scheuchzerio-Caricetea, (4) Molinio-Arrhenatheretea and other grassland classes, (5) Potamogetonetea and Lemnetea, (6) Papaveretea and other classes of annual ruderal vegetation, and (7) Epilobietea and other classes of perennial ruderal vegetation.

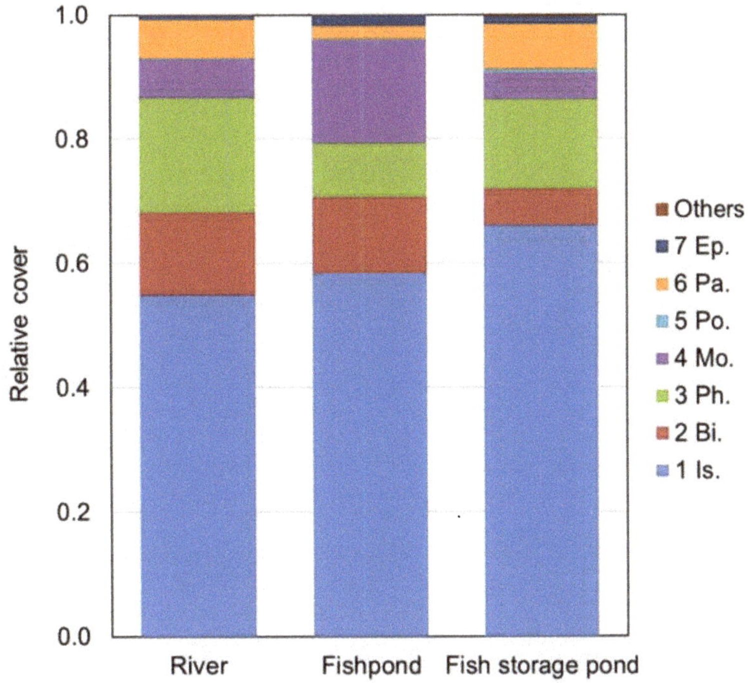

Figure 7. Relative cover (average of three 1 m^2 plots) of species groups according to vegetation classes (see Figure 6) in the three habitat types, rivers (n = 11), fishponds (n = 10), and fish storage ponds (n = 10).

3.6. Distribution Trends

The oldest dated *Cyperus fuscus* records are from 1833, which also provides the lower limit of the produced frequency chart (Figure 8). The curve of absolute (uncorrected) values shows important inter-annual fluctuations in the number of records during the whole study period (until 2015). These are, however, most remarkable after 1995 when the number of records strongly increased. Although the botanical activity was the highest in this period, the curve of the corrected species abundance still shows a clear trend in growing abundance values and, at the same time, low uncertainty expressed as 99% confidence intervals (Figure 8). The corrected values in some periods between the 1830s and 1910s are comparable or even higher than in the period after 1995 but uncertainty in the estimation of the corrected species abundance is high due to low numbers of the old records, particularly before the 1850s.

3.7. Climatic Conditions

All 19 bioclim variables differentiated significantly between the map quadrants with and without the records of *Cyperus fuscus* (Table 4). The differentiation was predominantly unilateral, i.e., the quadrants without *C. fuscus* mostly differed from quadrants occupied by the species, but not vice versa (Figure S3). Climatic variables related to temperature had larger effects than variables related to precipitation. The occurrences of *C. fuscus* correlated positively with temperature and negatively with precipitation. Summary charts showed that *C. fuscus* has the optimum in regions with about 600 mm annual precipitation and 8 °C annual mean temperature (Figure 9).

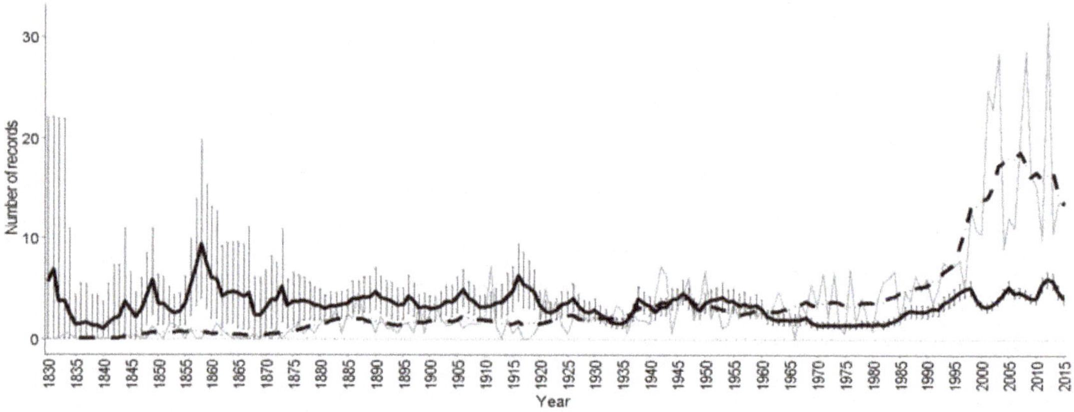

Figure 8. Numbers of individual records of *Cyperus fuscus* in the Czech Republic. Thin grey line–frequency curves; thick dashed line–11-year moving average; thick solid line–moving average corrected for botanical activity. Whiskers represent 99% confidence intervals of the means.

Table 4. Climatic variables and their significance in the canonical discriminant analysis (CDA). Biplot scores of explanatory variables represent the contribution of climatic variables to the first canonical axis separating presences and absences of *Cyperus fuscus* [77]. Climatic variables with positive biplot score values are positively correlated with *C. fuscus* occurrence, variables with negative values are positively correlated with *C. fuscus* absence. The significance was adjusted using the false discovery rate.

Climatic Variable	Description	Explained Variation [%]	*p* (adj.)	Biplot Score
bio10	Mean temperature of warmest quarter	19.1	<0.001	0.92
bio08	Mean temperature of wettest quarter	18.9	<0.001	0.91
bio05	Maximum temperature of warmest month	18.9	<0.001	0.91
bio01	Annual mean temperature	18.7	<0.001	0.91
bio11	Mean temperature of coldest quarter	15.2	<0.001	0.82
bio04	Temperature seasonality	13.6	<0.001	0.77
bio07	Temperature annual range	12.4	<0.001	0.74
bio06	Minimum temperature of coldest month	9.4	<0.001	0.65
bio02	Mean diurnal range	9.3	<0.001	0.64
bio12	Annual precipitation	6.7	<0.001	-0.54
bio14	Precipitation of driest month	6.6	<0.001	-0.54
bio17	Precipitation of driest quarter	6.4	<0.001	-0.53
bio19	Precipitation of coldest quarter	6.3	<0.001	-0.53
bio16	Precipitation of wettest quarter	5.0	<0.001	-0.47
bio18	Precipitation of warmest quarter	5.0	<0.001	-0.47
bio15	Precipitation seasonality	4.5	<0.001	0.44
bio13	Precipitation of wettest month	4.3	<0.001	-0.44
bio09	Mean temperature of driest quarter	4.2	<0.001	0.43
bio03	Isothermality	3.4	<0.001	0.39

3.8. Climatic Niche Modelling

The best maxent model (LQH, RM = 1) resulted in high AUC score 0.83 ± 0.0002 (mean ± variance) and low omission rate 0.11 ± 0.0004. The occurrences of *Cyperus fuscus* in the Czech Republic are concentrated in three regions: river valleys and surrounding lowlands in N and SE parts and fishpond basins in SW parts of the country's territory (Figure 10). The species clearly prefers regions with higher mean annual temperatures, which are rather poor to moderately rich in precipitation (Table 4, Figure 9). These regions are shown to be important for populations of *C. fuscus* already in the 19th century, with many further finds during the 20th and 21st century. In contrast, most localities in cold regions (where older occurrences are exceptional) were recorded in the 20th and, most importantly, 21st centuries (Figure 10).

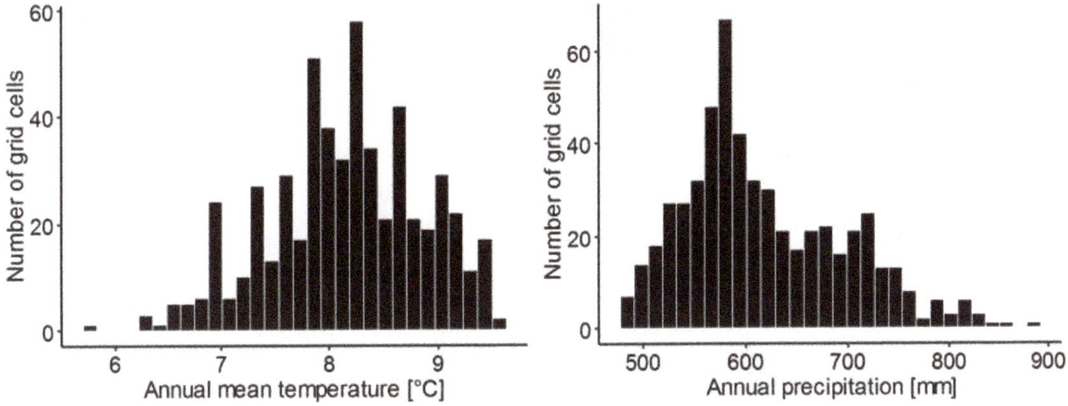

Figure 9. Histograms of the annual mean temperature and annual precipitation of the mapping grid cells in the Czech Republic occupied by *Cyperus fuscus*.

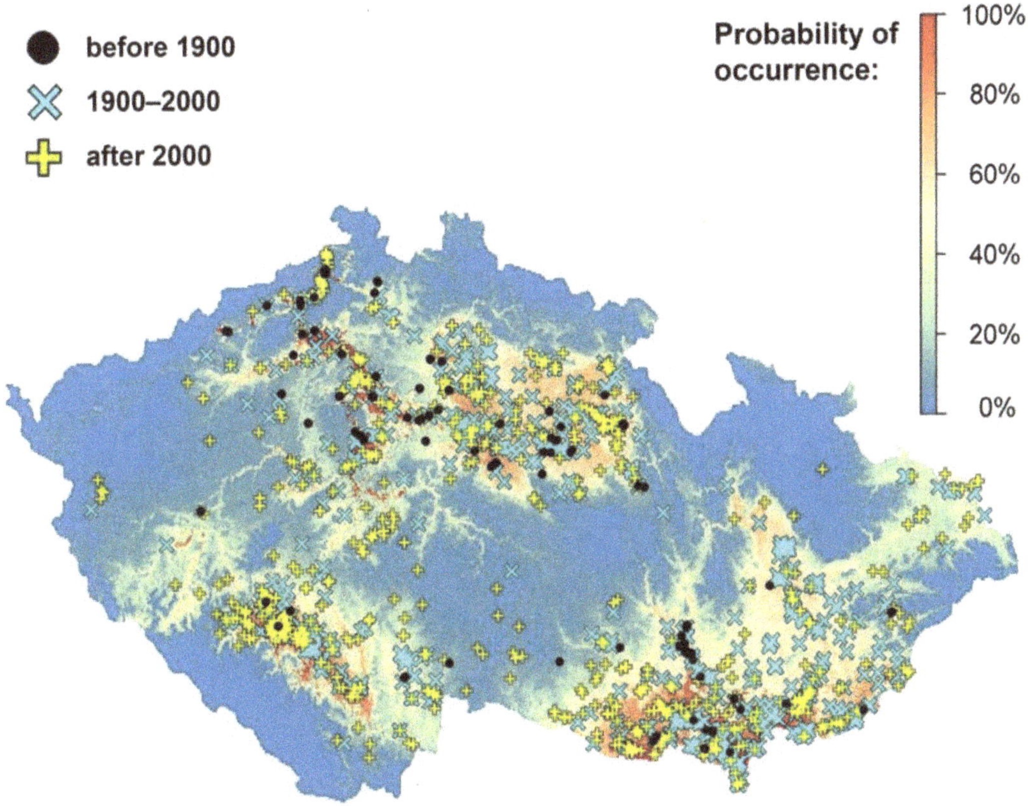

Figure 10. Predicted climatic suitability of habitats for *Cyperus fuscus*. The actual species' occurrences are plotted with black circles, turquoise x-marks, and yellow crosses.

4. Discussion

4.1. Vegetation and Cyperus fuscus Frequency Change over Time

As a large part of the relevés with *Cyperus fuscus* from the phytosociological databases lacked any environmental data, such as habitat type, type of substrate, or substrate moisture, we could only classify the relevés into the predefined syntaxa, describe the overall species composition of the vegetation, and analyze how it is influenced by the year of sampling and environmental variables expressed as Ellenberg indicator values (EIVs). We also analyzed the correlations between the cover of *C. fuscus* and the species number and cover of archaeophytes and neophytes.

Our results show that *Cyperus fuscus* grows in a broad range of non-forest vegetation types, with an optimum in the communities of the Isoëto-Nanojuncetea class. It is also frequent in the vegetation of the Bidentetea tripartitae and Phragmito-Magnocaricetea classes. Although the species of the latter two classes are usually taller and competitively stronger than small wetland annuals of the Isoëto-Nanojuncetea class, it seems that *C. fuscus* is rather tolerant of these conditions. This is probably related to its morphological plasticity: when soil moisture is high, the species can reach a height of up to several dozen centimeters [81]. Under sub-optimal substrate moisture not only *C. fuscus* but also potential competitors from the other wetland classes, particularly Bidentetea, decrease their size. Even the stands classified within the class Isoëto-Nanojuncetea were characterized by high frequency (Table 1) and sometimes also cover (data not displayed) of the species of the Bidentetea and Phragmito-Magnocaricetea classes. An important share of these rather nutrient demanding species points out the tolerance of *C. fuscus* to elevated nutrient amounts (particularly nitrogen, sometimes also phosphorus) in the soil. However, its optimum is in habitats with average amounts of nutrients, as shown by the correlation of *C. fuscus* cover and EIVs for nutrients (Figure 2 and Figure S1). As in other mudflat species, the optimum of *C. fuscus* may be modified by high competition pressure of nitrophilous species under the conditions of high nutrient amounts (see e.g., Šumberová et al. [82] for competition by *Leersia oryzoides* to small wetland annuals). Indeed, during our field data sampling we often encountered numerous populations of *C. fuscus* on deep sapropelic muddy soils, which are known as an extraordinary nutrient rich substrate. Moreover, sapropelic substrates also store high amounts of water, which have a positive effect on the growth of *C. fuscus* [24]. These populations usually occurred on sites with limited cover of taller herbs, which is in accordance with very high demands of *C. fuscus* for light (see the EIVs for light, Figure 2). On the other hand, *C. fuscus* populations also exist in nitrogen and phosphorus poor, but calcium rich fens or in stands of the rare association Cyperetum flavescentis. Such relevés, mainly historical ones, were very rare in our dataset; most of them came from the Alps [83], some of them from the Carpathians (Hájek, unpublished data). As the occurrence of *C. fuscus* in these regions is rather sporadic, poorly predictable, and the populations growing in small-scale gaps in perennial vegetation are usually rather small, we did not include these regions into our field trips. This part of the vegetation variability was captured thanks to the data from the phytosociological databases; however, we could miss a relevant part of the genetic variability of *C. fuscus*.

Cyperus fuscus is considered a thermophilous species, especially due to the high temperatures required for germination [24] and the concentration of its occurrence in warm regions [23]. As its spread related to climate warming is supposed [23], we expected to detect distribution and frequency changes of this species in the Czech Republic (herbarium and other occurrence data) and, at the same time, changes in species composition of the vegetation with *C. fuscus* towards higher share of thermophilous species. We confirmed an increasing number of localities of *C. fuscus* in the last two decades, whereby the species is more frequently found in regions where the habitat conditions should not be suitable for its occurrence according to climatic modelling (Figure 10). These occurrences of the species outside of its common range are probably related to the extraordinary hot summers since the 1990s. The populations recorded may be temporary or they could be stored for a couple of years in the soil seed bank, where they are waiting for a similar situation [84,85].

Although the growing number of localities may be partly related to high field research activity in the country, also the corrected values (Figure 8) show the increasing trend and at the same time the high data reliability (the higher peaks in the 19th and the beginning of the 20th centuries have lower reliability, as shown by broad confidence intervals). Therefore, it might be quite surprising that we could not detect any increase of *C. fuscus* cover in the recent relevés (Figure 2). It is probably due to the sampling of most of the relevés without any focus to record the stands of *C. fuscus*. Besides our own data from 2012, the relevés were collected randomly to document relevant wetland vegetation types. The relevés with the occurrence of *C. fuscus* (cover $\geq$ 1%) were selected subsequently from the databases. Moreover, for the study of species cover changes, permanent plots on selected sites would be an optimal method (see Šumberová et al. [10] for *Crassula aquatica*). For a thermophilous species, like *C. fuscus*, we would also expect a higher temperature optimum expressed as EIV (Figure 2). Similarly, the climatic optima derived from the mapping quadrants occupied by *C. fuscus* in the Czech Republic seem to be quite low for mean annual temperature and quite high for yearly sum of precipitation in comparison to the climate of the warmest parts of the Czech Republic (temperature: 9–10 °C, precipitation: 500–550 mm [86]). We suppose that the recent spread of *C. fuscus* into the colder regions, which is directly reflected in the number of mapping quadrants, is behind these patterns. Considering the EIVs, some species with rather low temperature demands accompany *C. fuscus*, particularly in colder regions, while other thermophytes are absent. Precipitation probably plays an important role in the recent distribution of *C. fuscus*. The warmest parts of the countries studied are also the regions with rather low precipitation. There, *C. fuscus* is usually concentrated in the large river floodplains with a suitable mesoclimate (particularly high air humidity and thus slower substrate desiccation). It seems that the frequent temperature extremes in colder regions recently allow the species to spread because of more favorable precipitation amounts and/or their distribution throughout the growing season.

Cyperus fuscus is also considered as basiphilous (particularly calciphilous) [87] (pp. 221–226) [88]; this relationship should be particularly remarkable in colder regions [89]. Our results did not show a significant relationship of *C. fuscus* to soil reaction, but if we considered even the insignificant result (p = 0.092; Figure 2), we could see that there is no sign of higher EIV for soil reaction in this species. In contrast, it seems to be rather generalist in the relationship to soil reaction. Like for temperature, some transitional status of a large part of the relevés, where in one and the same stand species with different habitat requirements (in this case for soil reaction and calcium content) are mixed, probably also plays a role here. These could be for instance the relevés of the Polygono-Eleocharitetum ovatae association, which include some species preferring acidic non-calcareous substrates (e.g., *Carex bohemica*) but usually lack a higher amount of basiphilous/calciphilous species. It is likely that high summer temperatures in the last two decades not only help *Cyperus fuscus* to cross the climatic limits in some regions, but also to cope with the conditions of lower calcium availability. However, an opposite explanation is also possible: the continual change of soil chemistry as a consequence of land use and management (e.g., fishpond manuring and liming, wastewater disposal) and climate change synergy [90]. These conditions would support the species with higher requirements for soil reaction and/or nutrients, but the acidophilous species with lower nutrient demands could still thrive as relics of former soil properties.

Although there was a high share of archaeophytes and neophytes compared to the number of native species in the whole dataset of the relevés with *Cyperus fuscus*, we did not find a significant negative influence of aliens on *C. fuscus* cover. In most of the relevés occurred at least one alien species, and in many of them, more than 10% of the total species number was formed by archaeophytes and/or neophytes. It is more than given by Chytrý et al. [91] for the littoral vegetation where also the communities of wetland annuals and perennials were included. However, the vegetation in the relevés from fish storage ponds and sandy or gravelly river deposits often exhibits transition to ruderal communities, as shown by the relatively high amount of ruderal species in our plots for complex research of *C. fuscus* (Figures 6 and 7); many of these species are aliens. On the other hand, the cover of alien species in most of the relevés was lower than 10%. As many of these species were annuals having typically similar or lower height than *C. fuscus*, their competitive ability is also likely to be similar or lower than that of *C. fuscus* (see species traits in e.g., *Lindernia dubia* or *Veronica peregrina* in the database PLADIAS, www.pladias.cz, accessed on 1 March 2021). Moreover, annual species like *Echinochloa crus-galli*, which are capable of forming dense and tall stands, often occur in low, prostrate forms in the habitats of *C. fuscus*; this growth form is probably related to management, such as mowing [82] or to natural disturbances, such as grazing by wild ducks, geese, and swans, as this species may constitute an important part of their diet [92]. These disturbances enable coexistence of species with various habitat requirements and competitive abilities within one stand. Some other aliens, particularly terrestrial ruderal weeds, may mark the margin of the ecological range of *C. fuscus*, such as sites with substrates that are too dry for *C. fuscus*. In such cases, the low cover of *C. fuscus* is not the result of competition by aliens.

Importantly, neophytes show an increase in their species numbers in the recent relevés, but the cover did not change with time. The cover of archaeophytes even decreased. Still, we wanted to know if there is some overlap of ecological demands (expressed as EIVs) of both groups of aliens and *Cyperus fuscus* at its optimum. We did not detect overlaps for basic factors, such as nutrients, moisture, and light, which would explain why neither the archaeophytes nor the neophytes have any influence on *C. fuscus*: There is obviously no competition for basic resources in most cases. The overlaps of EIVs for continentality, temperature, and soil reaction may be attributable to similar geographical distribution patterns of the relevés with high *C. fuscus* cover and with high archaeophyte/neophyte cover (or species number). As each alien species has its specific traits, we compared ecological requirements of *C. fuscus*, presented here, and of *Lindernia dubia*, an alien analyzed in an earlier study [93]. Optima expressed as EIVs of *C. fuscus* are much more like those of the native *L. procumbens* than those of *L. dubia*, which is less moisture and light and more temperature and nutrient demanding.

In fact, *Cyperus fuscus* thrives under high substrate moisture, i.e., on sites with waterlogged or, in the time after seedlings' establishment, even shallowly flooded substrate [24,81]. However, many relevés in our dataset are from sites, where the substrate desiccates rather early (e.g., fish storage pond bottoms or sandy river deposits); the moisture is still sufficient (although not optimal) for *C. fuscus* growth [27], but, at the same time, it also enables germination of terrestrial species. The presence of large numbers of terrestrial species in the relevés obviously pushed down the EIV value for moisture of *C. fuscus* in our analyses (compared to the corrected values in www.pladias.cz (accessed on 21 February 2021), where *C. fuscus* has the EIV 9 for moisture). Still, it is difficult to evaluate both the species' moisture range and optimum, as the populations of different habitat types seem to differ in their moisture requirements and flooding tolerance [27].

Although we have shown that there is a high number of alien as well as native ruderal species in the vegetation with *Cyperus fuscus*, the same applies for rare and threatened taxa. We did not analyze these species in more detail, because there are large differences among the national species lists of Austria [15], the Czech Republic [19], Poland [94], and Slovakia [95], not only in the particular species assessments, but also in the classification systems. However, there are several dozen species in the vegetation with *C. fuscus* (Table S5), which are on the Red Data List of at least one of the target countries. Most of them are characteristic species of the Isoëto-Nanojuncetea class. Among the species, which are under high threat pressures in most of the target countries, are for instance *Cyperus flavescens*, *C. michelianus*, *Elatine hydropiper*, *Lindernia procumbens* (protected by Bern Convention on the Conservation of European Wildlife and Natural Habitats and also by the Habitats Directive of the European Union), and *Pulicaria vulgaris*. Even species like *Eleocharis acicularis*, *E. ovata*, and *Limosella aquatica*, having high frequencies in our dataset (Table 1 and Table S5), are highly threatened in Austria [15]. These facts highlight *C. fuscus* as a potential indicator of habitat conditions suitable for a range of species, which are much more threatened than *C. fuscus* itself.

4.2. Genetics and Soil Seed Bank

In contrast to our expectation derived from previous publication records ($2n$ = 36 for the Czech Republic [32], $2n$ = 72 for Slovakia [33]), we found no variation of chromosome numbers and no substantial variation in genome sizes. The sampling of populations for flow cytometry was rather dense in the Czech Republic, eastern Austria, western Slovakia, and western Hungary and it included various habitat and vegetation types. The genome size obtained related to $2n$ = 36 chromosomes. The number of $2n$ = 72 chromosomes reported from a gravel pit near Lakšárska Nová Ves, Slovakia [33], could not be found in the samples used in the current study. We searched for *Cyperus fuscus* in this area but could not find it again there. The closest sampled population with similar ecological conditions (acidic sand) is from the dam reservoir Vodná nádrž Horná Studená voda (constructed on the site of a former large gravel pit) NW of Tomky, Slovakia (P. Kúr & K. Tremetsberger, 20.9.2013) and has $2n$ = 36 chromosomes. Plants from other gravel pits in Slovakia (NW of Dubnica nad Váhom and SE of Lednické Rovne) did not significantly deviate in genome size, so that our data give no evidence of the occurrence of polyploids in *C. fuscus.* One might hypothesize that the appearance of the cytotype with $2n$ = 72 chromosomes was only temporary, possibly due to a lower viability. The incidence of spontaneous autopolyploids may often not be detected. *Cyperus fuscus* with $2n$ = 36 chromosomes behaves genetically as a diploid with no more than two alleles per microsatellite locus per individual and allele dosage corresponding to a diploid [28,96]. Roalson [97] noted that peaks in haploid chromosome numbers in the genus *Cyperus* at 18, 36, and 54 could be interpreted as 1X, 2X, 3X pattern. Another possible interpretation is genetic/cytological diploidization. Uniform genome sizes and genetic indication of diploid-like behavior recommend *C. fuscus* as an ecological-genetic model plant.

Previously, no greater genetic differentiation than that between sampling plots within a given site was found between plants of *Cyperus fuscus* derived from seeds stored in the soil and plants derived from above-ground individuals [28]. Böckelmann et al. [28] also showed that *C. fuscus* immigrants have similar probabilities of germinating or entering the soil seed bank after arrival in river or anthropogenic sites. Here, we used Sørensen's similarity index as an indicator of species turnover between the soil and the yearly above-ground vegetation. The index did not differ significantly among the three studied habitat types (rivers, fishponds, and fish storage ponds), although there is a slight tendency for higher similarity in species composition at river sites. In contrast, in *Cardamine amara,* which is also found in habitats influenced by water movements, there was a slight tendency for lower similarity in species composition at riverbanks compared to retention basins and meadows [31]. We conclude that more natural sites along rivers and anthropogenic habitats disconnected from the river with high density of *Cyperus fuscus* do not seem to differ much in their dynamics between the soil seed bank and the above-ground population/vegetation, neither with respect to individual species (genetic diversity) nor with respect to the vegetation (species diversity).

The structural equation model revealed a negative effect (standardized regression weight = -0.291, $p = 0.013$) of plant community (vegetation) diversity (measured by the Shannon index) on within-population genetic diversity (H_S). A negative effect can be expected, because *Cyperus fuscus* is a small and short-lived pioneer species (intermediate between the r- and s-strategy types) with high demand for light [85,98]; it is specialized to the conditions encountered in mudflat habitats, which are not suitable for many species. Under conditions of high competition from many other species, fewer genotypes of *C. fuscus* may be able to survive, because a diverse community of competitors may limit the ability of the individual species to use different parts of the environment (i.e., the niche breadth is reduced) [41]. The effect of the habitat type on H_S, however, is not mediated by plant community diversity (indirect effect of the fishpond habitat mediated by Shannon diversity = 0.041; indirect effect of the fish storage pond habitat mediated by Shannon diversity = 0.036), because the direct effect of the habitat type on Shannon diversity is weak and not significant (Figure 5). The strong and significant effect of the habitat type on H_S must therefore be mediated by other mechanisms than plant community diversity, composition, or cover (Figures 6 and 7), such as differences in the strength of selection pressures, founder effects when anthropogenic habitats were created through initiation of fish farming activities, or different levels and patterns of immigration [28]. Despite that the vegetation of anthropogenic habitats harbors a great diversity of Isoëto-Nanojuncetea species (similar to that of riverine stands of *C. fuscus*), *C. fuscus* plants from riverbanks exceed those from anthropogenic habitats in terms of genetic diversity, adaptation to flooding (lower cost of low-oxygen escape strategy), and fitness [27,28]. This has important implications for conservation: It is necessary to preserve the dynamic riverine habitats if one wants to maintain all the adaptive potential of the species.

Supplementary Materials: The following are available online at https://www.mdpi.com/article/10.3390/w13091277/s1, Figure S1: Pairwise relationships between the *Cyperus fuscus* cover in the phytosociological relevés and selected independent variables, Figure S2: Scatterplot of species richness in the soil and corresponding relevé, Figure S3: Overlap between climatic conditions of map squares with and without the confirmed presence of *Cyperus fuscus* shown by the canonical discriminant analysis (CDA), Table S1: Localities with *Cyperus fuscus* selected for complex study, Table S2: Merged taxa, Table S3: List of species in the soil and corresponding relevé, Table S4: Chromosome numbers, ploidies, and 1C-values in populations of *Cyperus fuscus*, Table S5: Percentage synoptic table of the vegetation (diagnostic species) with *Cyperus fuscus* (full version).

Author Contributions: Conceptualization, K.T., K.-G.B., and K.Š.; Data curation, K.T., P.D., Z.K., J.B., Z.H., and K.Š.; Formal analysis, P.K., K.T., Z.K., and K.Š.; Funding acquisition, S.P., K.T., K.-G.B., Z.H., and K.Š.; Investigation, P.K., S.P., K.T., P.D., Z.K., J.B., K.-G.B., Z.H., A.M., and K.Š.; Writing—original draft, P.K., K.T., and K.Š.; Writing—review & editing, S.P., P.D., Z.K., J.B., K.-G.B., Z.H., and A.M. All authors have read and agreed to the published version of the manuscript.

Funding: This research was funded by the Austrian Agency for International Cooperation in Education and Research, grant number CZ 13/2012 (to K.T.), the Ministry of Education, Youth and Sports of the Czech Republic, grant number 7AMB12AT015 (to K.Š., Z.H. and S.P.), the Hochschuljubiläumsstiftung der Stadt Wien, grant number H-2488/2012 (to K.T.), and the Austrian Science Fund (FWF), grant number P24558-B16 (to K.G.B.). Elaboration of this paper was funded by the institutional support number RVO 67985939 of the Institute of Botany, Academy of Sciences of the Czech Republic (to K.Š., Z.H. and S.P.). P.D. was supported by the Technology Agency of the Czech Republic project number SS02030018–"Center for Landscape and Biodiversity (DivLand)". Open access funding was provided by the BOKU Vienna Open Access Publishing Fund.

Institutional Review Board Statement: Not applicable.

Informed Consent Statement: Not applicable.

Data Availability Statement: The new relevés used in this study (111 relevés from plots for complex investigation of *Cyperus fuscus* collected in 2012 and 37 additional relevés collected between 2000 and 2019; see Section 2.2.1) are available from the Austrian Vegetation Database—AVD [44] (relevé numbers 366393–366413) and the Czech National Vegetation Database—CNPD [45] (relevé numbers 447561–447566, 447572–447589, 447860–447962).

Acknowledgments: The authors are grateful to Kateřina Bubíková (Prague, Czech Republic) and custodians and contributors of phytosociological databases (see the Appendix A for the main contributors) of Austria (custodian Wolfgang Willner), the Czech Republic (Ilona Knollová), Poland (Grzegorz Swacha), Slovakia (Jozef Šibík), and the Gravel Bar Database (Veronika Kalníková) for providing vegetation relevés, Kateřina Bubíková (Prague, Czech Republic), Steffen Hameister, Elke Naumer-Bernhardt, and Klemens and Michaela Wernisch (Vienna, Austria), Flavia Landucci (Brno, Czech Republic), and Ildikó Varga (Budapest, Hungary) for sampling plant material for chromosome counts and/or flow cytometry, Hanna Weiss-Schneeweiss and Eva M. Temsch (Vienna, Austria) for using laboratory, help with chromosomal analyses and interpretation, and support with flow cytometric measurements, Markéta Chudomelová and Hana Fialová for their help with field data sampling, Reinhard Hössinger (Vienna, Austria) for help with the structural equation model, Michal Ducháček (Prague, Czech Republic), Zdeněk Kaplan (Průhonice, Czech Republic), and Richard Hrivnák (Bratislava, Slovakia) for their tips on *Cyperus fuscus* localities, Jiří Danihelka (Brno, Czech

Republic) for his advice to elaboration of the data on distribution trends, and Peter Poschlod (Regensburg, Germany) for his advice to literature. Fish farmers of a number of fish farms in the Czech Republic, particularly Rybářství Hluboká nad Vltavou Cz. Ltd., Blatenská ryba Ltd., Rybářství Chlumec nad Cidlinou Comp., Klatovské rybářství Comp., Líšno Comp., Rybářství Mariánské Lázně Ltd., Rybniční hospodářství Ltd., Štičí líheň Esox Ltd., Rybářství Doksy Ltd., Dvůr Lnáře Ltd., and Krajské školní hospodářství České Budějovice, kindly allowed us our research in complexes of fish storage ponds with restricted access to the general public and provided us with information about the management of their fishponds and fish storage ponds.

Conflicts of Interest: The authors declare no conflict of interest.

Appendix A

Main Sources of Vegetation Data and General Rules Applied for their Selection

Custodians of the national databases in Austria, the Czech Republic, Poland, and Slovakia, and the custodian of the Gravel Bar Vegetation Database were asked for all the available relevés with the occurrence of *Cyperus fuscus*. Our own data, not yet included in national databases, were also included (145 relevés). Altogether, 1206 relevés were summarized in the selection procedure.

During the selection procedure in Turboveg database software we used visual prescreening and filtering of the data to remove all the relevés lacking some important part of the header data or having outlier position within the dataset. We removed (1) all the relevés with the cover of the target species less than 1% (i.e., cover value r according to Braun-Blanquet cover scale)—these relevés mainly contained vegetation types with often accidental occurrence of *Cyperus fuscus*, vegetation mosaics, etc.; (2) all the relevés without coordinates and/or alternative precise location, such as the nearest settlement or locality name; (3) all the relevés collected in other than our four target Central European countries, i.e., AT, CZ, PL, and SK; and (4) all the relevés without plot size information or with plot size smaller than 0.5 m^2 or larger than 50 m^2.

In the final step of the selection procedure, we performed a sort of geographical stratification. To reduce oversampling, only max. three relevés for each site (i.e., point given by coordinates or locality name, when the coordinates were missing) were considered. For fish storage pond systems, each pond was considered separately, as there is usually quite a high variability among the individual ponds. If there were very many relevés from a single large fishpond or water reservoir, not the coordinates but the part of the water body (e.g., NE) was considered, but the distance of at least 0.5 km between the sites was required.

Exceptions from the number of the three relevés per site were allowed in the following cases: (1) the relevés were assigned by their author or by our expert opinion to different phytosociological classes; (2) the relevés have been collected in different periods, in intervals of at least five years. The final relevé number from all the databases after the selection procedure was 925.

(1) Czech Republic—CNPD (Czech National Phytosociological Database; originally 516 relevés)

The final relevé number after the selection procedure was 446. This number includes 107 published relevés and 339 unpublished relevés. Published relevés were mainly contributed by Jaroslav Rydlo (36, particularly from the papers [99] and [100]), L. Bartoňová (9, all of them in [101]), L. Malíková (5, all of them in [102]) and J. Vicherek (17, mainly in [103]). Unpublished relevés were mainly contributed by K. Šumberová (267), S. Hejný (9), J. Danihelka (12), J. Vicherek (12), F. Krahulec (7), Z. Hroudová (9), M. Chytrý (6), and V. Sedláček (5).

(2) Slovakia—CDF (Slovak Vegetation Database, also called Central Database of Phytosociological Samples; originally 147 relevés)

The final relevé number after the selection procedure was 96. From this number, altogether 55 relevés have been published and 41 unpublished. Published relevés were mainly collected by H. Oťaheľová (13, various papers, in each of them less than 5 relevés), J. Vicherek (6 relevés, all in [104]), and T. Homola (5 relevés in various papers). The other authors contributed less than 5 relevés (R. Hrivnák, D. Dítě, L. Šomšák, and others). Unpublished relevés were mainly contributed by M. Valachovič (10 relevés) and H. Oťaheľová (26), in this number also the manuscript notes by S. Hejný (14 relevés).

(3) Poland—PVD (Polish Vegetation Database; originally 337 relevés)

The final relevé number after the selection procedure was 218. In this number, altogether 157 relevés were published and 61 unpublished. All unpublished relevés were collected by Z. Kącki. Within the published relevés, the data from the papers by J. Borysiak ([105], 35 relevés), R. Krawczyk et al. ([106], 17 relevés), T. Macicka-Pawlik et al. ([107], 8 relevés), Z. Podbielkowski ([108], 6 relevés), A. Popiela ([109], 39 relevés), K. Spałek ([110], 6 relevés; [111], 10 relevés), and M. Zając & A. Zając ([112], 5 relevés) were most frequent.

(4) Austria—AVD (Austrian Vegetation Database; originally 50 relevés)

The final relevé number after the selection procedure was 32. In this number, 21 relevés were already published. Two main papers were authored by A. Traxler ([113], 11 relevés) and E. Aichinger ([83], 8 relevés). The 11 unpublished relevés belong to so-called AVL projects, e.g., FFH-Kartierung Donauauen, provided by several authors (usually the authors were not given in the header data).

(5) Gravel Bar Vegetation Database—GBVD (originally 11 relevés from IT, PL, BG, and RO)

In the two relevés from PL (= the only target country), *Cyperus fuscus* only occurred with the cover value "r", and therefore no relevé was used from this database.

References

1. Khan, F.A.; Ansari, A.A. Eutrophication: An ecological vision. *Bot. Rev.* **2005**, *71*, 449–482. [CrossRef]
2. Mooij, W.M.; Hülsmann, S.; De Senerpont Domis, L.N.; Nolet, B.A.; Bodelier, P.L.E.; Boers, P.C.M.; Pires, L.M.D.; Gons, H.J.; Ibelings, B.W.; Noordhuis, R.; et al. The impact of climate change on lakes in the Netherlands: A review. *Aquat. Ecol.* **2005**, *39*, 381–400. [CrossRef]
3. Usio, N.; Imada, M.; Nakagawa, M.; Akasaka, M.; Takamura, N. Effects of pond draining on biodiversity and water quality of farm ponds. *Conserv. Biol.* **2013**, *27*, 1429–1438. [CrossRef] [PubMed]
4. Petsch, D.K. Causes and consequences of biotic homogenization in freshwater ecosystems. *Int. Rev. Hydrobiol.* **2016**, *101*, 113–122. [CrossRef]
5. Zohary, T.; Ostrovsky, I. Ecological impacts of excessive water level fluctuations in stratified freshwater lakes. *Inland Waters* **2011**, *1*, 47–59. [CrossRef]
6. Dhir, B. Status of aquatic macrophytes in changing climate: A perspective. *J. Environ. Sci. Technol.* **2015**, *8*, 139–148. [CrossRef]
7. Short, F.T.; Kosten, S.; Morgan, P.A.; Malone, S.; Moore, G.E. Impacts of climate change on submerged and emergent wetland plants. *Aquat. Bot.* **2016**, *135*, 3–17. [CrossRef]
8. Mucina, L.; Bültmann, H.; Dierßen, K.; Theurillat, J.P.; Raus, T.; Čarni, A.; Šumberová, K.; Willner, W.; Dengler, J.; Gavilán García, R.; et al. Vegetation of Europe: Hierarchical floristic classification system of vascular plant, bryophyte, lichen, and algal communities. *Appl. Veg. Sci.* **2016**, *19* (Suppl. 1), 3–264. [CrossRef]
9. Deil, U. A review on habitats, plant traits and vegetation of ephemeral wetlands—A global perspective. *Phytocoenologia* **2005**, *35*, 533–706. [CrossRef]
10. Šumberová, K.; Ducháček, M.; Lososová, Z. Life-history traits controlling the survival of *Tillaea aquatica*: A threatened wetland plant species in intensively managed fishpond landscapes of the Czech Republic. *Hydrobiologia* **2012**, *689*, 91–110. [CrossRef]
11. Richert, E.; Achtziger, R.; Dajdok, Z.; Günther, A.; Heilmeier, H.; Hübner, A.; John, H.; Šumberová, K. Rare wetland grass *Coleanthus subtilis* in Central and Western Europe—Current distribution, habitat types, and threats. *Acta Soc. Bot. Polon.* **2016**, *85*, 1–16. [CrossRef]
12. Gazaix, A.; Klesczewski, M.; Bouchet, M.A.; Cartereau, M.; Molina, J.; Michaud, H.; Muller, S.D.; Pirsoul, L.; Gauthier, P.; Grillas, P.; et al. A history of discoveries and disappearances of the rare annual plant *Lythrum thesioides* M.Bieb.: New insights into its ecology and biology. *Bot. Lett.* **2020**, *167*, 201–211. [CrossRef]
13. Lansdown, R.V. *Cyperus Fuscus*. The IUCN Red List of Threatened Species 2013: eT164079A13545918. 2013. Available online: https://dx.doi.org/10.2305/IUCN.UK.2013-1.RLTS.T164079A13545918.en (accessed on 21 February 2021).
14. Stoltze, M.; Pihl, S. *Rødliste 1997 over Planter og Dyr i Danmark*; Red List 1997 of Plants and Animals in Denmark; Miljø- og Energiministeriet, Danmarks Miljøundersøgelser & Skov- og Naturstyrelsen: Copenhagen, Denmark, 1998.
15. Niklfeld, H.; Schratt-Ehrendorfer, L. (Eds.) Rote Liste gefährdeter Farn- und Blütenpflanzen (Pteridophyta und Spermatophyta) Österreichs. 2. Fassung. In *Rote Listen Gefährdeter Pflanzen Österreichs. Grüne Reihe Bundesministerium Umwelt, Jugend und Familie 10*; Austria Medien Service: Graz, Austria, 1999; pp. 33–130.
16. Cheffings, C.M.; Farrell, L. The Vascular Plant Red Data List for Great Britain. *Species Status* **2005**, *7*, 1–116.

17. Bornand, C.; Gygax, A.; Juillerat, P.; Jutzi, M.; Möhl, A.; Rometsch, S.; Sager, L.; Santiago, H.; Eggenberg, S. *Rote Liste Gefässpflanzen. Gefährdete Arten der Schweiz*; Bundesamt für Umwelt; Bern & Info Flora: Geneva, Switzerland, 2016.
18. Metzing, D.; Garve, E.; Matzke-Hajek, G.; Adler, J.; Bleeker, W.; Breunig, T.; Caspari, S.; Dunkel, F.G.; Fritsch, R.; Gottschlich, G.; et al. Rote Liste und Gesamtartenliste der Farn- und Blütenpflanzen (Trachaeophyta) Deutschlands. In *Rote Liste Gefährdeter Tiere, Pflanzen und Pilze Deutschlands. Band 7: Pflanzen*; Metzing, D., Hofbauer, N., Ludwig, G., Matzke-Hajek, G., Eds.; Bundesamt für Naturschutz: Bonn-Bad Godesberg, Germany, 2018; Volume 70, pp. 13–358.
19. Grulich, V.; Chobot, K. (Eds.) *Červený seznam cévnatých rostlin ČR*; The Red List of Vascular Plants of the Czech Republic; Agentura ochrany přírody a krajiny ČR: Praha, Czech Republic, 2017; Volume 35, pp. 75–132.
20. Van Landuyt, W.; Hoste, I.; Vanhecke, L. Rode Lijst van de vaatplanten van Vlaanderen en het Brussels Hoofdstedelijk Gewest [Red List of vascular plants of Flanders and the Brussels capital region]. In *Atlas van de Flora van Vlaanderen en het Brussels Gewest [Atlas of the Flora of Flanders and the Brussels Region]*; Van Landuyt, W., Hoste, I., Vanhecke, L., Van den Bremt, P., Vercruysse, W., De Beer, D., Eds.; Instituut voor Natuur- en Bosonderzoek, Nationale Plantentuin van België & Flo.Wer.: Brussel, Belgium, 2006; pp. 69–81.
21. *Cyperus fuscus* L. Estonian Red List of Threatened Species. eBiodiversity. Available online: http://vana.elurikkus.ut.ee/kirjeldus.php?id=4208&inst=&org=&lang=eng (accessed on 17 February 2021).
22. Colling, G. Red List of the Vascular Plants of Luxembourg. *Ferrantia* **2005**, *42*, 1–77.
23. Kaplan, Z.; Danihelka, J.; Štěpánková, J.; Ekrt, L.; Chrtek, J., Jr.; Zázvorka, J.; Grulich, V.; Řepka, R.; Prančl, J.; Ducháček, M.; et al. Distributions of vascular plants in the Czech Republic. Part 2. *Preslia* **2016**, *88*, 229–322.
24. Von Lampe, M. Wuchsform, Wuchsrhythmus und Verbreitung der Arten der Zwergbinsengesellschaften. *Diss. Bot.* **1996**, *266*, 1–353.
25. Ansong, M.; Pickering, C. Are weeds hitchhiking a ride on your car? A systematic review of seed dispersal on cars. *PLoS ONE* **2013**, *8*, e80275. [CrossRef]
26. Šumberová, K.; Ducháček, M. Analysis of plant soil seed banks and seed dispersal vectors: Its potential and limits for forensic investigations. *Forensic Sci. Int.* **2017**, *270*, 121–128. [CrossRef]
27. Böckelmann, J.; Tremetsberger, K.; Šumberová, K.; Grausgruber, H.; Bernhardt, K.G. Fitness and growth of the ephemeral mudflat species *Cyperus fuscus* in river and anthropogenic habitats in response to fluctuating water-levels. *Flora* **2017**, *234*, 135–149. [CrossRef]
28. Böckelmann, J.; Tremetsberger, K.; Šumberová, K.; Kohl, G.; Grausgruber, H.; Bernhardt, K.G. Genetic variation in an ephemeral mudflat species: The role of the soil seed bank and dispersal in river and secondary anthropogenic habitats. *Ecol. Evol.* **2020**, *10*, 3620–3635. [CrossRef]
29. Abernethy, V.J.; Willby, N.J. Changes along a disturbance gradient in the density and composition of propagule banks in floodplain aquatic habitats. *Plant Ecol.* **1999**, *140*, 177–190. [CrossRef]
30. Aponte, C.; Kazakis, G.; Ghosn, D.; Papanastasis, V.P. Characteristics of the soil seed bank in Mediterranean temporary ponds and its role in ecosystem dynamics. *Wetlands Ecol. Manage.* **2010**, *18*, 243–253. [CrossRef]
31. Koch, M.; Huthmann, M.; Bernhardt, K.G. *Cardamine amara* L. (Brassicaceae) in dynamic habitats: Genetic composition and diversity of seed bank and established populations. *Basic Appl. Ecol.* **2003**, *4*, 339–348. [CrossRef]
32. Krahulcová, A. Chromosome numbers in selected monocotyledons (Czech Republic, Hungary, and Slovakia). *Preslia* **2003**, *75*, 97–113.
33. Májovský, J. Index of chromosome numbers of Slovakian flora (Part 6). *Acta Fac. Rerum Nat. Univ. Comen. Bot.* **1978**, *26*, 1–42.
34. Montgomery, L.; Khalaf, M.; Bailey, J.P.; Gornall, R.J. Contributions to a cytological catalogue of the British and Irish flora, 5. *Watsonia* **1997**, *21*, 365–368.
35. Löve, Á. Chromosome number reports LXXXIX. *Taxon* **1985**, *34*, 727–730. [CrossRef]
36. Löve, Á.; Löve, D. Cyto-taxonomical studies on boreal plants. III. Some new chromosome numbers of Scandinavian plants. *Arkiv Bot.* **1944**, *31 A*, 1–22.
37. Levin, D.A. *The Role of Chromosomal Change in Plant Evolution*; Oxford University Press: Oxford, NY, USA, 2002.
38. Čertner, M.; Kúr, P.; Kolář, F.; Suda, J. Climatic conditions and human activities shape diploid-tetraploid coexistence at different spatial scales in the common weed *Tripleurospermum inodorum* (Asteraceae). *J. Biogeogr.* **2019**, *46*, 1355–1366. [CrossRef]
39. Hroudová, Z.; Zákravský, P. Ecology of two cytotypes of *Butomus umbellatus*. III. Distribution and habitat differentiation in the Czech and Slovak Republics. *Folia Geobot. Phytotax.* **1993**, *28*, 425–435. [CrossRef]
40. Vellend, M. Species diversity and genetic diversity: Parallel processes and correlated patterns. *Am. Nat.* **2005**, *166*, 199–215. [CrossRef] [PubMed]
41. Vellend, M.; Geber, M.A. Connections between species diversity and genetic diversity. *Ecol. Lett.* **2005**, *8*, 767–781. [CrossRef]
42. Van der Maarel, E. Transformation of cover-abundance values in phytosociology and its effect on community similarity. *Vegetatio* **1979**, *38*, 97–114.
43. Dengler, J.; Jansen, F.; Glöckler, F.; Peet, R.K.; De Cáceres, M.; Chytrý, M.; Ewald, J.; Oldeland, J.; Finckh, M.; Lopez-Gonzalez, G.; et al. The Global Index of Vegetation-Plot Databases (GIVD): A new resource for vegetation science. *J. Veg. Sci.* **2011**, *22*, 582–597. [CrossRef]
44. Willner, W.; Berg, C.; Heiselmayer, P. Austrian Vegetation Database. *Biodivers. Ecol.* **2012**, *4*, 333. [CrossRef]
45. Chytrý, M.; Rafajová, M. Czech National Phytosociological Database: Basic statistics of the available vegetation-plot data. *Preslia* **2003**, *75*, 1–15.

46. Kącki, Z.; Śliwiński, M. The Polish Vegetation Database: Structure, resources and development. *Acta Soc. Bot. Polon.* **2012**, *81*, 75–79. [CrossRef]
47. Valachovič, M. Centrálna Databáza Fytocenologických Zápisov (CDF) na Slovensku [Central Database of Phytosociological Relevés (CDF) in Slovakia]. In Proceedings of the Zborník Referátov zo 7. Zjazdu SBS pri SAV, Hrabušice, Slovakia, 21–25 June 1999; Leskovjanská, A., Ed.; Správa NP Slovenský raj, Spišská Nová Ves & SBS: Bratislava, Slovakia, 1999; pp. 75–77. (In Slovak).
48. Kalníková, V.; Kudrnovsky, H. Gravel Bar Vegetation Database. *Phytocoenologia* **2017**, *47*, 109–110. [CrossRef]
49. Hennekens, S.M.; Schaminée, J.H.J. TURBOVEG, a comprehensive data base management system for vegetation data. *J. Veg. Sci.* **2001**, *12*, 589–591. [CrossRef]
50. Chytrý, M.; Otýpková, Z. Plot sizes used for phytosociological sampling of European vegetation. *J. Veg. Sci.* **2003**, *14*, 563–570. [CrossRef]
51. Tichý, L. JUICE, software for vegetation classification. *J. Veg. Sci.* **2002**, *13*, 451–453. [CrossRef]
52. Euro+Med PlantBase. The Information Resource for Euro-Mediterranean Plant Diversity. Available online: https://www.emplantbase.org/home.html (accessed on 7 March 2021).
53. Kącki, Z.; Łysko, A.; Dajdok, Z.; Kobierski, P.; Krawczyk, R.; Nowak, A.; Rosadziński, S.; Popiela, A. Formalized classification of ephemeral wetland vegetation (Isoëto-Nanojuncetea class) in Poland (Central Europe). *PeerJ* **2013**. under review.
54. Chytrý, M.; Tichý, L. National vegetation classification of the Czech Republic: A summary of the approach. *Phytocoenologia* **2018**, *48*, 121–131. [CrossRef]
55. Ellenberg, H.; Weber, H.E.; Düll, R.; Wirth, W.; Werner, W.; Paulißen, D. Zeigerwerte von Pflanzen in Mitteleuropa. 2nd ed. *Scr. Geobot.* **1992**, *18*, 1–258.
56. Wood, S.N. Thin-plate regression splines. *J. R. Stat. Soc. Ser. B Stat. Methodol.* **2003**, *65*, 95–114. [CrossRef]
57. Wood, S.N. Stable and efficient multiple smoothing parameter estimation for generalized additive models. *J. Am. Stat. Assoc.* **2004**, *99*, 673–686. [CrossRef]
58. Wood, S.N. Fast stable restricted maximum likelihood and marginal likelihood estimation of semiparametric generalized linear models. *J. R. Stat. Soc. Ser. B Stat. Methodol.* **2011**, *73*, 3–36. [CrossRef]
59. Wood, S.N.; Pya, N.; Saefken, B. Smoothing parameter and model selection for general smooth models (with discussion). *J. Am. Stat. Assoc.* **2016**, *111*, 1548–1575. [CrossRef]
60. Wood, S.N. *Generalized Additive Models: An Introduction with R*, 2nd ed.; CRC Press, Taylor & Francis Group: New York, NY, USA, 2017.
61. Pyšek, P.; Danihelka, J.; Sádlo, J.; Chrtek, J., Jr.; Chytrý, M.; Jarošík, V.; Kaplan, Z.; Krahulec, F.; Moravcová, L.; Pergl, J.; et al. Catalogue of alien plants of the Czech Republic (2nd edition): Checklist update, taxonomic diversity and invasion patterns. *Preslia* **2012**, *84*, 155–255.
62. Tokarska-Guzik, B.; Dajdok, Z.; Zając, M.; Zając, A.; Urbisz, A.; Danielewicz, W.; Hołdyński, C. *Rośliny Obcego Pochodzenia w Polsce ze Szczególnym Uwzględnieniem Gatunków Inwazyjnych*; Generalna Dyrekcja Ochrony Srodowiska: Warszawa, Poland, 2012.
63. Sørensen, T. A method of establishing groups of equal amplitude in plant sociology based on similarity of species content and its application to analyses of the vegetation on Danish commons. *Biol. Skr. Dan. Vid. Selsk.* **1948**, *5*, 1–34.
64. R Core Team. *R: A Language and Environment for Statistical Computing*; R Foundation for Statistical Computing: Vienna, Austria, 2018; Available online: https://www.R-project.org/ (accessed on 10 December 2020).
65. Schwarzacher, T.; Heslop-Harrison, P. *Practical In Situ Hybridisation*, 2nd ed.; BIOS: Oxford, UK, 2000.
66. Schönswetter, P.; Suda, J.; Popp, M.; Weiss-Schneeweiss, H.; Brochmann, C. Circumpolar phylogeography of *Juncus biglumis* (Juncaceae) inferred from AFLP fingerprints, cpDNA sequences, nuclear DNA content and chromosome numbers. *Mol. Phylogenet. Evol.* **2007**, *42*, 92–103. [CrossRef]
67. Temsch, E.M.; Greilhuber, J.; Krisai, R. Genome size in liverworts. *Preslia* **2010**, *82*, 63–80.
68. Otto, F.J.; Oldiges, H.; Göhde, W.; Jain, V.K. Flow cytometric measurement of nuclear DNA content variations as a potential in vivo mutagenicity test. *Cytometry* **1981**, *2*, 189–191. [CrossRef]
69. Galbraith, D.W.; Harkins, K.R.; Maddox, J.M.; Ayres, N.M.; Sharma, D.P.; Firoozabady, E. Rapid flow cytometric analysis of the cell cycle in intact plant tissues. *Science* **1983**, *220*, 1049–1051. [CrossRef]
70. Thiers, B.M. *Index Herbariorum*; The New York Botanical Garden: Bronx, NY, USA; Available online: http://sweetgum.nybg.org/science/ih/ (accessed on 21 February 2021).
71. Wild, J.; Kaplan, Z.; Danihelka, J.; Petřík, P.; Chytrý, M.; Novotný, P.; Rohn, M.; Šulc, V.; Brůna, J.; Chobot, K.; et al. Plant distribution data for the Czech Republic integrated in the Pladias database. *Preslia* **2019**, *91*, 1–24. [CrossRef]
72. Petts, G.E. River regulation in Europe: An historical perspective. *Regul. Rivers: Res. Manag.* **1987**, *1*, 363–369.
73. Danihelka, J.; Niklfeld, H.; Šípošová, H. *Viola elatior*, *V. pumila* and *V. stagnina* in Austria, Czechia and Slovakia: A story of decline. *Preslia* **2009**, *81*, 151–171.
74. Fick, S.E.; Hijmans, R.J. WorldClim 2: New 1-km spatial resolution climate surfaces for global land areas. *Int. J. Climatol.* **2017**, *37*, 4302–4315. [CrossRef]
75. Hijmans, R. R Package Raster: Geographic Data Analysis and Modeling. Version 3.0-14/r3417. Available online: https://R-Forge.R-project.org/projects/raster/ (accessed on 10 December 2020).
76. Niklfeld, H. Bericht über die Kartierung der Flora Mitteleuropas. *Taxon* **1971**, *20*, 545–571. [CrossRef]

77. Šmilauer, P.; Lepš, J. *Multivariate Analysis of Ecological Data using CANOCO 5*, 2nd ed.; Cambridge University Press: Cambridge, UK, 2014.
78. Phillips, S.J.; Dudík, M. Modeling of species distribution with Maxent: New extensions and a comprehensive evaluation. *Ecography* **2008**, *31*, 161–175. [CrossRef]
79. Kass, J.M.; Vilela, B.; Aiello-Lammens, M.E.; Muscarella, R.; Merow, C.; Anderson, R.P. Wallace: A flexible platform for reproducible modeling of species niches and distributions built for community expansion. *Methods Ecol. Evol.* **2018**, *9*, 1151–1156. [CrossRef]
80. Fielding, A.; Bell, J. A review of methods for the assessment of prediction errors in conservation presence/absence models. *Environ. Conserv.* **1997**, *24*, 38–49. [CrossRef]
81. Pietsch, W. Beitrag zur Gliederung der europäischen Zwergbinsengesellschaften (Isoëto-Nanojuncetea Br.-BI. & Tx. 1943). *Vegetatio* **1973**, *28*, 401–438.
82. Šumberová, K.; Lososová, Z.; Horáková, V. Vegetation dynamics on exposed pond bottoms in the Českobudějovická basin (Czech Republic). *Phytocoenologia* **2005**, *35*, 421–448. [CrossRef]
83. Aichinger, E. *Vegetationskunde der Karawanken*; Gustav Fischer: Jena, Germany, 1933.
84. Poschlod, P.; Rosbakh, S. Mudflat species: Threatened or hidden? An extensive seed bank survey of 108 fish ponds in Southern Germany. *Biol. Conserv.* **2018**, *225*, 154–163.
85. Phartyal, S.S.; Rosbakh, S.; Ritz, C.; Poschlod, P. Ready for change: Seed traits contribute to the high adaptability of mudflat species to their unpredictable habitat. *J. Veg. Sci.* **2020**, *31*, 331–342. [CrossRef]
86. Tolasz, R. (Ed.) *Climate Atlas of Czechia*; Czech Hydrometeorological Institute & Palacký University: Prague & Olomouc, Czech Republic, 2007.
87. Hejný, S. *Ökologische Charakteristik der Wasser- und Sumpfpflanzen in den slowakischen Tiefebenen (Donau- und Theissgebiet)*; Verlag der Slowakischen Akademie der Wissenschaften: Bratislava, Slovakia, 1960.
88. Müller-Stoll, W.R.; Pietsch, W. Ökologische Untersuchungen über die Gesellschaft des Eleocharito-Caricetum bohemica auf wasserfrei gewordenen Teichböden in Zentraleuropa. *Verh. Zool. Bot. Ges. Österreich* **1985**, *123*, 51–70.
89. Šumberová, K.; Lososová, Z.; Fabšicová, M.; Horáková, V. Variability of vegetation of exposed pond bottoms in relation to management and environmental factors. *Preslia* **2006**, *78*, 235–252.
90. Francová, K.; Šumberová, K.; Janauer, G.A.; Adámek, Z. Effects of fish farming on macrophytes in temperate carp ponds. *Aquac. Int.* **2019**, *27*, 413–436. [CrossRef]
91. Chytrý, M.; Pyšek, P.; Tichý, L.; Knollová, I.; Danihelka, J. Invasions by alien plants in the Czech Republic: A quantitative assessment across habitats. *Preslia* **2005**, *77*, 339–354.
92. Farmer, J.A.; Webb, E.B.; Pierce, R.A.; Bradley, K.W. Evaluating the potential for weed seed dispersal based on waterfowl consumption and seed viability. *Pest Manag. Sci.* **2016**, *73*, 2592–2603. [CrossRef]
93. Šumberová, K.; Lososová, Z.; Ducháček, M.; Horáková, V.; Fabšičová, M. Distribution, habitat ecology, soil seed bank and seed dispersal of threatened *Lindernia procumbens* and alien *Lindernia dubia* (Antirrhinaceae) in the Czech Republic. *Phyton* **2012**, *52*, 39–72.
94. Zarzycki, K.; Szelag, Z. Red list of the vascular plants in Poland. In *Red List of Plants and Fungi in Poland*; Mirek, Z., Zarzycki, K., Wojewoda, W., Szelag, Z., Eds.; Polish Academy of Sciences & W. Szafer Institute of Botany: Kraków, Poland, 2006; pp. 11–20.
95. Eliáš, J.P.; Dítě, D.; Kliment, J.; Hrivnák, R.; Feráková, V. Red list of ferns and flowering plants of Slovakia, 5th edition (October 2014). *Biologia* **2015**, *70*, 218–228. [CrossRef]
96. Böckelmann, J.; Wieser, D.; Tremetsberger, K.; Šumberová, K.; Bernhardt, K.G. Isolation of nuclear microsatellite markers for *Cyperus fuscus* (Cyperaceae). *Appl. Plant Sci.* **2015**, *3*, 1500071. [CrossRef]
97. Roalson, E.H. A synopsis of chromosome number variation in the Cyperaceae. *Bot. Rev.* **2008**, *74*, 209–393. [CrossRef]
98. Landolt, E.; Bäumler, B.; Erhardt, A.; Hegg, O.; Klötzli, F.; Lämmler, W.; Nobis, M.; Rudmann-Maurer, K.; Schweingruber, F.H.; Theurillat, J.P.; et al. *Flora Indicativa*, 2nd ed.; Haupt: Bern, Switzerland, 2010.
99. Rydlo, J. Vodní makrofyta v oblasti soutoku Labe a Vltavy [Aquatic macrophytes in the area of the Labe and Vltava confluence]. *Muz. Souč. Ser. Natur.* **2006**, *21*, 25–70. (In Czech)
100. Rydlo, J. Vodní makrofyta ve stojatých vodách na Poděbradsku a Nymbursku [Aquatic macrophytes of stagnant waters in the Poděbrady and Nymburk regions]. *Muz. Souč. Ser. Natur.* **2005**, *20*, 11–134. (In Czech)
101. Bravencová, L.; Musil, Z.; Reiter, A. Flóra a vegetace obnaženého dna Znojemské a Vranovské údolí nádrže (střední Podyjí) [Flora and vegetation of the exposed bottom of the Znojmo and Vranov water reservoirs (middle Dyje basin)]. *Thayensia* **2007**, *7*, 153–173. (In Czech)
102. Malíková, L. Vliv Obhospodařování na Strukturu Makrovegetace. Vliv Obhospodařování Rybníků na Strukturu Vegetace a Flóry Obnažených den [Influence of Management on the Structure of Macrovegetation. Influence of Pond Management on the Structure of Vegetation and Flora of Exposed Bottoms]. Bachelor's Thesis, University of South Bohemia, České, Budějovice, 2000.
103. Vicherek, J. Poznámky k syntaxonomii asociace *Cyperus fuscus-Chenopodium glaucum* as. Klika 1935. *Spisy Přírod. Fak. Univ. J. E. Purkyně Brno* **1969**, *501*, 63–73. (In Czech)
104. Vicherek, J. Poznámky k cenologické afinitě *Myosurus minimus* L. [Remarks on the coenological affinity of *Myosurus minimus* L.]. *Preslia* **1968**, *40*, 387–396. (In Czech)

105. Borysiak, J. *Struktura Aluwialnej Roślinności Lądowej Środkowego i Dolnego Biegu Warty [The Structure of Alluvial Vegetation of the Middle and Lower Warta River]*; Wydawnictwo Naukowe UAM: Poznan, Poland, 1994. (In Polish)
106. Krawczyk, R.; Cwener, A.; Michalczuk, W.; Zubel, R. Ephemeral wetland communities of Isoëto-Nano-Juncetea class—New data from south-eastern Poland. *Biodiv. Res. Conserv.* **2016**, *42*, 41–54. [CrossRef]
107. Macicka-Pawlik, T.; Wilczyńska, W. Zbiorowiska roślinne starorzeczy w dolinie środkowego biegu Odry [Plant communities of oxbow lakes in the middle Odra river valley]. *Acta Univ. Wratislav. Pr. Bot.* **1996**, *64*, 73–120. (In Polish)
108. Podbielkowski, Z. Roślinność stawów rybnych woj. Warszawskiego [Vegetation of fishponds in the province of Warsaw]. *Monogr. Bot.* **1968**, *27*, 3–123. (In Polish) [CrossRef]
109. Popiela, A. Zbiorowiska z klasy *Isoëto-Nanojuncetea* na terenie Polski Zachodniej [Communities of the Isoëto-Nanojuncetea class in western Poland]. *Fragm. Flor. Geobot. Ser. Polon.* **1996**, *3*, 289–310. (In Polish)
110. Spałek, K. Bidentetum cernuae (Kobendza 1948) Slavnic 1951 w zbiornikach zaporowych na Śląsku Opolskim [Bidentetum cernuae (Kobendza 1948) Slavnic 1951 in dam reservoirs in Opole Silesia]. *Acta Bot. Siles.* **2008**, *3*, 137–144. (In Polish)
111. Spałek, K.; Nowak, A. Zbiorowiska namułkowe z klasy Isoëto-Nanojuncetea w zbiornikach zaporowych na Śląsku Opolskim [Mudflat communities of the class Isoëto-Nanojuncetea in dam reservoirs in Opole Silesia]. *Fragm. Flor. Geobot. Polon.* **2006**, *13*, 361–368. (In Polish)
112. Zając, M.; Zając, A. Zbiorowiska z klasy *Isoëto-Nanojuncetea* na dnach wysychających stawów w południowej części Kotliny Oświęcimskiej [Communities of the class Isoëto-Nanojuncetea on the bottoms of drying-up ponds in the southern part of the Oświęcim basin]. *Zesz. Nauk. Uniw. Jagiell., Pr. Bot.* **1988**, *17*, 155–160. (In Polish)
113. Traxler, A. Zwergbinsengesellschaften in Österreich. Diploma Thesis, University of Vienna, Vienna, Austria, 1990.

water

MDPI

Article

Distribution of Vascular Plant Communities in Slovenian Watercourses

Igor Zelnik *, Urška Kuhar, Matej Holcar, Mateja Germ and Alenka Gaberščik

Department of Biology, Biotechnical Faculty, University of Ljubljana, Jamnikarjeva 101, 1000 Ljubljana, Slovenia; ursska.kuhar@gmail.com (U.K.); matej.holcar@bf.uni-lj.si (M.H.); mateja.germ@bf.uni-lj.si (M.G.); alenka.gaberscik@bf.uni-lj.si (A.G.)
* Correspondence: igor.zelnik@bf.uni-lj.si

Abstract: Rivers and streams are heterogenous ecosystems that host a great number of vascular plant communities. The territory of Slovenia is highly diverse regarding geomorphologic, geologic, climatic, and edaphic conditions. We presumed that environmental variability will also affect the distribution of hygrophilous vascular plants in running waters and consequently the structure of plant communities they form. We analyzed macrophyte, spatial, and environmental parameters in 906 stretches of the watercourses occurring in the Dinaric, Pannonian, and Po lowland hydro-ecoregions. We determined 87 vascular plant taxa. The most abundant were *Myriophyllum spicatum*, *Phalaris arundinacea*, and *Potamogeton nodosus*. Submerged macrophytes presented about one third of total species abundance, while amphiphytes were somewhat less abundant. Canonical correspondence analysis (CCA) revealed that distance from the source explained 15.1% of the growth form type variability, and current velocity and latitude explained 4.1% each. With the assessed parameters, we explained 31.6% of the variability. When CCA was run with taxa, only 20.9% of their variability was explained with statistically significant parameters. We distinguished 25 different plant associations belonging to five classes and nine alliances. The majority of defined plant communities were distributed in different watercourses belonging to different hydro-ecoregions. Only seven communities had a narrower distribution range, three of them on karst poljes. Among them, the new association *Mentho aquaticae-Oenanthetum fistulosae* from the river Mali Obrh on the Loško polje was described in this contribution.

Keywords: macrophytes; growth forms; plant communities; distribution; rivers; streams; Slovenia

Citation: Zelnik, I.; Kuhar, U.; Holcar, M.; Germ, M.; Gaberščik, A. Distribution of Vascular Plant Communities in Slovenian Watercourses. *Water* **2021**, *13*, 1071. https://doi.org/10.3390/w13081071

Academic Editor: Massimo Zacchini

Received: 9 March 2021
Accepted: 10 April 2021
Published: 13 April 2021

Publisher's Note: MDPI stays neutral with regard to jurisdictional claims in published maps and institutional affiliations.

1. Introduction

Rivers, as well as streams, exhibit high spatial and temporal heterogeneity, which is a consequence of their interactions across four dimensions: longitudinal, lateral and vertical, and temporal [1]. These interactions enable the exchange of energy, matter, and organisms [2]. The strength of spatial interactions is changing along the river flow and due to human influences on water flow, riverbed shape, and its connection to the floodplains in catchments [3]. These interactions may result in different disturbances, such as increased current velocity, erosion, fluctuations of water level, siltation and sedimentation, and other pressures like changes in water quality (i.e., pH, temperature and concentrations of dissolved oxygen, nutrients, and toxic metals) [4,5]. Thus, rivers present a highly unpredictable habitat for organisms. The extent of the pressures exerted on the rivers depends on the properties of the catchment, namely slope, geology, land cover parameters, and human impact on river morphology. All these parameters may significantly alter riverine communities, including macrophytes [6–8].

Macrophyte communities respond to changes in their habitats and reflect the quality of the wider environment [9–13]. However, the relationship is also reversed, since longitudinal transport of dissolved nutrients and particulate matter in rivers depends on the type of macrophyte communities and their retention potential [14]. Different macrophyte communities vary significantly in their productivity and respond differently to nutrient concentrations [15]. Therefore, the relationships between nutrient level and community structure also vary significantly [16]. In addition to nutrients, current velocity is also considered as an important factor shaping macrophyte communities [17,18], affecting their distribution, species composition, biomass, and macrophyte metabolism [19,20]. In many cases, current velocity determines whether plants will successfully colonize and establish stable communities in a specific habitat [21–23]. Additionally, current velocity also affects plant communities indirectly, through the effects on the substrate and on erosion processes. Macrophyte communities are also the result of species competition, especially in terms of competition for light, nutrients, and space [24]. Competitive potential of macrophytes depends on their traits regarding growth dynamics, carbon assimilation, energy harvesting, and nutrient uptake [25]. Nutrient uptake by macrophytes in rivers is variable due to seasonality and their growth dynamics [26]. All these parameters may exert negative effects on macrophyte communities. In the case of European lowland watercourses, a decline in macrophyte species diversity and unification of communities was detected in recent decades. In many systems, fast-growing species with a wide ecological range are favored, which prevents the development of other more demanding species [27,28].

Globally, many vascular macrophyte species have broad distribution ranges, however, about 39% of the ca. 412 genera containing aquatic vascular macrophytes are endemic to a single biogeographic region [29]. In the case of 345 sites in central North America, temperature did not turn out to be an important factor affecting macrophyte distribution, although it may exert an indirect effect on waterbody properties [30]. For fluvial lakes, it was shown that the landscape morphometric parameters explained a great share of spatial distribution of submerged macrophyte communities [31], while in the case of eastern Pyrenean lakes water chemistry, altitude, vegetation cover of the catchment, and nutrient availability were the major environmental factors affecting macrophyte distribution [32]. Linear models indicated that richness of emergent species in ponds was negatively affected by total phosphorous, while the land cover factors affected submerged species richness [33].

The Slovenian territory is geographically highly diverse regarding geomorphologic, geologic, climatic, and edaphic conditions. Slovenia is also situated on the junction of four biogeographic regions, namely the Alpine, Dinaric, Pannonian, and (sub)Mediterranean regions [34–36]. The high environmental diversity of Slovenia results in high biodiversity [37]. For the purpose of aquatic ecosystem management, Slovenian territory was re-delineated to the Alpine, Dinaric, Pannonian, and Po regions [38]. Many watercourses, especially in the Dinaric region, are marked by an intermittent water regime, which generates physical disturbances due to extreme water level fluctuations and affects conditions in rivers, namely nutrient availability and occasionally even intermittence [39]. Studies of many authors [6,40,41] have revealed positive relationships between moderate water level changes and species diversity, as well as productivity.

Vegetation of aquatic ecosystems and distribution of aquatic plant communities have been studied with significantly lower intensity compared to terrestrial plant communities [42,43]. The result is a lack of knowledge in this research area, at least in some parts of Europe [44], which should be elaborated, especially due to high threats to these habitats. The above-mentioned facts are even more true in the case of running waters [42]. With this contribution, we would like to fill this gap for the part of central Europe, since our research area covers three hydro-ecoregions, namely the Po lowland, the Dinaric, and the Pannonian lowland, and represents an area reaching beyond the borders of the territory of Slovenia, which is also evident from Figure 1.

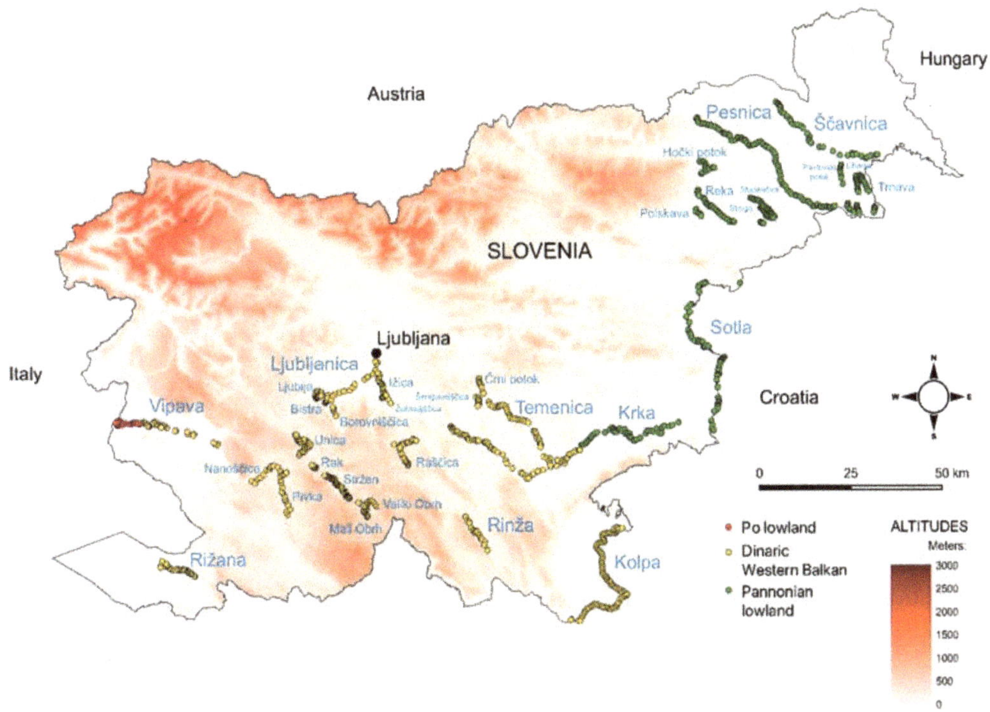

Figure 1. Map of Slovenia showing the distribution of surveyed sites. Circles of different colors indicate different hydro-ecoregions. The font size of the name roughly indicates the size of the specific watercourse.

In this study, we investigated 34 Slovenian watercourses belonging to the hydro-ecoregions the Po lowland, Dinarides, and the Pannonian lowland, in order to examine growth forms of macrophyte species and macrophyte communities in these watercourses, and to point out the potential regionality of these communities. We also expected that environmental variability would affect number of taxa, their abundance, and the occurrence of different growth forms.

2. Materials and Methods

2.1. Study Area

The geologic and climatic conditions of Slovenia are very diverse. Sedimentary rocks, of which limestone and dolomite predominate, cover 65.1% of the territory. Loose sediments (gravel, sand) are found in tectonic depressions and river valleys and contribute an additional 29.2%. Metamorphic rocks are mostly concentrated in northeast Slovenia accounting for 4.1%, while igneous rocks cover only 1.6% of the surface and are found along the Periadriatic fault and in the Pohorje Mountains [45]. Three types of climate occur in Slovenia, namely temperate humid climate with hot summers, temperate continental climate, and mountain climate, with the highest precipitation in the Julian Alps (in the northwest), which is at least three times the precipitation of the Pannonian northeastern Slovenia [46]. Slovenia is also the junction of four biogeographic regions, namely the Alpine, Dinaric, Pannonian, and (sub)Mediterranean regions [34–36].

For the purpose of this study, we investigated watercourses that belong to three hydro-ecoregions, namely Dinaric, Po lowland, and Pannonian lowland. The watercourses from the Alpine region were excluded, as vascular plants are scarce in this region. The majority of the Slovenian territory (about 16.500 km^2) is drained into the Black Sea, while 3.750 km^2 is drained into the Adriatic Sea [47]. Only two among the studied rivers (Vipava and Rižana) belong to the Adriatic catchment area. Altogether, we examined 1141 reaches of which 535 reaches belong to the hydro-ecoregion Dinarides, 352 reaches belong to the hydro-ecoregion Pannonian lowland, while only 19 belong to the hydro-ecoregion Po lowland. The watercourses Vipava (50), Rižana (32), Kolpa (96), Krka (108), Sotla (54), Temenica (51), Mali Obrh (12), Veliki Obrh (15), Stržen (42), Rak (8), Nanoščica (12), Pivka (18), Unica (29), Ljubljanica (22), Ižica (23), Ljubija (10), Bistra (7), Borovniščica (7), Želimeljščica (3), Rinža (18), Rašica (14), Polskava (12), Črni potok (13), Šentpavelščica (9), Hočki potok (14), Reka (21), Studenčnica (11), Struga (22), Pesnica (85), Ščavnica (39), Pavlovski potok (16), Libanja (17), and Trnava (16) were examined (Figure 1).

2.2. Macrophyte Data Set

Surveys were carried out in the years 2002–2006 along the entire stream courses. New sampling stretches were determined according to major changes in habitat characteristics and macrophyte distribution; 100 m stretches were determined on ca. 500 m intervals. A similar approach, but with 1000 m stretches, was used in the macrophyte survey of the Danube River [48]. Macrophyte surveys were performed from the bank of the stream, by wading in the stream or from a boat, using a rake with hooks. We recorded submerged, floating, and emergent vascular plants, bryophytes, charophytes, and filamentous algae. Macrophyte species abundance was estimated as a relative plant biomass using a five-degree scale, namely 1–very rare, 2–rare, 3–commonly present, 4–frequent, and 5–predominant, as proposed by Kohler and Janauer [49]. These values were transformed by the function x^3, as suggested by Schneider and Melzer [50]. The plants that were sampled in the vegetative phenological phase, which prevented identification to the species level, were only recorded on the genus level. We differentiated the following growth forms: pleustophytes (pl–free-floating on or below the water surface), submerged anchored species (sa–rooted in the sediment), floating-leaved species (fl–rooted in the sediment, forming exclusively or additionally leaves floating on the water surface), amphiphytes (am–having the ability to produce terrestrial and aquatic growth forms, or aquatic and aerial leaves), and helophytes (he–anchored in the water-saturated sediment, with plant assimilation areas permanently in the air) [51]. Stretches that were colonized only by bryophytes and stoneworts, and stretches with very low abundance of vascular species, namely less than 4 vascular plant species with lowest abundance values, were excluded from the analyses. A set of 906 sampled stretches was analyzed. Species names followed the nomenclature of *Flora Europaea* [52]. For the purpose of community analyses, the ordinal values of the Kohler-scale were transformed into quantitative values ("quantities") [53]. We equalized the transformed values as percentage cover-abundance values according to Braun-Blanquet [54], as previously tested and used by other authors [55,56]. We also used these "adjusted" values to examine if the clusters obtained by cluster analysis meet the definitions of specific associations, which are mostly defined by presence and/or abundance of diagnostic species. These values for dominant species are presented in Table A1 in Appendix A.

2.3. Environment Assessment

The assessment of environmental conditions in the selected watercourses was performed in the same stretches as the survey of macrophytes. We applied a modified Riparian, Channel, and Environmental (RCE) Inventory, proposed by Petersen [9] that comprised 11 parameters, each describing four levels of environmental gradient. The parameters include land-use type beyond the riparian zone, characteristics of the riparian zone (width, completeness, and vegetation type), and morphology of the stream channel (bank structure and undercutting, occurrence of retention structures, sediment accumulation, type of stream bottom and detritus, and dynamics of the flow, i.e., occurrence of riffles, pools, and meanders) [7,57]. Each parameter includes four categories comprising quality gradient, coded numerically from 1 to 4; 1 represented good, close to natural condition, while quality gradient values from 2 to 4 indicated worsening of environmental conditions or differences either due to regional characteristics or due to the longitudinal character of the river. We also assessed current velocity according to Janauer et al. [58] using a 4-level scale (1–no visible current, 2–slow, >0 to 30 cm s^{-1}, 3–medium, 35 to 65 cm s^{-1}, and 4–fast current, >70 cm s^{-1}).

Maps were made using QGIS version 3.16 [59].

2.4. Data Analyses and Classification of Aquatic Plant Communities

Diversity was measured as the number of species and the Shannon–Wiener diversity index (H′), which was calculated based on mean cover values with PAST, version 2.17c [60]. Correlations between diversity, abundance of macrophyte communities, and environmental parameters were calculated with Spearman correlation on ranks using PAST, version 2.17c [60]. Similarity of macrophyte communities between the stretches was also calculated with the aforementioned program. Ward's method was used as linkage method and Euclidean distance was used as similarity index [61], which provided clustering with lowest levels of chaining [62]. This enabled the comparison of similarity among the assemblages from different locations and their classification into plant communities. We used the approach based on the dominant species, which is also diagnostic for specific associations. This approach is frequently used for aquatic vegetation [63]. Dominance-based classification for species-poor plant communities is most similar to the approaches in traditional phytosociology [64].

Classification of plant communities was based on the definitions proposed by Šumberová and co-workers [65–68], Zaliberová and Škodová [69], and Landucci et al. [70]. Nomenclature of higher syntaxa (classes, alliances) followed Mucina et al. [71], while nomenclature of associations was in accordance with Šumberová and co-workers [65–68], Landucci et al. [70], and Zaliberová and Škodová [69].

2.5. The Influence of Environmental Factors on Macrophyte Community Composition

Firstly, detrended correspondence analysis (DCA) was performed, which helped us to evaluate whether the gradients in the matrix of macrophyte species data were linear or unimodal, and choose the appropriate direct gradient analysis for further analyses. If the eigenvalue for the first axis was at least 0.4 and gradient length exceeded 3 standard deviations, respectively, as suggested by ter Braak and Verdonschot [72], canonical correspondence analysis (CCA) was performed, while in other cases redundancy analysis would have been conducted (RDA). Species abundance values were log(x + 1)-transformed. Results of these analyses enabled the assessment of the relationships between environmental parameters and distribution and abundance of macrophytes and their growth forms. In addition to 12 of the above-mentioned environmental parameters, we also tested the relationships between macrophytes and presence and abundance of filamentous algae, as well as selected geographical parameters, i.e., longitude and latitude, distance from the source, slope, and altitude.

We used forward selection, where 999 permutations were performed in every round to rank the relative importance of explanatory variables and to avoid co-linearity [73]. Only the parameters with significance $p < 0.05$ (or $p \leq 0.001$ in the case of growth forms) were considered for further analyses. All analyses were performed using CANOCO for Windows 4.5 program package [74,75].

3. Results

3.1. Macrophyte Taxa and Their Growth Forms

In the surveyed stretches of selected watercourses, we determined a total of 87 vascular plant taxa that belonged to different growth forms. When comparing total abundance of vascular macrophytes in the surveyed watercourses, submerged macrophytes prevailed with 36.0%, followed by amphiphytes (29.9%). A somewhat lower share (18.2%) was attributed to helophytes and floating-leaved plants (12.9%), while pleustophytes, namely all the free-floating vascular plant taxa, had the lowest share (3.1%). The most abundant species was *Myriophyllum spicatum* (12.0%), followed by *Phalaris arundinacea* (6.7%), and *Potamogeton nodosus* (6.6%), while the abundance of the species *Berula erecta, P. perfoliatus*, alien species *Elodea canadensis, P. crispus, P. pectinatus, Nuphar luteum, P. natans*, and *Schoenoplectus lacustris* contributed 3–4% each. Another 76 species altogether contributed about a half (51.1%) to the total relative abundance. Relative abundances of macrophyte species in stretches of the surveyed watercourses are presented in Figure 2.

Figure 2. Relative abundance of macrophyte species presented as size of the circles in the stretches of the surveyed watercourses.

Correlation analyses revealed positive relationships between species diversity and retention structures, sediment, and detritus deposition, while there were negative relationships between filamentous algae abundance and current velocity and slope. In addition, a negative relationship was also obtained between species diversity and distance from the source, while a positive relationship was found between species diversity and altitude (Table 1).

Table 1. Spearman correlation coefficients between diversity, abundance, and assessed environmental variables. S-W–Shannon–Wiener diversity index; ***—very highly significant ($p < 0.001$), **—highly significant ($p < 0.01$), *—significant ($p < 0.05$).

Variable	No. of Taxa	S-W	Abundance
Distance from the source	−0.314 ***	−0.304 ***	−0.215 ***
Altitude	0.383 ***	0.363 ***	0.175 ***
Slope	−0.141 ***	−0.132 ***	−0.167 ***
Land use	0.084 **	0.097 **	0.115 ***
Width of riparian zone	ns	ns	ns
Completeness of riparian zone	0.123 ***	0.131 ***	0.133 ***
Vegetation of riparian zone	0.072 *	0.078 *	0.131 ***
Retention structures	0.188 ***	0.169 ***	0.104 **
Sediment deposition	0.196 ***	0.206 ***	0.184 ***
Riverbed bottom	0.207 ***	0.180 ***	0.202 ***
Flow dynamics	0.120 ***	0.129 ***	0.121 ***
Presence of detritus	0.216 ***	0.189 ***	0.192 ***
Current velocity	−0.365 ***	−0.322 ***	−0.386 ***
Presence of filamentous algae	−0.205 ***	−0.191 ***	ns

3.2. The Relationships between Growth Forms and Environmental Factors

The CCA ordination plot (Figure 3) showing the relationships between the abundance of growth forms and significant environmental parameters in different stretches is presented in Figure 3. This analysis revealed that distance from the source explained 15.1% of the growth form type variability, current velocity and latitude explained 4.1% each, the presence of filamentous algae 2.7%, altitude and river bottom 1.4% each, and sediment deposition explained less than 0.5%. All these parameters were very highly significant ($p \leq 0.001$). Altogether, with the environmental parameters, we explained 31.6% of the variability of presence and abundance of plant growth form types. Submerged plants and floating-leaved plants were related to distance from the source, pleustophytes to current velocity, while amphiphytes and helophytes were related to slope and river bottom.

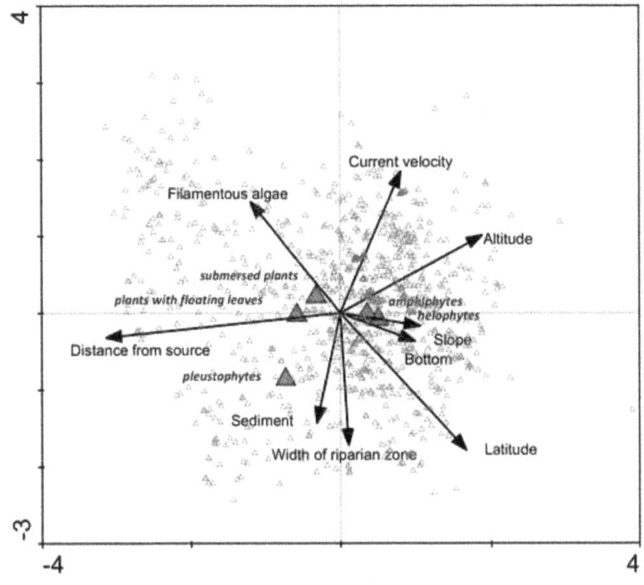

Figure 3. CCA (canonical correspondence analysis) plot showing the relationships between the abundance of different growth forms and significant environmental parameters detected in different stretches of the examined rivers.

3.3. Relationships between Species and Environmental Factors

CCA revealed that all the tested environmental parameters significantly affected the variability of species presence and abundance ($p \leq 0.001$), however, the share of explained variance was rather low, altogether 20.9% (Figure 4). Geographical parameters were revealed to be the most important. Distance from the source explained 4.9%, altitude 3.8%, longitude 2.4%, latitude 1.7%, current velocity 1.1%, and the presence of filamentous algae and width of riparian zone 0.95% of the variability each. The remaining parameters explained altogether an additional 5.5% of the variability of species presence and abundance.

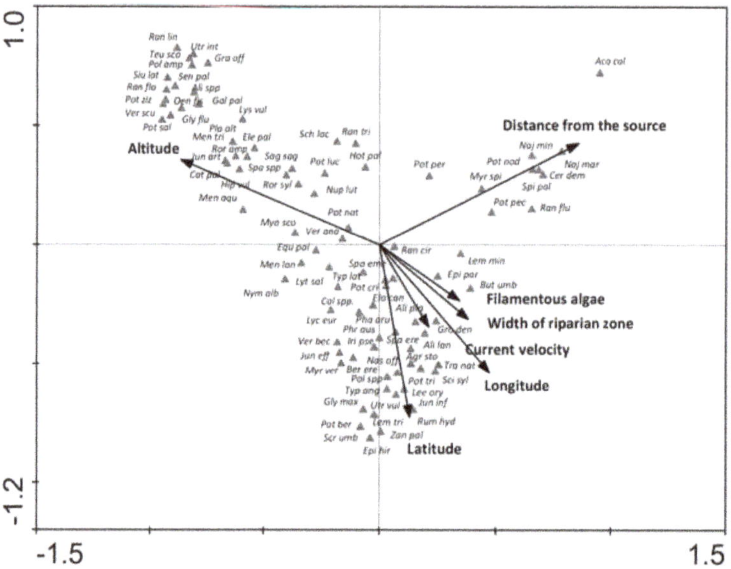

Figure 4. CCA plot showing the relationships between species presence and abundance, and significant environmental parameters detected in different stretches of the examined rivers. Only the parameters explaining at least 0.95% of the variability of species presence and abundance are presented. Abbreviated species names: *Acorus calamus* (*Aco cal*), *Agrostis stolonifera* agg. (*Agr sto*), *Alisma lanceolatum* (*Ali lan*), *Alisma plantago-aquatica* (*Ali pla*), *Alisma* spp. (*Ali spp*), *Berula erecta* (*Ber ere*), *Butomus umbellatus* (*But umb*), *Callitriche* spp. (*Cal spp.*), *Caltha palustris* (*Cat pal*), *Ceratophyllum demersum* (*Cer dem*), *Eleocharis palustris* (*Ele pal*), *Elodea canadensis* (*Elo can*), *Epilobium hirsutum* (*Epi hir*), *Epilobium parviflorum* (*Epi par*), *Equisetum palustris* (*Equ pal*), *Galium palustre* (*Gal pal*), *Glyceria fluitans* (*Gly flu*), *Glyceria maxima* (*Gly max*), *Gratiola officinalis* (*Gra off*), *Groenlandia densa* (*Gro den*), *Hippuris vulgaris* (*Hip vul*), *Hottonia palustris* (*Hot pal*), *Iris pseudacorus* (*Iri pse*), *Juncus articulatus* (*Jun art*), *Juncus effusus* (*Jun eff*), *Juncus inflexus* (*Jun inf*), *Leersia oryzoides* (*Lee ory*), *Lemna minor* (*Lem min*), *Lemna trisulca* (*Lem tri*), *Lycopus europaeus* (*Lyc eur*), *Lysimachia vulgaris* (*Lys vul*), *Lythrum salicaria* (*Lyt sal*), *Mentha aquatica* (*Men aqu*), *Mentha longifolia* (*Men lon*), *Menyanthes trifoliata* (*Men tri*), *Myosotis scorpioides* agg. (*Myo sco*), *Myriophyllum spicatum* (*Myr spi*), *Myriophyllum verticillatum* (*Myr ver*), *Najas marina* (*Naj mar*), *Najas minor* (*Naj min*), *Nasturtium officinale* (*Nas off*), *Nuphar luteum* (*Nup lut*), *Nymphaea alba* (*Nym alb*), *Oenanthe fistulosa* (*Oen fis*), *Phalaris arundinacea* (*Pha aru*), *Phragmites australis* (*Phr aus*), *Plantago altissima* (*Pla alt*), *Polygonum amphibium* (*Pol amp*), *Polygonum* spp. (*Pol spp*), *Potamogeton berchtoldii* (*Pot ber*), *Potamogeton crispus* (*Pot cri*), *Potamogeton lucens* (*Pot luc*), *Potamogeton natans* (*Pot nat*), *Potamogeton nodosus* (*Pot nod*), *Potamogeton pectinatus* (*Pot pec*), *Potamogeton perfoliatus* (*Pot per*), *Potamogeton salicifolius* (*Pot sal*), *Potamogeton trichoplyllus* (*Pot tri*), *Potamogeton x zizii* (*Pot ziz*), *Ranunculus circinatus* (*Ran cir*), *Ranunculus flammula* (*Ran fla*), *Ranunculus fluitans* (*Ran flu*), *Ranunculus lingua* (*Ran lin*), *Ranunculus trichophyllus* (*Ran tri*), *Rorippa sylvatica* (*Ror amp*), *Rorippa amphibia* (*Ror syl*), *Rumex hydrolapathum* (*Rum hyd*), *Sagittaria sagittifolia* (*Sag sag*), *Schoenoplectus lacustris* (*Sch lac*), *Scirpus sylvaticus* (*Sci syl*), *Scrophularia umbrosa* (*Scr umb*), *Senecio paludosus* (*Sen pal*), *Sium latifolium* (*Siu lat*), *Sparganium emersum* (*Spa eme*), *Sparganium erectum* agg. (*Spa ere*), *Sparganium* spp. (*Spa spp.*), *Spirodela polyrhiza* (*Spi pol*), *Teucrium scordium* (*Teu sco*), *Trapa natans* (*Tra nat*), *Typha angustifolia* (*Typ ang*), *Typha latifolia* (*Typ lat*), *Utricularia intermedia* (*Utr int*), *Utricularia vulgaris* (*Utr vul*), *Veronica anagallis-aquatica* (*Ver ana*), *Veronica beccabunga* (*Ver bec*), *Veronica scutellata* (*Ver scu*), *Zannichellia palustris* (*Zan pal*).

3.4. *Macrophyte Communities*

The similarity of macrophyte communities is presented in Figure 5, showing clustering of stretches. We distinguished 34 different clusters that were the basis for the definition of 25 associations belonging to five classes and nine alliances. Synoptic table shows the relative importance of different taxa (dominant and subdominant), expressed as average values of cover-abundance (Table A1), while plant communities based on these clusters are presented in Table 2. Overall, we determined 25 dominant taxa and 44 subdominant taxa.

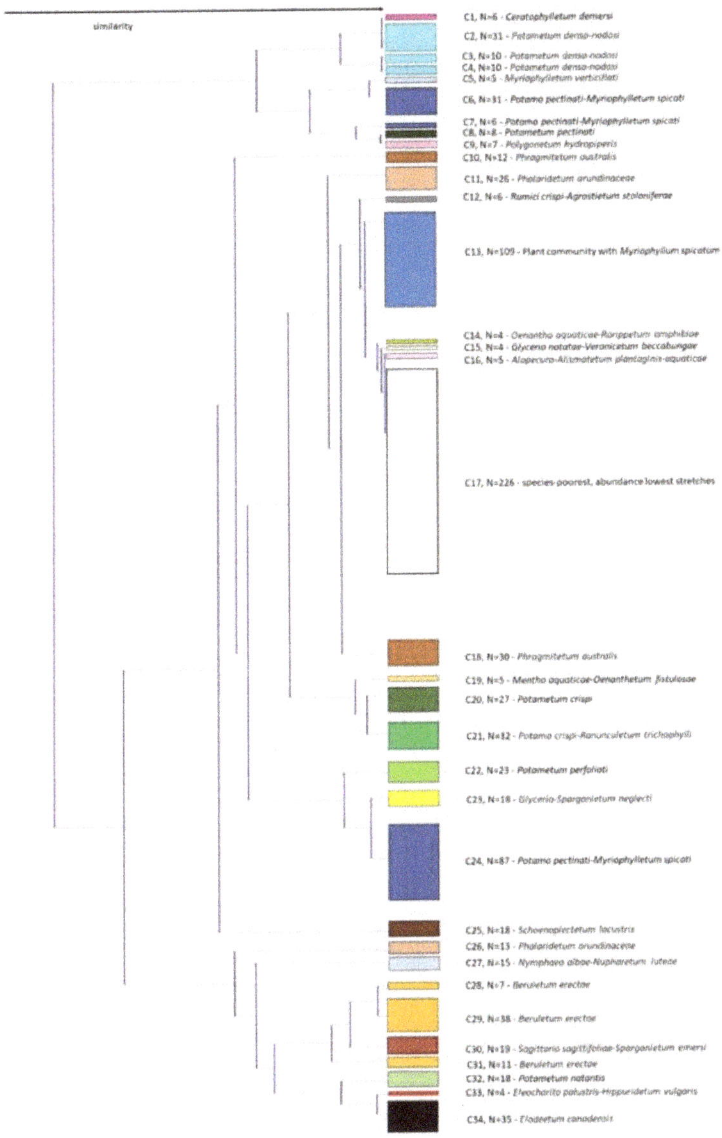

Figure 5. Similarity of the macrophyte stands, recorded in the studied watercourses using Ward's linkage method and Euclidean distance similarity index.

Table 2. Clusters classified into plant communities, their formal definitions as well as dominant and subdominant species.

Cluster Number	N	Plant Communities	Formal Definition	
			Coverage (%) of Dominant Taxa; Subdominant	NOT
1	6	*Ceratophylletum demersi* Corillion 1957	*Ceratophyllum demersum* >50%	*Potamogeton crispus* >25% *Potamogeton natans* >25% *Potamogeton nodosus* >25% *Ceratophyllum submersum* >25% *Nuphar luteum* >25% *Phragmites australis* >25% *Sagittaria sagittifolia* >25% *Sparganium erectum* >25%
2	31	*Potametum denso-nodosi* de Bolós 1957	*Potamogeton nodosus* >25%; codominant *C. demersum*	
3	10	*Potametum denso-nodosi* de Bolós 1957	*Potamogeton nodosus* >25%; subdominant *C. demersum*	*Sagittaria sagittifolia* > 25% *Sparganium emersum* >25%
4	10	*Potametum denso-nodosi* de Bolós 1957	*Potamogeton nodosus* >25%	
5	5	*Myriophylletum verticillati* Gaudet ex Šumberová in Chytrý 2011	*Myriophyllum verticillatum* >25%	
6	31	*Potamo pectinati-Myriophylletum spicati* Rivas Goday 1964	*Myriophyllum spicatum* >50%	*Butomus umbellatus* >25% *Nuphar luteum* >25% *Sparganium emersum* >25%
7	6	*Potamo pectinati-Myriophylletum spicati* Rivas Goday 1964	*Myriophyllum spicatum* >50%	*Potamogeton pectinatus*
8	8	*Potametum pectinati* Carstensen ex Hilbig 1971	*Potamogeton pectinatus* >50%	*Zannichellia palustris* >5%
9	7	*Polygonetum hydropiperis* Passarge 1965	*Polygonum hydropiper/mite* >50% codom.: *Myriophyllum spicatum*, *Phalaris arundinacea*	*Chenopodium glaucum* *Ranunculus sceleratus* *Urtica dioica* >25%
10	12	*Phragmitetum australis* Savič 1926	*Phragmites australis* >50%	
11	26	*Phalaridetum arundinaccae* Libbert 1931	*Phalaris arundinacea* >25% AND (groups *Bidens frondosa*/*Cirsium oleraceum*/*Urtica dioica*)	
12	6	*Rumici crispi-Agrostietum stoloniferae* Moor 1958	*Rorippa sylvestris* AND *Agrostis stolonifera* agg. >25% OR *Bidens tripartita* >50%	
13	109	Community with *Myriophyllum spicatum*	dominant *Myriophyllum spicatum*	*low dominance (15%)–Phalaris arundinacea, Potamogeton nodosus*
14	4	*Oenantho aquaticae-Rorippetum amphibiae* Lohmeyer 1950	*Rorippa amphibia* >25%; *Alisma plantago-aquatica*	*Glyceria maxima* cover >25% *Phalaris arundinacea*>25% *Phragmites australis*>25% *Sagittaria sagittifolia*>25% *Schoenoplectus lacustris* >25%*Sparganium erectum* >25%
15	4	*Glycerio notatae-Veronicetum beccabungae* Landucci et al. 2020	*Veronica beccabunga*	
16	5	*Alopecuro-Alismatetum plantaginis-aquaticae* Bolbrinker 1984	*Alisma plantago-aquatica* >25%	*Alisma lanceolatum* >25% *Glyceria maxima* >25% *Sparganium emersum* >25% *Sparganium erectum* >25%

Table 2. *Cont.*

Cluster Number	N	Plant Communities	Formal Definition	
			Coverage (%) of Dominant Taxa; Subdominant	**NOT**
17	226	species-poorest. abundance lowest stretches	*Phalaris arundinacea. Lytrum salicaria, Myosotis scorpioides*	
18	30	*Phragmitetum australis* Savič 1926	*Phragmites australis*	*low dominance (22%); Nuphar luteum*
19	5	*Mentho aquaticae-Oenanthetum fistulosae* ass. nova	*Oenanthe fistulosa* >50%	
20	27	*Potametum crispi* von Soó 1927	*Potamogeton crispus* >25%	*Nuphar luteum*>25% *Potamogeton natans*>25% *Phragmites australis*>25%
21	32	*Potamo crispi-Ranunculetum trichophylli* Imchenetzky 1926	*Ranunculus trichophyllus* >25%; *Potamogeton crispus*	*Glyceria notata*>25% *Juncus articulatus*>25%
22	23	*Potametum perfoliati* Miljan 1933	*Potamogeton perfoliatus* >25%	-
23	18	*Glycerio-Sparganietum neglecti* Koch 1926	*Sparganium erectum agg.* >25%	*Typha latifolia* >25%
24	87	*Potamo pectinati-Myriophylletum spicati* Rivas Goday 1964	dominant *Myriophyllum spicatum*	
25	18	*Schoenoplectetum lacustris* Chouard 1924	*Schoenoplectus lacustris* >25%	-
26	13	*Phalaridetum arundinaceae* Libbert 1931	*Phalaris arundinacea* >25% AND (groups *Bidens frondosa*/*Cirsium oleraceum*/*Urtica dioica*)	
27	15	*Nymphaeo albae-Nupharetum luteae* Nowiński 1927	*Nuphar luteum* >25%	*Nymphaea alba* >5% *Typha latifolia*>25%
28	7	*Beruletum erectae* Roll 1938	*Berula erecta* >5%; co-dominant *Callitriche* spp.	*Potamogeton perfoliatus*, *Sparganium emersum*
29	38	*Beruletum erectae* Roll 1938	*Berula erecta* >5%; subdom. *Callitriche* spp.	
30	19	*Sagittario sagittifoliae-Sparganietum emersi* Tüxen 1953	*Sagittaria sagittifolia* >25% OR *Sparganium emersum* >25%	*Butomus umbellatus* >25% *Glyceria maxima* >25% *Potamogeton gramineus* >25% *Rorippa amphibia* cover >25% *Sparganium erectum* >25%
31	11	*Beruletum erectae* Roll 1938	*Berula erecta* >5%	
32	18	*Potametum natantis* Hild 1959	*Potamogeton natans* >25%	*Callitriche palustris* s. l. >25% *Eleocharis palustris* >25% *Glyceria fluitans* >25% *Hippuris vulgaris* >25% *Nuphar luteum* >25% *Phragmites australis* >25% *Potamogeton nodosus* >25% *Sparganium emersum* >25% *Sparganium erectum* >25%
33	4	*Eleocharito palustris-Hippuridetum vulgaris* Passarge 1964	*Hippuris vulgaris* >25%	-
34	35	*Elodeetum canadensis* Nedelcu 1967	*Elodea canadensis* >50%	*Equisetum fluviatile* >25% *Potamogeton natans* >25% *Schoenoplectus lacustris* >25% *Sparganium erectum* >25%

The majority of the defined communities are distributed in various watercourses belonging to different hydro-ecoregions (Figures 1 and 6). However, 7 out of 25 communities had a narrower distribution range (Figure 6). Among these were the association *Myriophylletum verticillati*, found only in the river Struga (NE Slovenia), the associations *Potamo crispi-Ranunculetum trichophylli* and *Schoenoplectetum lacustris*, which were mostly limited to watercourses in karst poljes, the association *Polygonetum hydropiperis* that occurred only in 7 stretches in the lower part of the river Ščavnica (NE Slovenia), the association *Eleocharito palustris-Hippuridetum vulgaris* that was one of the rarest and occurred only in the river Ižica in central Slovenia, the association *Glycerio notatae-Veronicetum beccabungae*, which was recorded only in 4 stretches of the upper part of the river Temenica in central Slovenia, and the newly described association *Mentho aquaticae-Oenanthetum fistulosae* that was recorded only in 5 stretches of the river Mali Obrh on Loško polje, which is also a karstic polje.

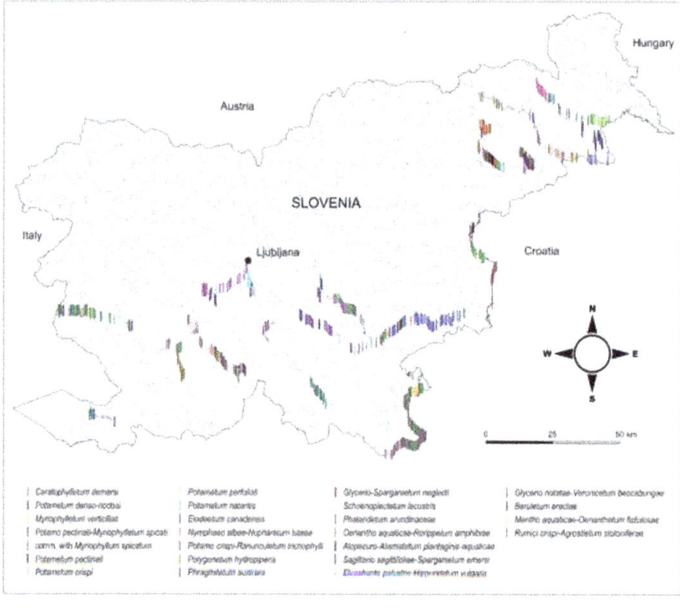

Figure 6. Plant communities in stretches of the surveyed watercourses.

Association *Ceratophylletum demersi* Corillion 1957

This was the only community dominated by pleustophytes. The association is defined with the characteristic species *C. demersum*, which had the highest abundance and was accompanied by *M. spicatum* and *P. nodosus*. Species richness was high. The distribution of this association was narrow, since it was found only in the six most lentic stretches of the rivers Krka and Struga (Figures 1 and 6).

Association *Potametum denso-nodosi* de Bolós 1957

This vegetation type is dominated by the species *P. nodosus*, which has both natant and submerged leaves, but the former may be absent in running water. Clusters 2 and 3 encompassed the majority of stretches except from the Krka River, where *P. nodosus* was co-dominant with *C. demersum* and *Najas marina*. Cluster 4 encompassed stands with the most prominent dominance of characteristic species. They were recorded in three watercourses from three hydrological regions (Temenica, Pavlovski potok, and Vipava) (Figures 1 and 6). Three types of this community were recorded, which were classified into three clusters.

Association *Myriophylletum verticillati* Gaudet ex Šumberová in Chytrý 2011

This vegetation type is dominated by *M. verticillatum*. These stands were the most species-rich and were only present in the river Struga.

Association Potamo pectinati-Myriophylletum spicati Rivas Goday 1964

The far most abundant species in this community was *M. spicatum*, which was also the most common macrophyte in the studied watercourses. This association was found in 31 stretches in watercourses from all three regions of Slovenia (Figures 1 and 6). A smaller cluster was recorded in the lower part of the river Ščavnica in the most eastern margin of the study area and differed from the larger cluster due to high abundance of the amphiphyte *Butomus umbellatus*, which was present in *f. aquatica* and *f. terrestris*. The third largest group of 87 stretches was classified into a sub-optimally developed form of the association *Potamo pectinati-Myriophylletum spicati*. In these stands, the characteristic species *M. spicatum* was the most abundant species, which had about two-fold the cover of the type described below, but still lower abundance than what is required for classification into association (≥50%). The second most abundant species was *P. nodosus*. This was the species-poorest vegetation type (4.2). The stands had the same distribution pattern as the stands of optimally developed association. Based on these indicators, we classified this group into the association *Potamo pectinati-Myriophylletum spicati*. This was the most frequently recorded association.

Community with Myriophyllum spicatum

We classified this second largest cluster (Figure 5) into a suboptimal form of the above-described association, since the characteristic species did not reach the abundance for classification into association. The subdominant species were *Phalaris arundinacea* and *P. nodosus*, with significantly lower abundances.

Association *Potametum pectinati* Carstensen ex Hilbig 1971

This association is dominated by *P. pectinatus*. This plant association occurred only in eight stretches in the rivers Krka, Borovniščica, Ižica, and Stržen. Nonetheless, the characteristic and dominant species *P. pectinatus* was very common.

Association *Potametum crispi* von Soó 1927

The dominant species in this association is *P. crispus* (Table 2, 36%), which is present in all 27 stretches and is about three times more abundant than *M. spicatum* (13%). It was determined in the rivers Krka, Sotla, Temenica, Hočki potok, Polskava, Pavlovski potok, and even Vipava. The average number of species was low (7.8).

Association *Potametum perfoliati* Miljan 1933

This plant association is defined by the dominance of the characteristic species *P. perfoliatus*, which was two times more abundant than the second most abundant species *M. spicatum* (22%). Species richness was moderate (average 10.0). The third most abundant species in these 23 stretches was *P. crispus*. It was detected in the lower flow of Kolpa and in Rak, Vipava, Pesnica, Rinža, and Ižica (Figures 1 and 6).

Association *Potametum natantis* Hild 1959

This plant association is defined by the characteristic and dominant species with floating leaves, *P. natans*, which was the most abundant (51%). The sub-dominant species were *N. luteum* and *P. lucens*. Species richness and total abundance were moderate. It was detected in 18 stretches in Krka, Ljubljanica and along the lower flow of the river Ižica.

Association *Elodeetum canadensis* Nedelcu 1967

This community is dominated by the invasive alien species *E. canadensis*. It developed in Krka, Temenica, Ljubljanica, Ižica, and Stržen. Species richness was moderate.

Association Nymphaeo albae-Nupharetum luteae Nowiński 1927

This community is dominated by the large-sized floating-leaved macrophyte *N. luteum*, which often develops only submerged leaves in running waters. The distribution of this association in surveyed rivers is limited to the river Stržen in the Cerkniško polje and to the river Ščavnica in the Pannonian region, while single stretches were recorded in the river Vipava (Po lowland) and in the river Rinža (Dinarides).

Association Potamo crispi-Ranunculetum trichophylli Imchenetzky 1926

These stands are dominated by *Ranunculus trichophyllus*. It occurs in oligo-mesotrophic to eutrophic water bodies, such as alluvial pools and ditches. It is often accompanied by *Eleocharis acicularis*. It was distributed in 32 stretches in the rivers on karst poljes (Veliki Obrh, Stržen, Rak, Unica), which have an extremely fluctuating water tables, as they sometimes even run dry, and in two stretches of the river Vipava.

Association *Polygonetum hydropiperis* Passarge 1965

This vegetation type is usually dominated by *Polygonum hydropiper* or *P. mite*. It occurred only in seven stretches in the lower part of the river Ščavnica.

Association *Phragmitetum australis* Savič 1926

In the riverbeds of the surveyed watercourses, this association was sub-optimally developed. The distribution of these stands is ubiquitous, as *P. australis* is the dominant species. It occurred in 42 stretches in watercourses of the coastal part of Slovenia (Rižana), Po lowland (Vipava) in central Slovenia, as well as in the Pannonian region.

Association Glycerio-Sparganietum neglecti Koch 1926

The community dominated by *Sparganium erectum* agg. was detected in 18 stretches and included stands dominated by the species *S. erectum*, *S. neglectum*, and *S. microcarpum*.

Association Schoenoplectetum lacustris Chouard 1924

This community is dominated by the tall-growing species *Schoenoplectus lacustris* that forms sparse stands. Its distribution was limited to the rivers on karst poljes (Mali Obrh, Stržen, Unica), with extremely fluctuating water tables.

Association *Phalaridetum arundinaceae* Libbert 1931

Stands of this association are dominated by *Phalaris arundinacea*. They were found in watercourses in all three regions.

Association Oenantho aquaticae-Rorippetum amphibiae Lohmeyer 1950

This association is dominated by *Rorippa amphibia*, which often forms dense monospecific stands. It was recorded in the river Stržen and in two Pannonian rivers.

Association Alopecuro-Alismatetum plantaginis-aquaticae Bolbrinker 1984

This association includes stands with the characteristic and dominant species *Alisma plantago-aquatica*. It was found in five stretches in Sotla, Hočki potok, and Pavlovski potok.

Association Sagittario sagittifoliae-Sparganietum emersi Tüxen 1953

This association includes stands dominated by *Sagittaria sagittifolia* and *S. emersum*, respectively, which are both characteristic and dominant species of this association. Stands of this association were recorded in the river Ljubljanica and its tributary in central Slovenia, and in the river Ščavnica in the northeastern edge of the study area.

Association Eleocharito palustris-Hippuridetum vulgaris Passarge 1964

This vegetation type is represented by stands of *Hippuris vulgaris*. This association was the rarest among the studied associations since it only occurred in the river Ižica in central Slovenia.

Association Glycerio notatae-Veronicetum beccabungae Landucci et al. 2020

Stands dominated by *Veronica beccabunga* were classified to the association *Glycerio notatae-Veronicetum beccabungae*, which was defined in the year 2020. The subdominant species were *Juncus effusus*, *P. australis*, and *Agrostis stolonifera*. These stands were among the species-poorest (average 4.3) in our study. This association was rare, since it was only recorded in four stretches of the upper part of the river Temenica in central Slovenia.

Association *Beruletum erectae* Roll 1938

This association is dominated by the amphiphyte macrophyte *Berula erecta*. We found this community in three different forms considering floristic composition, one with the co-dominant *Callitriche* spp., the most common with the subdominant *Callitriche* spp., and the third, where *B. erecta* was the most dominant species. This association was recorded in 18 different watercourses all over Slovenia in all the regions, from the coastal part and Po lowland to the most eastern watercourses in the Pannonian lowland.

Association Mentho aquaticae-Oenanthetum fistulosae ass. nova

This association has not yet been described, so we define it for the first time here in this contribution. This new description is also the consequence of its limited distribution, as it was only recorded in five stretches of the river Mali Obrh. This river flows on the Loško polje, which is a karst polje, a feature found only in the Dinaric region within Europe, for which exceptional water level fluctuations are characteristic, as sometimes the entire area may even run dry. Stands of this community are dominated by the characteristic species *Oenanthe fistulosa*, accompanied with numerous amphiphytes from the class *Phragmito-Magnocaricetea*, which represent the majority in number and abundance. This was the third richest plant community in our study.

Holotypus hoc loco: Oenanthe fistulosa 63%, Mentha aquatica 38%, Juncus articulatus 38%, Phalaris arundinacea 15%, Alisma spp. 15%, Rorippa amphibia 15%, Caltha palustris 3%, Potamogeton lucens 15%, Schoenoplectus lacustris 15%, Lycopus europaeus 3%, Sium latifolium 3%. (GK Coordinates: 459,376, 61,043; altitude: 570.2 m a.s.l.; slope: 0.58; current velocity: 1; length: 100 m).

Association Rumici crispi-Agrostietum stoloniferae Moor 1958

This association occurs in sites that are frequently dry. In fact, they are mostly inundated at moderate or high-water levels. This association was recorded in the river Mali Obrh in karst polje and in two streams in the Pannonian lowland.

Species-poorest sections with lowest abundance of macrophytes

This group of stretches could not be classified into any plant communities due to the low abundance and various ecology of the present species. This was by far the largest group of stretches (Figure 5), which encompassed one quarter of the total set of the recorded stretches. This group was among the species-poorest groups (6.5 species on average) with a very low abundance value. The most common species in this type were marsh species (helophytes) *P. arundinacea*, *Lytrum salicaria*, and *Myosotis scorpioides*, found sporadically close to the banks within these stretches regardless of the region. This type is scattered along the entire set of the studied watercourses.

4. Discussion

4.1. Species Presence and Abundance

High heterogeneity of sub-catchments of the surveyed watercourses resulted in high heterogeneity of environmental parameters and consequently high vascular plant diversity and low total share of species with a wide ecological valence. This environmental heterogeneity provides high habitat diversity, as shown in the case of macrophyte diversity in ponds [76]. The meta-analysis of aquatic macrophytes on a global scale revealed that the majority of macrophyte species are distributed within narrow ranges, and only 1.21% of the total 3457 macrophyte species are widely distributed [77]. The most abundant species in the surveyed Slovenian watercourses were *M. spicatum*, *P. arundinacea*, *P. nodosus*, *B. erecta*, *P. perfoliatus*, alien species *E. canadensis*, *P. crispus*, *P. pectinatus*, *N. luteum*, *P. natans*, and *S. lacustris*. All of these species are known to have wide ecological ranges [78,79]. Some of them even became invasive in different parts of the world. The presence and abundance of these invasive species depend on their potential for introduction in certain environments and environmental limitations to their distributions [80]. Among the most abundant species detected in this study was the alien species *E. canadensis*. However, it does not exhibit an invasive character in the surveyed watercourses, since the environmental diversity of these watercourses prevents its spreading [81]. In addition, *E. canadensis* is indicative of impacted lowland sites [82], as also revealed in this study.

In the surveyed watercourses, we detected a relatively low share of submerged plants. This is possibly due to the low average stream order of these watercourses and thus shading of riverbeds with riparian vegetation. Species with amphibious character are an especially important component of karst watercourses, where extreme water level fluctuations present a factor of permanently changing environmental conditions. These changes do not only present stress for plants, but also disturbances that result in dieback of submerged species. Thus, only well-adapted amphibious species and cosmopolites may withstand these changes [39]. In the case of northern European rivers, less than half of aquatic plant species exhibited an amphibious character [83]. Among the more abundant species found in the surveyed watercourses, *P. arundinacea*, *B. erecta*, *N. luteum*, and *S. lacustris* may develop different growth forms in addition to numerous low-growing amphibious species like *M. aquatica*, *R. amphibia*, *V. anagallis-aquatica*, and *M. scorpioides* agg., which may also contribute significantly to plant abundance. Multiple environmental changes during water level fluctuations like CO_2 concentration, light intensity and quality, temperature, and osmotic stress may trigger the development of a new, better-adapted growth form or leaves [84]. In addition, these fluctuations also affect environmental conditions for ecosystem processes, especially the exchange of substances between sediments and water, and mineralization rates [85], and thus the availability of nutrients in water [86]. Changes are most pronounced during floods, when nutrients are released from the terrestrial systems into water. Due to intensive mineralization rates, the sediments of intermittent water bodies usually contain a lower proportion of organic matter and consequently a lower nutrient content [87], which may also affect species composition. Among growth forms, we detected a somewhat lower share of helophytes and floating-leaved plants, while the lowest share was seen for free-floating vascular plant taxa. This was a consequence of the environmental properties of these watercourses that mainly present tributaries of larger rivers, such as rivers Sava and Drava.

4.2. The Relationships between Macrophyte Presence and Abundance, and Environmental Factors

In our study, spatial parameters explained the greatest share of species presence and abundance variability. This is expected, since according to the research of Tapia Grimaldo et al. [88], spatial variables may be used as a proxy for various environmental variables that are important for defining macrophyte communities. The significant contribution of distance from the source to the explained variance is partly a consequence of the river continuum effect [89] that is based on discharge gradient, current velocity, flow pattern, bed, bank, and community structure along the river course. Considering single parameters (marginal effects), altitude was of equal importance. Since we excluded the alpine region from our research, the most pronounced altitudinal gradient in the surveyed watercourses was seen for the karst intermittent watercourses that flow through different poljes along the Idrija fault which are located at different altitudes and host different plant species [39]. Kennedy et al. [90] showed that the main environmental parameters for shaping macrophyte community composition and diversity are altitude and stream order, along with some chemical parameters.

Correlation analysis between the environmental parameters and species diversity and abundance showed highly significant results (Table 1). Altitude has been shown to be a strong predictor of aquatic macrophyte diversity irrespective of geographical location [91] and is usually negatively related with species diversity [42]. However, in our study, we found positive correlations between macrophyte species diversity and altitude. The strong negative relationship with current velocity found in our study is in accordance with the results, gathered in the survey of middle-sized streams in Latvia, where the highest macrophyte diversity was determined in streams with low water velocity and gravelly substrate, while the species-poor macrophyte communities were characteristic for streams with high water velocity and sandy substrate [92]. The most sensitive macrophytes to high water velocities are macrophytes that are anchored to the sediment, while macrophytes with a free-floating growth form are the least sensitive [51]. The latter may also be passively transported along the river flow, while the former might be mechanically damaged or loaded with sediment. Low to moderate velocities have a beneficial effect on growth, as they increase the availability of carbon dioxide and nutrients, while higher speeds represent mechanical stress and thus have a negative effect [93]. This is in accordance with the intermediate disturbance hypothesis, which presumes that maximum species richness occurs under moderate disturbance regime [94]. In addition, the current supports plant communities by transport of plant propagules along the watercourse [24].

4.3. *Plant Communities*

Syntaxonomic scheme of the studied plant communities

1. Class *Lemnetea* O. de Bolòs et Masclans 1955
    1.1. Alliance *Stratiotion* Den Hartog et Segal 1964
        1 Association *Ceratophylletum demersi* Corillion 1957
2. Class *Potamogetonetea* Klika in Klika et Novák 1941
    2.1. Alliance *Potamogetonion* Libbert 1931
        2 Association *Potametum denso-nodosi* de Bolós 1957
        3 Association *Myriophylletum verticillati* Gaudet ex Šumberová in Chytrý 2011
        4 Association *Potamogetono pectinati-Myriophylletum spicati* Rivas Goday 1964
        5 Community with *Myriophyllum spicatum*
        6 Association *Potametum pectinati* Carstensen ex Hilbig 1971
        7 Association *Potametum crispi* Soó 1927
        8 Association *Potametum perfoliati* Miljan 1933
        9 Association *Potametum natantis* Hild 1959
        10 Association *Elodeetum canadensis* Nedelcu 1967
    2.2. Alliance *Nymphaeion albae* Oberdorfer 1957
        11 Association *Nymphaeo albae-Nupharetum luteae* Nowiński 1927
    2.3. *Ranunculion aquatilis* Passarge ex Theurillat in Theurillat et al. 2015
        12 Association *Potamo crispi-Ranunculetum trichophylli* Imchenetzky 1926
3. Class *Bidentetea tripartitae* Tüxen et al. ex von Rochow 1951
    3.1. Alliance *Bidention tripartitae* Nordhagen ex Klika et Hadač 1944
        13 Association *Polygonetum hydropiperis* Passarge 1965
4. Class *Phragmito-Magnocaricetea* Klika in Klika et Novák 1941
    4.1. Alliance *Phragmition communis* Koch 1926
        14 Association *Phragmitetum australis* Savič 1926
        15 Association *Glycerio-Sparganietum neglecti* Koch 1926
        16 Association *Schoenoplectetum lacustris* Chouard 1924
        17 Association *Phalaridetum arundinaceae* Libbert 1931
    4.2. Alliance *Eleocharito palustris-Sagittarion sagittifoliae* Passarge 1964
        18 Association *Oenantho aquaticae-Rorippetum amphibiae* Lohmeyer 1950
        19 Association *Alopecuro-Alismatetum plantaginis-aquaticae* Bolbrinker 1984
        20 Association *Sagittario sagittifoliae-Sparganietum emersi* Tüxen 1953
        21 Association *Eleocharito palustris-Hippuridetum vulgaris* Passarge 1964
    4.3. Alliance *Glycerio-Sparganion* Br.-Bl. et Sissingh in Boer 1942
        22 Association *Glycerio notatae-Veronicetum beccabungae* Landucci et al. 2020
        23 Association *Beruletum erectae* Roll 1938
        24 Association *Mentho aquaticae-Oenanthetum fistulosae* ass. nova
5. Class *Molinio-Arrhenatheretea* Tx. 1937
    5.1. Alliance *Potentillion anserinae* Tx. 1947
        25 Association *Rumici crispi-Agrostietum stoloniferae* Moor 1958

The surveyed stretches differed greatly regarding species composition and abundance, and we classified these assemblages into 25 different associations. Most of the determined associations are widespread in the surveyed watercourses, with the exception of 7 of them that occur more locally, 3 out of these in karst poljes.

The association *Ceratophylletum demersi* was the only community dominated by pleustophytes, occurring in the studied watercourses, and was detected in six stretches. It usually occurs in waterbodies with no or only slow current, in eutrophic to hypertrophic water, which can also reach high turbidity [65]. However, in lakes, C. *demersum* may also play an important role in stabilization and maintenance of a clear-water state at high P concentrations [95].

The association *Potametum denso-nodosi* occurs in watercourses with low flow velocity or in artificial channels, mostly with water depths of 30–60 cm, as was also the case in our study. These habitats are usually frequently disturbed, while sediments are mainly mineral [66]. The dominant species is *P. nodosus*. This species is widespread and develops both natant and submerged leaves [79].

The association *Myriophylletum verticillati* occurs in mesotrophic to eutrophic clear water, e.g., in channels and lentic sections of streams [66]. *M. spicatum* and *M. verticillatum* thrive in waters with high calcium contents [96].

The association *Potamo pectinati-Myriophylletum spicati* was the species-poorest vegetation type, which indicated the competitive character of the dominant species *M. spicatum* with a broad ecological range [81]. Large-branched specimens that produce dense stands and dispersion by fragmentation present a competitive advantage over smaller species [97]. Thus, it is most common in places with regular disturbances. It also tolerates turbid waters and even desiccation of its habitat for a short period [66].

Community with *M. spicatum* formed the second largest cluster of stretches. This community was sub-optimally developed, since the characteristic species did not reach the level of abundance for classification into association. The subdominant species were *P. arundinacea* and *P. nodosus*, with significantly lower abundances.

The association *Potametum pectinati* thrives in eutrophic and hypertrophic waters with high turbidity and was mostly found in the middle and lower parts of watercourses. This association is dominated by *P. pectinatus*, which has high phenotypic plasticity [79]. Its reproduction is generally vegetative. Therefore, tubers play an important role in survival of its populations [98].

The association *Potametum crispi* was found in different parts of Slovenia, mainly in the watercourses that are not very deep. The community is typical for waters loaded with nutrients [66].

The association *Potametum perfoliati* was defined by the dominance of the characteristic species *P. perfoliatus*, which is indicative of a wide range of nutrient conditions and pH [99].

The association *Potametum natantis* was only found in few stretches of three lowland rivers. This species-poor vegetation type is dominated by *P. natans*, which forms dense stands with broad leaves, floating on the water surface [79]. It mostly occurs in mesotrophic, but also in oligotrophic or eutrophic water bodies, which are 20–100 cm deep, with no or slow current, e.g., in alluvial pools, channels, and lentic sections of streams [66].

The association *Elodeetum canadensis* is dominated by the invasive alien species *Elodea canadensis*. It was found only in about 4% of the surveyed stretches. This species prefers deeper eutrophic water bodies like channels and lentic sections of streams. It was not found in karst watercourses, since it does not tolerate drying [39]. Although this vegetation type is considered as species-poor, species richness in our case was moderate (10.7), which indicated low invasiveness of the characteristic species in the studied watercourses [81]. They mostly occurred in central Slovenia.

The association *Nymphaeo albae-Nupharetum luteae* is characteristic for the river Stržen in the Cerkniško polje (Dinarids) and for the river Ščavnica in the Pannonian region. It tolerates high water level fluctuations, which mark the water regime in the Cerkniško polje [100]. According to Šumberová [66], it usually develops in mesotrophic to eutrophic water of alluvial pools, channels, and lentic sections of streams. This community is dominated by the large-sized floating-leaved macrophyte *Nuphar luteum*, which mainly develops submerged leaves in running waters. This association was mostly found in the river Stržen in the Cerkniško polje, and in the river Ščavnica in the Pannonian region.

The association *Potamo crispi-Ranunculetum trichophylli* is the only association from this alliance. It encompasses the vegetation of aquatic plants, that is mainly present in rivers on karst poljes with an extremely fluctuating water table [101].

The association *Polygonetum hydropiperis* is usually dominated by *P. hydropiper* or *P. mite*. It colonizes the banks of rivers and streams, seasonally flooded pools in river floodplains, ditches, and channels, where soils tend to dry out quickly [67]. This association was rare in the studied watercourses.

Monospecific stands of the association *Phragmitetum australis* consist of tall and dense shoots of *P. australis* [102]. This species has a broad ecological range, occurring in oligotrophic to eutrophic, acidic to basic sites [68], having low regional bioindicator potential [103]. It is widely distributed from 70° N to the tropics [104]. It occurs in the littoral, along the watercourses, channels, and in deltas, in up to 2-m-deep waters [100]. In the surveyed riverbeds, this association was sub-optimally developed. It was abundant in the riparian zone, which was, however, not the subject of our research.

The association *Glycerio-Sparganietum neglecti* includes stands dominated by the species *S. erectum*, *S. neglectum*, and *S. microcarpum*. In some floras, these species are considered as subspecies of *S. erectum*, while in others, they are treated as independent species [70,105]. They have a very similar morphology, distribution, and habitats, which are overlapping, so they had been frequently misidentified or not distinguished from each other. Therefore, it was not possible to define three different associations [106]. They usually thrive in shallow mesotrophic to eutrophic water bodies, in ditches and channels with slowly running water, as well as in lentic sections on lower river courses. The dominant *S. erectum* agg. develops aerial and long submerged floating leaves, which are tolerant to moderate current velocity. Usually, optimal water level is 10–60 cm, but sites may dry out for short periods in summer [43,68].

The association *Schoenoplectetum lacustris* is dominated by the tall-growing species *S. lacustris* that favors mesotrophic to eutrophic water, constantly wet conditions and is usually found deeper than *P. australis* [70]. It often colonizes a transitional zone between the open water and reed vegetation. The distribution of this association was limited to rivers on karst poljes.

The association *Phalaridetum arundinaceae*, which is dominated by *P. arundinacea*, previously classified to the association *Rorippo-Phalaridetum* [68]. This association occurred on stream banks of the middle reaches of watercourses with high current velocity and is characterized by species tolerant of disturbance. Landucci et al. [70] classified stands dominated by *P. arundinacea* into one single association *Phalaridetum arundinaceae*, which was assigned to the alliance *Phragmition communis* since the floristic differences between these associations are too small.

The association *Alopecuro-Alismatetum plantaginis-aquaticae* was only found in five stretches. It usually occurs in shallow water in sites, which are seasonally dry. Habitats are typically mesotrophic to eutrophic and inundated by up to 50 cm. However, occurrences on wet and occasionally inundated soil are also common [68]. In our study, this association was found in the watercourses of the Pannonian area with such character.

The association *Sagittario sagittifoliae-Sparganietum emersi* includes stands dominated by *S. sagittifolia* and *S. emersum*, respectively, which are both characteristic and dominant species of this association. In rivers, both species produce different leaf forms, namely submerged or emergent, or even partly floating leaves on the water surface. This association favors up to 120 cm deep eutrophic waters [68]. In summer, the water table can decrease and expose the bottom. However, these dominant species can regenerate from their underground organs.

The association *Eleocharito palustris-Hippuridetum vulgaris* is represented by stands of *H. vulgaris*. In stretches with low current velocity, this species forms dense stands [107]. In deep waters, it appears in the submerged form, while in shallow waters, it occurs in the emergent form. It thrives in calcium-rich waters and is often found in cool waters. In general, the number of localities is low and is decreasing due to eutrophication and spread of competitively stronger plants [68]. This association was among the rarest and occurred only in the river Ižica in central Slovenia.

The association *Glycerio notatae-Veronicetum beccabungae* is dominated by *V. beccabunga* and was defined in 2020 due to the previously invalidly published names [70]. The subdominant species were *Juncus effusus, P. australis,* and *A. stolonifera*. These stands were rare and species-poor.

The association *Beruletum erectae,* which is dominated by the amphiphyte *B. erecta,* was found in 56 stretches of 18 watercourses in all the surveyed regions. The dominant species and consequently the association thrive in waters rich in nutrients and calcium, and can be found in different forms, namely in emergent stands in shallow water or on wet banks, in stands with leaves, partly floating on the surface in slow streams, and in submerged stands with strongly branched stems in fast streams [68], as was also the case in our study.

The association *Mentho aquaticae-Oenanthetum fistulosae* is a species-rich, newly defined association, limited to five stretches in the river Mali Obrh on Loško polje. Stands of this community are dominated by the low stature amphibious species *O. fistulosa,* which thrived in the company of numerous other amphiphytes.

The flood-meadow association *Rumici crispi-Agrostietum stoloniferae* is species-poor because of the dominance of the taxon *A. stolonifera* agg., which forms dense stands [69]. This association occurred only in six stretches in sites that are frequently dry, but mostly inundated at moderate or high-water levels.

5. Conclusions

High heterogeneity of the surveyed watercourses, their catchments, and environmental parameters resulted in high diversity of vascular plants and consequently in high diversity of plant communities they form. We found high heterogeneity of aquatic vegetation and low total share of species with a wide ecological valence. We determined a total of 87 vascular plants among which 36% of total abundance belonged to submerged macrophytes, 30% to amphiphytes, 18% to helophytes, 3% to floating-leaved plants, and only 3% to pleustophytes. The most abundant species was *M. spicatum* with a 12% share, followed by *P. arundinacea* and *P. nodosus*. Spatial parameters explained the highest share of species presence and abundance variability. The examined river stretches host 25 different associations belonging to 5 classes and 9 alliances, which are mostly distributed in different hydro-ecoregions. Seven plant associations had local distribution, and three of them were only found on karst poljes. This distribution pattern reveals specific environmental conditions in these karst poljes and lack of regionality elsewhere.

Author Contributions: Conceptualization, A.G. and I.Z.; methodology, M.H., A.G., U.K. and I.Z.; investigation: A.G., M.G. and U.K.; validation: M.H., A.G., M.G. and U.K., I.Z.; writing—original draft preparation M.H., A.G. and I.Z.; writing—review and editing, A.G. and I.Z.; visualization, A.G., M.H. and I.Z.; supervision, A.G. and I.Z.; project administration, A.G. and U.K.; funding acquisition, A.G. All authors have read and agreed to the published version of the manuscript.

Funding: This research was funded by Ministry of Higher Education, Science and Technology-Research programs Biology of plants (P1-0212) and Young researchers (3311-04-831246) (Slovenia).

Institutional Review Board Statement: Not applicable.

Informed Consent Statement: Not applicable.

Data Availability Statement: Data are stored within the documentation of P1-0212 Research program. Data are available on special request.

Acknowledgments: The authors thank to Mateja Grašič for reviewing the manuscript.

Conflicts of Interest: The authors declare no conflict of interest. The funders had no role in the design of the study; in the collection, analyses, or interpretation of data; in the writing of the manuscript, or in the decision to publish the results.

Appendix A

Table A1. Synoptic table of the clusters classified into plant communities. Their dominant (orange) and subdominant (yellow) species are presented in average values of cover-abundance values (%). 0 stands for values <0.5.

Cluster Number	1	2	3	4	5	6	7	8	9	10	11	12	13	14	15	16	17	18	19	20	21	22	23	24	25	26	27	28	29	30	31	32	33	34
number of stretches	6	31	10	10	5	31	6	8	7	12	26	6	109	4	4	5	226	30	5	27	32	23	18	87	18	13	15	7	38	19	11	18	4	35
dominant species of the plant associations																																		
Ceratophyllum demersum	63	37	36	.	.	.	.	5	.	.	.	.	.	.	.	3	0	.	.	1	.	.	2	0	.	.	.	.	1	.	.	.	.	.
Potamogeton nodosus	22	31	63	66	.	16	39	13	6	.	0	.	4	.	.	1	1	4	.	4	.	9	15	18	.	.	6	.	.	1	.	.	.	3
Myriophyllum verticillatum	3	.	.	2	63	.	.	2	.	.	.	.	1	.	.	.	0	.	.	3	0	.	.	0	.	3	.	.	1	.	.	0	.	2
Myriophyllum spicatum	26	27	22	23	39	64	63	21	42	9	3	.	15	.	.	.	1	11	.	13	4	22	19	33	5	3	4	.	3	2	.	12	13	7
Potamogeton pectinatus	18	24	29	.	15	2	59	63	39	.	0	.	0	.	.	14	0	.	.	0	1	6	6	1	.	.	.	.	1	0	.	8	.	1
Polygonum hydropiper/mite	1	.	.	1	2	2	5	6	45	6	2	1	0	.	.	26	0	1	.	2	.	.	8	0	.	2	2	.	0	1	1	.	.	1
Phragmites australis	9	1	1	8	13	2	.	4	.	63	3	3	1	8	13	1	1	22	.	1	0	2	0	1	.	3	3	1	4	1	5	.	.	2
Phalaris arundinacea	7	2	0	20	33	7	28	12	42	2	38	7	5	2	.	9	4	4	19	11	6	9	13	2	5	69	5	.	9	9	15	2	.	7
Agrostis stolonifera agg.	.	.	.	.	.	.	22	.	13	.	5	42	1	4	8	.	2	2	3	2	.	2	3	0	.	2	2	.	2	4	1	.	.	1
Rorippa amphibia	.	.	.	.	.	0	8	.	0	.	3	.	0	44	.	.	1	4	9	.	5	0	0	.	2	4	6	.	.	0	.	0	.	.
Veronica beccabunga	1	.	.	2	4	.	.	1	.	.	2	.	0	.	38	.	1	.	.	.	0	.	4	0	.	2	.	.	1	.	.	.	.	1
Alisma plantago-aquatica	1	.	.	1	.	2	8	1	5	5	1	2	0	10	.	29	1	2	.	3	.	0	6	1	.	3	2	.	1	2	0	.	.	2
Oenanthe fistulosa	.	.	.	.	.	.	.	.	.	.	.	.	.	.	.	.	0	.	63	.	4	.	.	.	3	5	.	.	.	.	.	.	.	.
Potamogeton crispus	3	0	0	7	.	2	19	0	18	6	1	.	3	.	.	1	1	2	.	36	13	13	9	2	.	0	0	0	3	5	0	3	8	9
Ranunculus trichophyllus	3	7	12	.	.	.	.	2	.	.	.	3	1	.	.	.	1	1	.	2	27	3	2	2	1	0	5	.	2	7	.	1	5	2
Potamogeton perfoliatus	5	8	2	.	.	6	4	10	8	1	1	.	2	.	.	.	1	1	.	0	7	43	.	3	15	.	0	31	0	5	.	13	1	9
Sparganium erectum agg.	6	0	3	8	5	2	13	10	.	11	3	11	1	.	.	3	1	5	.	4	.	1	41	2	.	.	2	.	3	10	15	.	.	1
Schoenoplectus lacustris	3	11	5	4	.	1	.	7	.	.	3	.	1	2	.	.	1	2	4	.	12	2	1	1	57	.	5	.	1	1	.	15	.	1
Nuphar luteum	.	.	.	12	.	0	23	5	.	3	.	.	0	.	.	.	1	17	.	.	0	1	.	0	4	.	65	.	.	7	.	19	.	6
Callitriche spp.	1	.	.	2	2	1	1	7	32	0	0	.	0	.	.	6	1	.	1	4	4	0	5	.	.	10	.	38	15	19	4	6	27	5
Berula erecta	1	0	0	3	17	0	1	11	.	8	3	1	1	.	.	.	1	3	.	4	.	1	4	0	.	5	4	38	24	13	65	7	27	3
Sparganium emersum	8	6	.	8	8	4	5	11	16	.	.	.	0	.	.	.	1	3	.	1	.	1	1	0	.	3	7	29	1	47	.	18	1	10
Butomus umbellatus	.	.	.	.	.	.	46	.	13	.	1	.	0	.	.	.	1	0	.	.	0	2	.	.	.	.	0	.	.	.	.	.	.	.
Potamogeton natans	9	1	.	.	8	1	.	21	.	11	.	.	0	.	.	.	0	7	.	2	3	9	.	1	1	1	.	0	.	15	3	51	10	13

Table A1. *Cont.*

Cluster Number	1	2	3	4	5	6	7	8	9	10	11	12	13	14	15	16	17	18	19	20	21	22	23	24	25	26	27	28	29	30	31	32	33	34
number of stretches	6	31	10	10	5	31	6	8	7	12	26	6	109	4	4	5	226	30	5	27	32	23	18	87	18	13	15	7	38	19	11	18	4	35
dominant species of the plant associations																																		
Hippuris vulgaris	.	.	.	.	.	.	.	.	.	.	.	.	.	.	.	.	1	1	.	.	7	.	.	.	1	.	4	0	1	.	.	7	63	2
Elodea canadensis	1	0	2	0	8	3	.	9	.	.	.	.	0	.	.	.	1	.	.	10	.	2	2	0	.	.	.	.	2	5	3	15	38	49
subdominant taxa	.	.	.	.	.	.	.	.	.	.	.	.	.	.	.	.	.	.	.	.	.	.	.	.	.	.	.	.	.	.	.	.	.	.
Najas marina	.	14	34	.	.	1	5	0	.	.	.	.	0	.	.	.	0	.	.	0	.	.	1	0	.	.	.	.	.	.	.	.	.	.
Zannichellia palustris	.	.	.	10	8	.	.	.	.	.	.	.	0	.	.	.	0	.	.	.	.	0	.	0	.	5	.	.	2	.	0	.	.	.
Potamogeton berchtoldii	.	.	.	.	26	2	.	2	.	.	1	.	1	.	.	3	0	.	.	1	.	.	11	0	.	2	.	.	3	.	1	.	.	1
Iris pseudacorus	9	0	1	.	13	1	28	9	8	3	6	1	1	.	8	2	1	2	.	4	0	1	4	1	1	2	1	3	5	3	4	1	.	4
Lythrum salicaria	.	.	.	2	8	1	5	1	5	7	10	.	1	1	.	2	3	2	1	3	3	2	1	1	12	2	1	.	3	1	2	0	.	2
Leersia oryzoides	.	.	.	2	.	1	.	6	.	.	5	.	1	.	.	.	0	.	.	.	.	.	8	0	.	.	.	.	.	.	.	.	.	.
Ranunculus circinatus	1	1	11	.	.	1	.	5	.	.	.	9	0	.	.	.	0	.	5	2	0	.	.	0	1	1	.	.	0	2	.	1	.	4
Sium latifolium	.	.	.	.	.	.	.	.	.	.	.	.	.	5	.	.	0	1	1	.	2	0	.	.	1	.	2	.	.	.	.	.	.	.
Nasturtium officinale	6	1	.	.	36	1	.	9	.	.	0	.	0	.	8	.	1	0	.	7	1	0	1	0	.	11	.	.	16	0	3	5	5	5
Juncus effusus	.	.	.	.	.	.	.	.	.	0	3	.	0	.	23	3	0	0	.	2	.	.	2	0	.	.	.	.	.	.	0	.	.	.
Sparganium spp.	.	.	.	9	.	3	.	.	.	.	1	3	1	.	8	.	1	3	1	4	21	4	.	2	16	.	3	.	1	.	.	.	.	2
Mentha aquatica	3	.	.	3	20	0	.	.	.	5	2	3	0	4	.	1	2	3	29	1	13	1	3	0	6	2	7	.	7	0	14	0	4	1
Myosotis scorpioides agg.	10	.	.	2	10	0	.	9	.	5	1	3	0	2	.	.	3	1	14	5	11	2	2	0	2	2	5	18	4	0	2	8	2	3
Alisma spp.	.	.	.	.	.	0	.	.	.	.	1	3	0	2	.	.	1	0	15	.	4	1	.	0	5	2	0	.	.	.	.	.	.	.
Polygonum amphibium	.	.	.	.	.	.	.	.	.	.	4	1	.	2	.	.	1	1	7	.	2	0	.	.	3	14	6	.	.	.	.	.	.	.
Potamogeton lucens	.	.	.	.	.	.	.	2	.	.	.	.	2	.	.	.	0	5	4	.	1	4	.	1	6	.	9	10	1	13	.	23	1	4
Veronica anagallis-aquatica	3	9	3	0	.	1	.	7	.	3	0	.	0	.	.	.	1	1	4	9	11	2	3	0	.	8	0	6	12	0	13	13	23	6
Sagittaria sagittifolia	.	.	.	.	.	.	.	5	.	.	.	.	.	.	.	.	0	1	.	.	0	4	.	1	.	.	0	.	.	0	.	19	.	0
Avg. nr. of species	12.5			6.8	19.6	5.3	13.7	12.5	12.9	8.2	8.3	7.3	4.5	7.8	4.3	.	6.5	8.8	14.0	7.8	14.6	10.0	11.8	4.2	13.6	9.3	11.5	6.7	8.2	8.8	9.6	11.5	.	10.7

References

1. Ward, J.V. The Four-Dimensional Nature of Lotic Ecosystems. *J. N. Am. Benthol. Soc.* **1989**, *8*, 2–8. [CrossRef]
2. Settele, J.; Scholes, R.; Betts, R.A.; Bunn, S.; Leadley, P.; Nepstad, D.; Overpeck, J.T.; Angel Taboada, M.; Adrian, R.; Allen, C.; et al. 2014: Terrestrial and inland water systems. In *Climate Change 2014: Impacts, Adaptation, and Vulnerability: Part A: Global and Sectoral Aspects. Contribution of Working Group II to the Fifth Assessment Report of the Intergovernmental Panel on Climate Change*; Cambridge University Press: Cambridge, UK, 2014; pp. 1–32.
3. Stanford, J.A. Rivers in the landscape: Introduction to the special issue on riparian and groundwater ecology. *Freshw. Biol.* **1998**, *40*, 402–406. [CrossRef]
4. Petts, G.E. Rivers: Dynamic components of catchment ecosystems. In *The River's Handbook: Hydrological and Ecological Principles*; Calow, P., Petts, G.E., Eds.; Blackwell Science: Oxford, UK, 1994; pp. 3–22.
5. Boulton, A.J.; Brock, M.A. *Australian Freshwater Ecology: Processes and Management*; Gleneagles Publishing: Adelaide, Australia, 1999; p. 118.
6. Baattrup-Pedersen, A.; Larsen, S.E.; Riis, T. Long-term effects of stream management on plant communities in two Danish lowland streams. *Hydrobiology* **2002**, *481*, 33–45. [CrossRef]
7. Kuhar, U.; Gregorc, T.; Renčelj, M.; Šraj-Kržič, N.; Gaberščik, A. Distribution of macrophytes and condition of the physical environment of streams flowing through agricultural landscape in north-eastern Slovenia. *Limnologica* **2007**, *37*, 146–154. [CrossRef]
8. Kuhar, U.; Germ, M.; Gaberščik, A. Macrophytes of watercourses in the Slovenian Danube Basin. In *Macrophytes of the River Danube Basin*; Janauer, G.A., Gaberščik, A., Květ, J., Germ, M., Exler, N., Eds.; Academia: Praha, Czech Republic, 2018; pp. 234–251.
9. Petersen, R.C. The RCE: A Riparian, Channel, and Environmental Inventory for small streams in the agricultural landscape. *Freshw. Biol.* **1992**, *27*, 295–306. [CrossRef]
10. Szoszkiewicz, K.; Ciecierska, H.; Kolada, A.; Schneider, S.C.; Szwabinska, M.; Ruszczynska, J. Parameters structuring macrophyte communities in rivers and lakes–results from a case study in North-Central Poland. *Knowl. Manag. Aquat. Ecosyst.* **2014**, *415*, 8. [CrossRef]
11. Kuhar, U.; Germ, M.; Gaberščik, A.; Urbanič, G. Development of a River Macrophyte Index (RMI) for assessing river ecological status. *Limnologica* **2011**, *41*, 235–243. [CrossRef]
12. Haury, J. Patterns of macrophyte distribution within a Breton brook compared with other study scales. *Landsc. Urban Plan.* **1995**, *31*, 349–361. [CrossRef]
13. Haslam, S. The evaluation of river pollution using vegetation in the Maltese islands. *Fresenius Environ. Bull.* **2000**, *9*, 347–351.
14. Verschoren, V.; Schoelynck, J.; Cox, T.; Schoutens, K.; Temmerman, S.; Meire, P. Opposing effects of aquatic vegetation on hydraulic functioning and transport of dissolved and organic particulate matter in a lowland river: A field experiment. *Ecol. Eng.* **2017**, *105*, 221–230. [CrossRef]
15. Bakker, E.; Van Donk, E.; Declerck, S.; Helmsing, N.; Hidding, B.; Nolet, B. Effect of macrophyte community composition and nutrient enrichment on plant biomass and algal blooms. *Basic Appl. Ecol.* **2010**, *11*, 432–439. [CrossRef]
16. Vestergaard, O.; Sand-Jensen, K. Alkalinity and trophic state regulate aquatic plant distribution in Danish lakes. *Aquat. Bot.* **2000**, *67*, 85–107. [CrossRef]
17. Haslam, S.M. *River Plants: The Macrophytic Vegetation of Watercourses*; Cardigan, Forrest Text: Ceredigion, UK, 2006; p. 450.
18. Franklin, P.; Dunbar, M.; Whitehead, P. Flow controls on lowland river macrophytes: A review. *Sci. Total. Environ.* **2008**, *400*, 369–378. [CrossRef]
19. Chambers, P.A.; Prepas, E.E.; Hamilton, H.R.; Bothwell, M.L. Current Velocity and Its Effect on Aquatic Macrophytes in Flowing Waters. *Ecol. Appl.* **1991**, *1*, 249–257. [CrossRef]
20. Dodds, W.K.; Biggs, B.J.F. Water Velocity Attenuation by Stream Periphyton and Macrophytes in Relation to Growth Form and Architecture. *J. North Am. Benthol. Soc.* **2002**, *21*, 2–15. [CrossRef]
21. Biggs, B.J.F. Hydraulic habitat of plants in streams. *Regul. Rivers Res. Manag.* **1996**, *12*, 131–144. [CrossRef]
22. Riis, T.; Biggs, B.J.F. Hydrologic and hydraulic control of macrophyte establishment and performance in streams. *Limnol. Oceanogr.* **2003**, *48*, 1488–1497. [CrossRef]
23. Riis, T.; Suren, A.M.; Clausen, B.; Sand-Jensen, K. Vegetation and flow regime in lowland streams. *Freshw. Biol.* **2008**, *53*, 1531–1543. [CrossRef]
24. Fox, A.M. Macrophytes. In *The River's Handbook: Hydrological and Ecological Principles*; Calow, P., Petts, G.E., Eds.; Blackwell Science: Oxford, UK, 1992; pp. 216–233.
25. Murphy, K.J.; Rørslett, B.; Springuel, I. Strategy analysis of submerged lake macrophyte communities: An international example. *Aquat. Bot.* **1990**, *36*, 303–323. [CrossRef]
26. Riis, T.; Tank, J.L.; Reisinger, A.J.; Aubenau, A.; Roche, K.R.; Levi, P.S.; Baattrup-Pedersen, A.; Alnoee, A.B.; Bolster, D. Riverine macrophytes control seasonal nutrient uptake via both physical and biological pathways. *Freshw. Biol.* **2019**, *65*, 178–192. [CrossRef]

27. Baattrup-Pedersen, A.; Larsen, S.E.; Riis, T. Composition and richness of macrophyte communities in small Danish streams-influence of environmental factors and weed cutting. *Hydrobiologia* **2003**, *495*, 171–179. [CrossRef]
28. Baattrup-Pedersen, A.; Riis, T. Macrophyte diversity and composition in relation to substratum characteristics in regulated and unregulated Danish streams. *Freshw. Biol.* **1999**, *42*, 375–385. [CrossRef]
29. Chambers, P.A.; Lacoul, P.; Murphy, K.J.; Thomaz, S.M. Global diversity of aquatic macrophytes in freshwater. In *Freshwater Animal Diversity Assessment*; Springer: Berlin, Germany, 2008; pp. 9–26.
30. Pip, E. Water temperature and freshwater macrophyte distribution. *Aquat. Bot.* **1989**, *34*, 367–373. [CrossRef]
31. Léonard, R.; Legendre, P.; Jean, M.; Bouchard, A. Using the landscape morphometric context to resolve spatial patterns of submerged macrophyte communities in a fluvial lake. *Landsc. Ecol.* **2007**, *23*, 91–105. [CrossRef]
32. Gacia, E.; Ballesteros, E.; Camarero, L.; Delgado, O.; Palau, A.; Riera, J.L.; Catalan, J. Macrophytes from lakes in the eastern Pyrenees: Community composition and ordination in relation to environmental factors. *Freshw. Biol.* **1994**, *32*, 73–81. [CrossRef]
33. Manzo, L.; Grech, M.; Epele, L.; Kutschker, A.; Miserendino, M. Macrophyte regional patterns, metrics assessment and ecological integrity of isolated ponds at Austral Patagonia (Argentina). *Sci. Total. Environ.* **2020**, *727*, 138617. [CrossRef] [PubMed]
34. Wraber, M. Pflanzengeographische Stellung und Gliedernung Sloweniens. *Vegetatio* **1969**, *17*, 176–199. [CrossRef]
35. Sket, B. Oblikuje se današnje živalstvo. In *Fauna of Slovenia*; Sket, B., Gogala, M., Kuštor, V., Eds.; Tehniška založba Slovenije: Ljubljana, Slovenia, 2003; pp. 41–55.
36. Illies, J. *Limnofauna Europaea: Eine Zusammenstellung Aller Die Europäischen Binnengewässer bewohnenden Mehrzelligen Tierarten Mit Angaben Über Ihre Verbreitung Und Ökologie*; Fischer: Stuttgart, Germany, 1978.
37. Mršić, N. *Biotic Diversity in Slovenia: Slovenia-the "Hot Spot" of Europe*; Ministrstvo za Okolje in Prostor, Uprava RS za Varstvo Narave: Ljubljana, Slovenia, 1997; p. 129.
38. Urbanic, G. Redelineation of European inland water ecoregions in Slovenia. *Rev. Hydrobiol.* **2008**, *1*, 17–25.
39. Šraj-Kržič, N.; Germ, M.; Urbanc-Berčič, O.; Kuhar, U.; Janauer, G.A.; Gaberščik, A. The quality of the aquatic environment and macrophytes of karstic watercourses. *Plant Ecol.* **2007**, *192*, 107–118. [CrossRef]
40. Hawes, I.; Riis, T.; Sutherland, D.; Flanagan, M. Physical Constraints to Aquatic Plant Growth in New Zealand Lakes. *J. Publ. Title* **2003**, *41*, 44–52.
41. Van Geest, G.J.; Coops, H.; Roijackers, R.M.M.; Buijse, A.D.; Scheffer, M. Succession of aquatic vegetation driven by reduced water-level fluctuations in floodplain lakes. *J. Appl. Ecol.* **2005**, *42*, 251–260. [CrossRef]
42. Alahuhta, J.; Lindholm, M.; Baastrup-Spohr, L.; García-Girón, J.; Toivanen, M.; Heino, J.; Murphy, K. Macroecology of macrophytes in the freshwater realm: Patterns, mechanisms and implications. *Aquat. Bot.* **2021**, *168*, 103325. [CrossRef]
43. Jenačković, D.D.; Zlatković, I.D.; Lakušić, D.V.; Ranđelović, V.N. The assessment of seasonal variability in emergent macrophyte communities. *Biologia* **2016**, *71*, 287–297. [CrossRef]
44. Landucci, F.; Řezníčková, M.; Šumberová, K.; Chytrý, M.; Aunina, L.; Biţă-Nicolae, C.; Bobrov, A.; Borsukevych, L.; Brisse, H.; ČARNI, A.; et al. WetVegEurope: A database of aquatic and wetland vegetation of Europe. *Phytocoenologia* **2015**, *45*, 187–194. [CrossRef]
45. Hrvatin, M.; Tičar, J.; Zorn, M. Rocks and Tectonic Structure of Slovenia. In *The Geography of Slovenia*; Perko, D., Ed.; Springer Nature Switzerland AG: Cham, Switzerland, 2020; pp. 23–34.
46. Komac, B.; Pavšek, M.; Topole, M. Climate and Weather of Slovenia. In *The Geography of Slovenia*; Perko, D., Ed.; Springer Nature Switzerland AG: Cham, Switzerland, 2020; pp. 71–89.
47. Kolbezen, M.; Pristov, J.; Bat, M.; Klemenc, B.; Hrček, D. *Surface Streams and Water Balance of Slovenia*; Ministrstvo za Okolje in Prostor: Ljubljana, Slovenia, 1998; p. 98.
48. Janauer, G.; Exler, N.; Anačkov, G.; Barta, V.; Berczik, Á.; Boža, P.; Dinka, M.; Georgiev, V.; Germ, M.; Holcar, M.; et al. Distribution of the Macrophyte Communities in the Danube Reflects River Serial Discontinuity. *Water* **2021**, *13*, 918. [CrossRef]
49. Köhler, A.; Janauer, G.A. Zur Methodik der Untersuchung von aquatischen Makrophyten in Fließgewässern. *Handbuch Angewandte Limnologie: Grundlagen-Gewässerbelastung-Restaurierung-Aquatische Ökotoxikologie-Bewertung-Gewässerschutz* **2004**, *8*, 1–22. [CrossRef]
50. Schneider, S.; Melzer, A. The Trophic Index of Macrophytes (TIM)—A New Tool for Indicating the Trophic State of Running Waters. *Int. Rev. Hydrobiol.* **2003**, *88*, 49–67. [CrossRef]
51. Janauer, G.A.; Gaberscik, A.; Kvet, J.; Germ, M.; Exler, N. *Macrophytes of the River Danube Basin*; Academia: Praha, Czech Republic, 2018; p. 408.
52. Kurtto, A.; Sennikov, A.; Lampinen, R. *Atlas Florae Europaeae (AFE)—Distribution of Vascular Plants in Europe*; The Committee for Mapping the Flora of Europeand Societas Biologica Fennica Vanamo: Helsinki, Finland, 2013; Available online: https://www.luomus.fi/en/atlas-florae-europaeae-afe-distribution-vascular-plants-europe (accessed on 9 February 2021).
53. Schaumburg, J.; Schranz, C.; Foerster, J.; Gutowski, A.; Hofmann, G.; Meilinger, P.; Schneider, S.; Schmedtje, U. Ecological classification of macrophytes and phytobenthos for rivers in Germany according to the water framework directive. *Limnologica* **2004**, *34*, 283–301. [CrossRef]
54. Braun-Blanquet, J. *Pflanzensoziologie, Grundzüge der Vegetationskunde*, 3rd ed.; Springer: Berlin, Germany, 1964; p. 631.
55. Lukács, B.A.; Tóthmérész, B.; Borics, G.; Várbíró, G.; Juhász, P.; Kiss, B.; Müller, Z.; G-Tóth, L.; Erős, T. Macrophyte diversity of lakes in the Pannon Ecoregion (Hungary). *Limnologica* **2015**, *53*, 74–83. [CrossRef]

56. Pall, K.; Gecheva, G.; Soaru-Minea, A.; Lukacs, B.S.P. *Intercalibration of the National Classifications of Ecological Status for Eastern Continental Lakes*; Publications Office of the European Union: Rue Mercier, Luxembourg, 2018; p. 81.
57. Germ, M.; Gaberščik, A.; Urbanc-Berčič, O. The wider environmental assessment of river ecosystems. *Acta. Biol. Slov.* **2000**, *43*, 13–19.
58. Janauer, G.A.; Exler, N.; Schmidt-Mumm, U. The harmonised method for the macrophyte and habitat survey in the MIDCC-project: River Danube, floodplain waters and tributaries. In *Macrophytes of the River Danube Basin*; Janauer, G.A., Gaberščik, A., Květ, J., Germ, M., Exler, N., Eds.; Academia: Praha, Czech Republic, 2018; pp. 14–34. ISBN 978-80-200-2743-6.
59. QGIS. Welcome to the QGIS project! Available online: http://www.qgis.org/ (accessed on 25 September 2020).
60. Hammer, Ø.; Harper, D.A.T.; Ryan, P.D. PAST: Paleontological statistics software package for education and data analysis. *Palaeontol. Electron.* **2001**, *4*, 1–9.
61. Murtagh, F.; Legendre, P. Ward's Hierarchical Agglomerative Clustering Method: Which Algorithms Implement Ward's Criterion? *J. Classif.* **2014**, *31*, 274–295. [CrossRef]
62. Zelnik, I. Vegetation of the Meadows from the Order Molinietalia W. Koch 1926 and Contact Sites in Slovenia. Ph.D. Thesis, University of Ljubljana, Ljubljana, Slovenia, 2005.
63. Hrivnák, R.; Bubíková, K.; Oťaheľová, H.; Šumberová, K. Formalised classification of aquatic vegetation in Slovakia. *Phytocoenologia* **2019**, *49*, 107–133. [CrossRef]
64. Landucci, F.; Tichý, L.; Šumberová, K.; Chytrý, M. Formalized classification of species-poor vegetation: A proposal of a consistent protocol for aquatic vegetation. *J. Veg. Sci.* **2015**, *26*, 791–803. [CrossRef]
65. Šumberová, K. Vegetation of free floating aquatic plants (Lemnetea). In *Vegetation of the Czech Republic. 3, Aquatic and Wetland Vegetation*; Chytrý, M., Ed.; Academia: Praha, Czech Republic, 2011; pp. 43–99.
66. Šumberová, K. Vegetation of aquatic plants rooted in the bottom (Potametea). In *Vegetation of the Czech Republic. 3, Aquatic and Wetland Vegetation*; Chytrý, M., Ed.; Academia: Praha, Czech Republic, 2011; pp. 100–247.
67. Šumberová, K.; Lososová, Z. Vegetation of annual nitrophilous wetland herbs (Bidentetea tripartitae). In *Vegetation of the Czech Republic. 3, Aquatic and Wetland Vegetation*; Chytrý, M., Ed.; Academia: Praha, Czech Republic, 2011; pp. 347–384.
68. Šumberová, K.; Hájková, P.; Chytrý, M.; Hroudová, Z.; Sádlo, J.; Hájek, M.; Hrivnák, R.; Navrátilová, J.; Hanáková, P.; Ekrt, L.; et al. Marsh vegetation (Phragmito-Magno-Caricetea). In *Vegetation of the Czech Republic. 3, Aquatic and Wetland Vegetation*; Chytrý, M., Ed.; Academia: Praha, Czech Republic, 2011; pp. 385–579.
69. Zaliberová, M.; Škodová, I. Flood-meadows. In *Plant Communities of Slovakia. 5. Grassland Vegetation*; Hegedüšová, K., Škodová, I., Eds.; Veda: Bratislava, Slovakia, 2014.
70. Landucci, F.; Šumberová, K.; Tichý, L.; Hennekens, S.; Aunina, L.; Biță-Nicolae, C.; Borsukevych, L.; Bobrov, A.; Čarni, A.; De Bie, E.; et al. Classification of the European marsh vegetation (Phragmito-Magnocaricetea) to the association level. *Appl. Veg. Sci.* **2020**, *23*, 297–316. [CrossRef]
71. Mucina, L.; Bültmann, H.; Dierßen, K.; Theurillat, J.; Raus, T.; Čarni, A.; Šumberová, K.; Willner, W.; Dengler, J.; García, R.G.; et al. Vegetation of Europe: Hierarchical floristic classification system of vascular plant, bryophyte, lichen, and algal communities. *Appl. Veg. Sci.* **2016**, *19*, 3–264. [CrossRef]
72. Ter Braak, C.J.F.; Verdonschot, P.F.M. Canonical correspondence analysis and related multivariate methods in aquatic ecology. *Aquat. Sci.* **1995**, *57*, 255–289. [CrossRef]
73. Hudon, C.; Gagnon, P.; Jean, M. Hydrological factors controlling the spread of common reed (Phragmites australis) in theSt. Lawrence River (Québec, Canada). *Écoscience* **2005**, *12*, 347–357. [CrossRef]
74. Ter Braak, C.J.; Smilauer, P. *CANOCO Reference Manual and CanoDraw for Windows User's Guide: Software for Canonical Community Ordination*, version 4.5; Research Output of Wageningen University and Research Staff: Ithaca, NY, USA, 2002.
75. Furnas, E.R. Available online: http://www.microcomputerpower.com/ (accessed on 2 March 2021).
76. Fernández-Aláez, M.; García-Criado, F.; García-Girón, J.; Santiago, F.; Fernández-Aláez, C. Environmental heterogeneity drives macrophyte beta diversity patterns in permanent and temporary ponds in an agricultural landscape. *Aquat. Sci.* **2020**, *82*, 1–12. [CrossRef]
77. Murphy, K.; Efremov, A.; Davidson, T.A.; Molina-Navarro, E.; Fidanza, K.; Betiol, T.C.C.; Chambers, P.; Grimaldo, J.T.; Martins, S.V.; Springuel, I.; et al. World distribution, diversity and endemism of aquatic macrophytes. *Aquat. Bot.* **2019**, *158*, 103127. [CrossRef]
78. Maberly, S.C.; Haslam, S.M. River Plants of Western Europe: The Macrophytic Vegetation of the Watercourses of the European Economic Community. *J. Ecol.* **1988**, *76*, 1248. [CrossRef]
79. Preston, C. *Pondweeds of Great Britain and Ireland*; Botanical Society of the British Isles: London, UK, 1995; p. 352.
80. Brainard, A.S.; Luzadis, V.A.; Schulz, K.L. Drivers of species richness, biomass, and dominance of invasive macrophytes in temperate lakes. *Biol. Invasions* **2021**, *23*, 1069–1085. [CrossRef]
81. Kuhar, U.; Germ, M.; Gaberščik, A. Habitat characteristics of an alien species Elodea canadensis in Slovenian watercourses. *Hydrobiologia* **2010**, *656*, 205–212. [CrossRef]
82. O'Hare, M.T.; Baattrup-Pedersen, A.; Nijboer, R.; Szoszkiewicz, K.; Ferreira, T. Macrophyte communities of European streams with altered physical habitat. *Hydrobiologia* **2006**, *566*, 197–210. [CrossRef]
83. Janauer, G.; Dokulil, M. Macrophytes and Algae in Running Waters. In *Biological Monitoring of Rivers: Applications and Perspectives*; John, W., Ed.; Sons Ltd.: Chichester, UK, 2006; pp. 89–109.

84. Li, G.; Hu, S.; Hou, H.; Kimura, S. Heterophylly: Phenotypic Plasticity of Leaf Shape in Aquatic and Amphibious Plants. *Plants* **2019**, *8*, 420. [CrossRef]
85. Boulton, A.J.; Brock, M.A.; Robson, B.J.; Ryder, D.S.; Chambers, J.M.; Davis, J.A. *Australian Freshwater Ecology: Processes and Management*; Wiley-Blackwell: Hoboken, NJ, USA, 2014; p. 386.
86. Nilsen, E.T.; Orcutt, D.M. Water limitation. In *The Physiology of Plants under Stress. Abiotic Factors*; John Wiley and Sons: New York, NY, USA, 1996; pp. 322–361.
87. Brink, F.V.D.; Van Der Velde, G.; Bosman, W.; Coops, H. Effects of substrate parameters on growth responses of eight helophyte species in relation to flooding. *Aquat. Bot.* **1995**, *50*, 79–97. [CrossRef]
88. Grimaldo, J.T.; O'Hare, M.T.; Kennedy, M.P.; Davidson, T.A.; Bonilla-Barbosa, J.; Santamaría-Araúz, B.; Gettys, L.; Martins, S.V.; Thomaz, S.M.; Murphy, K.J. Environmental drivers of freshwater macrophyte diversity and community composition in calcareous warm-water rivers of America and Africa. *Freshw. Biol.* **2017**, *62*, 1511–1527. [CrossRef]
89. Vannote, R.L.; Minshall, G.W.; Cummins, K.W.; Sedell, J.R.; Cushing, C.E. The River Continuum Concept. *Can. J. Fish. Aquat. Sci.* **1980**, *37*, 130–137. [CrossRef]
90. Kennedy, M.P.; Lang, P.; Grimaldo, J.T.; Martins, S.V.; Bruce, A.; Hastie, A.; Lowe, S.; Ali, M.M.; Sichingabula, H.; Dallas, H.; et al. Environmental drivers of aquatic macrophyte communities in southern tropical African rivers: Zambia as a case study. *Aquat. Bot.* **2015**, *124*, 19–28. [CrossRef]
91. Alahuhta, J.; Lindholm, M.; Bove, C.P.; Chappuis, E.; Clayton, J.; De Winton, M.; Feldmann, T.; Ecke, F.; Gacia, E.; Grillas, P.; et al. Global patterns in the metacommunity structuring of lake macrophytes: Regional variations and driving factors. *Oecologia* **2018**, *188*, 1167–1182. [CrossRef]
92. Grinberga, L. Environmental factors influencing the vegetation in middle-sized streams in Latvia. *Ann. Bot.* **2011**, *1*, 37–44. [CrossRef]
93. Madsen, J.D.; Chambers, P.A.; James, W.F.; Koch, E.W.; Westlake, D.F. The interaction between water movement, sediment dynamics and submersed macrophytes. *Hydrobiologia* **2001**, *444*, 71–84. [CrossRef]
94. Connell, J.H. Diversity in Tropical Rain Forests and Coral Reefs. *Science* **1978**, *199*, 1302–1310. [CrossRef]
95. Mjelde, M.; Faafeng, B. Ceratophyllum demersum hampers phytoplankton development in some small Norwegian lakes over a wide range of phosphorus concentrations and geographical latitude. *Freshw. Biol.* **1997**, *37*, 355–365. [CrossRef]
96. Hutchinson, G.E. The chemical ecology of three species of myriophyllum (Angiospermae, Haloragaceae) 1,2. *Limnol. Oceanogr.* **1970**, *15*, 1–5. [CrossRef]
97. Grace, B.J.; Wetzel, R.G. The Production Biology of Eurasian Watermilfoil (Myriophyllum spicatum L.): A Review. *J. Aquat. Plant Manag.* **1978**, *16*, 1–11.
98. Van Wijk, R. Ecological studies on Potamogeton pectinatus L. I. General characteristics, biomass production and life cycles under field conditions. *Aquat. Bot.* **1988**, *31*, 211–258. [CrossRef]
99. Germ, M.; Kuhar, U.; Gaberščik, A. Abundance and Diversity of Taxa Within the Genus Potamogeton in Slovenian Watercourses. In *Natural and Constructed Wetlands*; Springer International Publishing: Cham, Switzerland, 2016; pp. 283–291.
100. Dolinar, N.; Regvar, M.; Abram, D.; Gaberščik, A. Water-level fluctuations as a driver of Phragmites australis primary productivity, litter decomposition, and fungal root colonisation in an intermittent wetland. *Hydrobiologia* **2015**, *774*, 69–80. [CrossRef]
101. Gaberščik, A.; Grašič, M.; Abram, D.; Zelnik, I. Water Level Fluctuations and Air Temperatures Affect Common Reed Habitus and Productivity in an Intermittent Wetland Ecosystem. *Water* **2020**, *12*, 2806. [CrossRef]
102. Martinčič, A. Pteridophytes and Spermatophytes. In *Vanishing Lake–Monography on lake Cerknica (Slovenian with English Summary)*; Gaberščik, A., Ed.; Društvo ekologov Slovenije: Ljubljana, Slovenia, 2002; pp. 76–80.
103. Jenačković, D.D.; Zlatković, I.D.; Lakušić, D.V.; Ranđelović, V.N. Macrophytes as bioindicators of the physicochemical characteristics of wetlands in lowland and mountain regions of the central Balkan Peninsula. *Aquat. Bot.* **2016**, *134*, 1–9. [CrossRef]
104. Clevering, A.O.; Brix, H.; Lukavská, J. Geographic variation in growth responses in Phragmites australis. *Aquat. Bot.* **2001**, *69*, 89–108. [CrossRef]
105. Píšová, S.; Fér, T. Intraspecific differentiation of Sparganium erectum in the Czech Republic. *PRESLIA* **2020**, *92*. [CrossRef]
106. Landucci, F.; Republic, C.; Gigante, D.; Venanzoni, R.; Chytrý, M. Wetland vegetation of the class Phragmito-Magno-Caricetea in centralItaly. *Phytocoenologia* **2013**, *43*, 67–102. [CrossRef]
107. Misson, G.; Macor, A.; Boscutti, F.; Casolo, V. Ecological characterisation of Hippuris vulgaris populations growing in spring Water Rivers. *Phyton (Horn)* **2016**, *56*, 209–224. [CrossRef]

water MDPI

Article

Distribution of the Macrophyte Communities in the Danube Reflects River Serial Discontinuity

Georg A. Janauer [1], Norbert Exler [2], Goran Anačkov [3], Veronika Barta [4], Árpád Berczik [5], Pal Boža [3], Mária Dinka [5], Valeri Georgiev [6], Mateja Germ [7], Matej Holcar [7], Richard Hrivnák [8], Ružica Igić [3], Siniša Ozimec [9], Anca Sârbu [10], Brigitte Schmidt [4], Udo Schmidt-Mumm [4], Wolfgang Schütz [11], Katalin Sipos [12], Elemér Szalma [13], Jasenka Topić [14], Sonya Tsoneva [6], Milan Valachovič [8], Vladimir Valchev [6], Dragana Vukov [3], Igor Zelnik [7,*] and Alenka Gaberščik [7]

1 Department of Functional and Evolutionary Ecology, Division of Limnology, University of Vienna, Althanstrasse 14, A-1090 Vienna, Austria; georg.janauer@univie.ac.at
2 FDG Austria, DI Norbert Exler, Bentheimstrasse 9/25, A-1210 Vienna, Austria; norbert@exler.at
3 Department of Biology and Ecology, Faculty of Sciences, University of Novi Sad, Trg Dositeja Obradovića 3, 21000 Novi Sad, Serbia; goran.anackov@dbe.uns.ac.rs (G.A.); boza.pal@dbe.uns.ac.rs (P.B.); ruzica.igic@dbe.uns.ac.rs (R.I.); dragana.vukov@dbe.uns.ac.rs (D.V.)
4 Department of Limnology and Oceanography, University of Vienna, Althanstrasse 14, A-1090 Vienna, Austria; veronika.barta@aon.at (V.B.); brigitte.schmidt@univie.ac.at (B.S.); uschmidt@drei.at (U.S.-M.)
5 Centre for Ecological Research, Danube Research Institute, Hungarian Academy of Sciences, Karolina ut 29, H-1113 Budapest, Hungary; berczik.arpad@okologia.mta.hu (Á.B.); dinka.maria@outlook.hu (M.D.)
6 Institute of Biodiversity and Ecosystem Research at the Bulgarian Academy of Sciences, Acad. G. Bonchev Str., Bl. 23, 1113 Sofia, Bulgaria; Valeri.Georgiev@iber.bas.bg (V.G.); s.tsoneva@gmail.com (S.T.); vladomura@gmail.com (V.V.)
7 Department of Biology, Biotechnical Faculty, University of Ljubljana, Jamnikarjeva 101, 1000 Ljubljana, Slovenia; mateja.germ@bf.uni-lj.si (M.G.); matej.holcar@bf.uni-lj.si (M.H.); alenka.gaberscik@bf.uni-lj.si (A.G.)
8 Institute of Botany, Slovak Academy of Sciences, Dúbravská cesta 9, SK-845 23 Bratislava, Slovakia; richard.hrivnak@savba.sk (R.H.); milan.valachovic@savba.sk (M.V.)
9 Faculty of Agrobiotechnical Sciences Osijek, Josip Juraj Strossmayer University of Osijek, Vladimira Preloga 1, HR-31000 Osijek, Croatia; sinisa.ozimec@fazos.hr
10 Department of Botany and Microbiology, Faculty of Biology, University of Bucharest, Intrarea Portocalelor 1–3, Sector 6, 060101 Bucharest, Romania; anchusa24@yahoo.com
11 Büro fur Ökologie und Vegetationskunde, Im Jägeracker 28, D-79312 Emmendingen, Germany; wolf.schuetz@gmx.de
12 WWF Hungary, Almos vezert utja 69/A, H-1141 Budapest, Hungary; katalin.sipos@wwf.hu
13 Faculty of Education, University of Szeged Juhasz Gyula, Boldogasszony sgt. 6, H-6721 Szeged, Hungary; szalma@jgypk.u-szeged.hu
14 Sisačka cesta, 2. odvojak 45a, HR-10020 Zagreb, Croatia; jasenka.topic@gmail.com
* Correspondence: igor.zelnik@bf.uni-lj.si; Tel.: +386-1-320-3339

Citation: Janauer, G.A.; Exler, N.; Anačkov, G.; Barta, V.; Berczik, Á.; Boža, P.; Dinka, M.; Georgiev, V.; Germ, M.; Holcar, M.; et al. Distribution of the Macrophyte Communities in the Danube Reflects River Serial Discontinuity. *Water* **2021**, *13*, 918. https://doi.org/10.3390/w13070918

Academic Editor: Arantza Iriarte

Received: 26 January 2021
Accepted: 23 March 2021
Published: 27 March 2021

Abstract: The Danube is the second-longest river in Europe that is subjected to various man-made alterations, including those related to hydro-power plants. We surveyed and analyzed the presence and abundance of macrophytes in the main channel from 2582 river kilometers (rkm) to 171 rkm. We also assessed selected habitat parameters in the sampled river stretches. Sixty-eight different plant species were recorded along the entire course. Among neophytes, we found *Elodea nuttallii*, *E. canadensis*, *Vallisneria spiralis* and *Azolla filiculoides*. Based on similarity analysis, we distinguished 15 plant communities, most of which were defined as associations, which were classified into 5 alliances and represented three vegetation classes, namely vegetation of rooted hydrophytes Potamogetonetea, the vegetation of pleustophytes Lemnetea and vegetation of marshes Phragmitetea. The number and abundance of plant species, as well as plant communities recorded in single stretches, varied along the course. Canonical correspondence analysis (CCA) revealed that environmental parameters explained 21% of plant species composition. CCA runs with neophytes explained 41% of the variance, and current velocity, water transparency, species number and bank structure were significant variables. The present study revealed that the free-running sections of the river are poor in number and abundance of plant species, whereas impounded reaches mainly show an opposite result.

Keywords: Danube; large river; macrophytes; neophytes; plant communities; distribution

1. Introduction

The Danube is the second-longest river in Europe, forming the second largest river basin on the continent. The river basin, covering an area of 807.827 square kilometers [1], is recognized as a hotspot of European freshwater biodiversity [2]. The Danube River connects the territories and interests of at least 18 different countries. It flows through Germany, Austria, the Slovak Republic, Hungary, Croatia, Serbia, Romania, Bulgaria and Ukraine and gathers the water draining from catchments in Switzerland, Italy, the Czech Republic, Poland, Slovenia, Bosnia and Herzegovina, Montenegro, Albania and Moldova. As any other river, the Danube defines the quality of the landscapes and their ecological systems [3,4] and, at the same time, is influenced by their quality and human activities [5].

Lotic ecosystems change in time and space. The modern characterization of rivers has a history going back to about 1960, starting with Odum [6], Illies and Botoseanu [7], as well as with Hawkes [8]. An important step was taken by Vannote et al. [9], who described a continuous gradient of discharge, current velocity, flow pattern, bed and bank structure, dependent on hydrological and hydraulic replacements, showing adjustments in biota and their populations along the river course, named as the river continuum concept (RCC). Soon thereafter, Ward and Stanford [10] published the serial discontinuity concept, which extends the scope of the RCC, focusing on man-made alterations in rivers, especially hydro-power plants (HPP) and other structural modifications. A decade later, Ward and Stanford [11] extended the discontinuity model to rivers with floodplains. This relates well with large rivers like the Danube (e.g., [12]). Moog and co-workers [13] defined ten Danube section types along its course that based on the Danube River typologies regarding the catchment geology, slope [14], geomorphological landscape features and anthropogenic impacts, as well as geomorphological regions and the ecoregions of Illies [15].

The Danube River had been a crucial pathway of Central Europe regarding invasions, settlement, trade and navigation in the Danube basin for more than 2500 years [16]. These multiple socio-economic impacts exerted various pressures to the river [1] that resulted in increased pollution [17,18] and is a significant loss of habitats and species [19], which rank the Danube among the world's top 10 rivers at risk [1]. The main pressures are pollution by industrial and urban wastewater and runoff from agricultural landscapes, overfishing, urban development, dams and bank erosion [16]. In spite of these pressures, the river still shows a high ecological potential [19], harboring a variety of different groups of organisms, including aquatic macrophytes [20].

Aquatic macrophytes comprise a variety of taxonomic groups, from macroscopic algae (e.g., stoneworts), to bryophytes, pteridophytes and angiosperms [21]. They are variable regarding their anatomical and physiological traits, life-history and their tolerance to different physical, chemical and biotic stressors [22]. Macrophytes play a fundamental role in the structure and functioning of river ecosystems that is also the case in large rivers [23]. Macrophytes are involved in energy flow, nutrient cycling and also provide habitats for a large variety of organisms [24–27]. Macrophytes function as "biological engineers", causing significant morphological and ecological modifications of the river [28] by significantly altering sedimentation processes and flow patterns in streams [22,29,30]. They also absorb nutrients and other substances from water and sediment and thus affect its quality [31]. The function of macrophytes strongly depends on their growth form and abundance [32]. The presence and abundance of macrophytes in the river bed depend on water and sediment chemistry, water depth, flow velocity, and substrate characteristics [24,33–36], but less on water temperature [37]. Macrophytes are valuable indicators of water and sediment quality [38–41]. Since they respond to the changes in nutrient concentration, they can contribute significantly to the assessment of the river ecosystems [42]. Variability of macrophyte communities can, therefore, be explained by a variety of environmental parameters, mainly by altitude, flow, riverbed substrate, conductivity and alkalinity, but these communities do not necessarily reflect specific ecoregional distribution [35,43].

The knowledge of macrophyte diversity, their geographic distribution, their structural properties and habitat preferences is of crucial importance for science and enables the definition of the reference conditions for the assessment of the ecological status according to the Water Framework Directive [40]. These data may also enable the prediction of future changes in species composition due to human impacts. Various parameters that shape macrophyte communities may have synergistic and antagonistic effects [44]. For instance, Demars and Edwards [45] claim that one cannot separate the effect of nutrients from other environmental variables, especially along a steep environmental gradient, and they suggested that the response of macrophytes to nutrient enrichment should be studied in homogeneous groups defined by factors, such as alkalinity and slope.

The estimation of biotic communities in large rivers is very hard due to the complexity and spatial and temporal heterogeneity along the river [23,46], as well as due to its depth and width. This is also the case in the Danube River that varies significantly in slope [14], discharge [47], substrate, current velocity, bank structure and water transparency [48]. River regulation included straightening of the river channel by transforming meanders into oxbows, mostly with reduced connection to the main river channel. Artificial bank reinforcements guaranteed secure and technically stable navigation channels, and river bed incision was reduced where run-of-river hydro-power plants (HPP) were installed. Thus, the present status of the Danube is, unambiguously, far from being natural. However, according to Ward and Stanford [10], the most extensive technical interference with the river course and its biota are dams and impoundments of the HPPs. Besides significantly changing habitats, HPPs can have strong edge effects both upstream and downstream [49]. In addition, almost the entire length of the Danube suffers from substantially trimmed floodplain areas.

In this contribution, we analyze the presence and abundance of macrophytes, including neophytes, in the main channel of the Danube River from 2582 river kilometers (rkm) in Baden-Württemberg (SW Germany) to 171 rkm in the delta in Romania. Regarding various alternations along the river, we hypothesized that the distribution, diversity and abundance of macrophytes and their communities along the course will not be in accordance with the river continuum concept and regionality but will more likely reveal serial discontinuity. Regarding our research questions, we presume that these communities will be influenced by habitat parameters in the sampled river stretches.

2. Materials and Methods

2.1. Site Description

The locations of towns, HPPs and backlogs of reservoirs expressed as the river kilometers (rkm) are defined by Linz Rowing Club (ISTER) [50], HPP-Bavaria [51], and the Danube Commission [52]. The river section in Baden-Württemberg (Germany) is interrupted by 24 HPPs, and another 21 HPPs in the Bavarian section. The only Bavarian reach in "near-natural" conditions is between rkm 2328 and 2225 upstream HPP Passau. Then the Danube enters Austria, where 10 HPPs break the continuum in the main channel. However, one of Austria's river sections with no HPPs is the constrained Wachau Valley, a natural reserve. This river part with free-running conditions is located between HPP Melk (rkm 2038) and the backlog (rkm 2002) of the reservoir of HPP Altenwörth (rkm 1981). Passing the Danube downstream to Vienna through several impoundments, the HPP Freudenau at rkm 1921 is the last one in the Austrian section. The following free-running section ends where the backlog of the Čunovo Reservoir (in Slovakia) is clearly detectable at rkm 1859, downstream of Bratislava. Close to the head of Čunovo Reservoir, water is diverted to the Old Danube (former main channel), which represents the Slovak/Hungarian state border. The whole area is part of the large inland delta, with floodplain water bodies in both countries. The Čunovo Reservoir also feeds the diversion of water to the HPP Gabčikovo at rkm 1820. The next "free-running" river section starts close to Gönjü (rkm 1791), already in Hungary and continues for about 625 km to a point 15 rkm downstream Belgrade in Serbia. At this point in the river, the backlog of the Iron Gate I impoundment (length: 211 rkm) begins at mean discharge. The dam of Iron Gate I is located at rkm 943. The Iron Gate II HPP is located at rkm 864, as the very last obstacle in the Danube, enabling a free run towards the delta, close to the city of Galaţi (Galatz).

2.2. Macrophyte Survey

The survey was performed in the main channel of the Danube River from 2582 rkm in Baden-Württemberg (SW Germany) to 171 rkm in Romania in survey units of one-km length. Survey units are matched with the indications of the river kilometers of 1000 m length along the river. In the Danube channel, aquatic plants usually do not colonize river parts much deeper than about 1.5 m [20], where the river has a high potential for macrophyte colonization, even covering the whole river width in some locations. Downstream from the city of Ulm, the river widens, and the left and right bank were surveyed separately, all the way to the delta. The width of the vegetated belt depends on morphological parameters of the riverbed and banks, including human interventions. The field surveys were conducted by boat using the rake and grapnel and/or aqua-scope in deeper water during the peak vegetation period in summer in years from 2002 to 2004. The whole river was surveyed by moving the boat(s) in a zig-zag pattern in narrow reaches upstream of Ulm, whereas the near-bank areas were surveyed by slowly moving along the banks. While proceeding downstream, the observed species were recorded in the species list. At the end of each rkm, benchmarked by the navigation signs, the final estimate of abundance was noted down.

The presence and abundance of aquatic macrophytes and macrophyte assemblages, respectively, were assessed in survey units of one-km length in the main channel of the Danube. The relative abundance was evaluated using a five-degree scale as proposed by Kohler and Janauer [53] using 5-level scale, namely 1 = very rare; 2 = rare; 3 = common; 4 = frequent; 5 = abundant/predominant. Plants that were sampled in the phenological phase, preventing identification to the species level, were recorded as a genus. Species names were unified according to *Flora Europaea* [54].

2.3. Assessment of Habitat Parameters

During regular surveys habitat, the parameters listed below were assessed along with a survey of the macrophyte vegetation. Bank structure, sediment type, flow class and land use type in the areas adjacent to the river are described according to CORINE System [55] with some adaptations regarding the spatial aspects of the river surveys. We used a 1–5 or 1–6 scale to describe specific environmental gradients of environmental parameters. The categories for the selected parameters were the following: (a) bank structure—refers to the upper littoral, extending over the water level during mean discharges (1—concrete or other artificial embankment material, 2—large rocks used for bank stabilization and river regulation, 3—gravel, 4—sand, 5—fine inorganic material, 6—floating mats), (b) sediment type (1—artificial material, e.g., concrete, asphalt, etc., 2—solid rock, 3—gravel, 4—sand, 5—fine inorganic material, 6—detritus or other organic material), (c) flow class (1—no flow/stagnant, 2—low flow, just visible to ca. 30 $cm \cdot s^{-1}$; 3—medium flow, 35 to 65 $cm \cdot s^{-1}$; 4—high flow, >70 $cm \cdot s^{-1}$), (d) land use type (1—artificial surfaces, 2—agricultural areas, 3—open spaces with little or no vegetation (gravel bars, sands, riverbanks, bare rock), 4—scrubs, 5—forests, 6—wetlands). In addition, (e) water transparency (transparency ratio, the relationship between total water depth and Secchi depth) was determined.

2.4. Data Treatment

The relative abundance values [53] were transformed by function x^3 [42], which describes best the relationship between the five degrees of estimation and the actual quantity of the macrophytes [53]. The ordinal values of abundances were transformed into quantitative values ("quantities"), which provided a set of species-data suitable for statistical analyses [56]. It was shown that transformed values highly correlate with average values of percentage cover-abundance classes of plant species according to Braun-Blanquet [57], and were even used as substitutes of these percentage cover-abundance values (e.g., [58,59]). Thus, we also used the transformed values as percentage cover-abundance values according to Braun-Blanquet [57] to examine if the clusters obtained by cluster analysis meet the definitions of specific associations, which are mostly defined by the presence and/or abundance of diagnostic species. These "adjusted" values for dominant species are presented in Table 1.

Table 1. The description of different clusters regarding the number of stretches/plant assemblages (N) mean Secchi depth, mean number of species, dominant species, adjusted mean abundance of dominant species, species frequency and the type of association.

Cluster No.	N	Mean Secchi Depth (cm)	Mean No. of Species	Dominant Species	Adj. Mean Abundance Values	Species Frequency (%)	Association
1	86	126 ± 45	4.5 ± 1.7	*Myr spi* *Pot per* *Ali pla*	32.8 6.5 3.5	100 78 78	*Potamo pectinati-Myriophylletum spicati* Rivas Goday 1964
2	48	72 ± 19	3.6 ± 2.4	*But umb* *Agr sto* *Spa ere*	28.7 4.8 2.1	100 42 23	*Butometum umbellati* Philippi 1973
3	60	66 ± 23	1.8 ± 1.2	*Pot acu* *Aco cal*	25.1 1.8	100 100	*Potametum acutifolii* Segal ex Šumberová et Hrivnák in Chytrý 2011
4	5	66 ± 2	3.8 ± 2.5	*Zan pal* *But umb* *Pot pec*	41.8 3.4 2.0	100 60 60	*Parvo-Potamo-Zannichellietum pedicellatae* Soó 1947
5	74	48 ± 28	3.3 ± 2.2	*Pot pec* *Elo nut* *Pot cri* *Pha aru*	8.4 1.6 1.5 1.4	97 19 34 32	Species-poor and abundance low community with *Potamogeton pectinatus*
6	1345	70 ± 25	1.5± 2.2	*But umb* *Pha aru* *Pot per*	0.42 0.36 0.34	21 15 6	Species-poorest and abundance lowest sections.
7	219	89 ± 40	4.7 ± 2.1	*Myr spi* *Cer dem* *Pot per* *Pot cri*	6.6 3.7 3.4 3.3	86 42 52 61	suboptimal community *with Myriophyllum spicatum*
8	17	/	5.4 ± 1.0	*Pot cri* *Myr spi* *Pot pus*	26.4 9.1 7.5	100 100 94	*Potametum crispi* Soo 1927
9	5	116 ± 18	12.0 ± 4.1	*Pot pec* *Pot per* *Cer dem*	63.3 63.2 16.1	100 100 80	*Potametum perfoliati* Miljan 1933
10	29	78 ± 21	8.1 ± 3.9	*Pot pec* *Pot per* *Myr spi*	65.1 14.7 6.8	100 85 45	*Potametum pectinati* Carstensen ex Hilbig 1971
11	29	100 ± 21	10.0 ± 2.3	*Pot nat* *Pot per* *Pot gra*	26.2 22.1 17.3	100 97 97	*Potametum natantis* Hild 1959

Table 1. *Cont.*

Cluster No.	N	Mean Secchi Depth (cm)	Mean No. of Species	Dominant Species	Adj. Mean Abundance Values	Species Frequency (%)	Association
12	12	96 ± 17	11.5 ± 1.3	*Pot nat* *Pot pec* *Tra nat* *Pot per*	29.8 25.5 20.7 17.8	100 100 83 75	*Potametum natantis* Hild 1959and *Trapetum natantis* Kárpáti 1963 (6 out of 12)
13	19	130 ± 10	12.2 ± 2.8	*Cer sub* *Spi pol* *Pot pec* *Pot luc* *Lem min*	21.3 18.0 5.9 4.7 2.1	89 100 95 95 89	*Potamo-Ceratophylletum submersi* Pop 1962
14	33	84 ± 23	11.8 ± 2.9	*Pot luc* *Pot pec* *Cer dem* *Spi pol*	23.5 16.1 12.7 8.7	100 97 97 100	*Potametum lucentis* Hueck 1931
15	27	69 ± 23	6.0 ± 3.2	*Pot pec* *Pot nat* *Pot per*	31.6 3.8 3.7	100 25.9 44.4	*Potametum pectinati* Carstensen ex Hilbig 1971
16	7	123 ± 19	9.6 ± 3.3	*Pot per* *Cer dem* *Pot gra*	62.3 20.8 6.6	100 86 43	*Potametum perfoliati* Miljan 1933
17	44	109 ± 32	7.2 ± 2.8	*Pot per* *Pot pec* *Myr spi*	26.1 26.2 13.8	93 75 75	*Potametum perfoliati* Miljan 1933
18	10	68 ± 9	12.4 ± 2.9	*Pot pec* *Pot luc* *Cer dem* *Pot gra*	47.4 41.9 23.4 17.8	100 100 100 100	*Potametum lucentis* Hueck 1931
19	18	78 ± 7	12.9 ± 2.5	*Pot pec* *Cer dem* *Pot gra* *Pot luc*	32.7 29.6 27.2 19.4	100 100 100 94	*Potametum graminei* Lang 1967
20	23	70 ± 9	8.8 ± 3.8	*Cer dem* *Spi pol* *Pot luc*	30.6 13.5 7.8	100 57 26	*Ceratophylletum demersi* Corillion 1957
21	10	125 ± 18	13.2 ± 2.5	*Spi pol* *Cer dem* *Cer sub* *Wol arr*	56.8 29.3 19.8 14.5	100 90 90 60	*Lemno-Spirodeletum polyrhizae* Koch 1954 (6) and *Ceratophylletum demersi* Corillion 1957 (4)

2.5. Similarity of Macrophyte Communities

The similarity of the macrophyte assemblages between the stretches was calculated with PAST, version 2.17c [60]. Ward's method was used as linkage method, and Euclidean distance was used as similarity index [61], which have been used by others with macrophyte data [62,63] and which provided the clustering with the lowest level of chaining [64]. The aim of the clustering method was a comparison of similarity between the records and help in their classification to the defined plant communities. We used an approach based on the dominant species, which are also diagnostic species of the associations and is the most frequently used approach for aquatic vegetation [65]. Dominance-based classification for species-poor plant communities is most similar to the approaches in traditional phytosociology [66]. For the classification of the individual clusters to the described aquatic plant communities, we used the definitions found in [67].

2.6. The Influence of Environmental Factors on the Composition of the Macrophyte Community

First, a detrended correspondence analysis (DCA) was performed, which helped us to reveal whether the gradients in the matrix of macrophyte species data are linear or unimodal and which direct gradient analysis to choose for further analyses. When the eigenvalue for the first axis in DCA was <0.4 and/or gradient length was <3 standard deviations, Redundancy analysis (RDA) was applied, while in the case when these values were higher, we applied Canonical correspondence analysis (CCA) [68]. The results of these analyses enabled us to assess the relationship between environmental parameters and the composition and abundance of macrophytes. The influence of the environmental factors on the distribution of neophyte species was tested with canonical correspondence analysis. We used forward selection, where 999 permutations were performed in every round to rank the relative importance of explanatory variables and to avoid co-linearity [69]. Only parameters with $p < 0.05$ were considered in further analyses. In the first run, the multivariate analysis was performed for all recorded species, while in the second run, we selected only species having total abundances >1%. In addition, we performed multivariate analysis to explore the relationship between the neophyte species and environmental parameters. All analyses were performed using CANOCO for Windows 4.5 program package (Microcomputer Power: Ithaca, NY, USA) [68].

3. Results

3.1. Species Presence and Abundance

We recorded 68 different plant species along the entire studied length. The most common and abundant taxon was *Potamogeton pectinatus* with 14.5%, followed by *P. perfoliatus* with 11.6% and *Myriophyllum spicatum* with 10.1% (Figure 1). Among the most abundant species presenting more than 2% of total abundance were seven species of the genus *Potamogeton*. On the whole, we recorded 11 species of this genus, namely *P. pectinatus, P. coloratus, P. pussilus, P. trichophyllus, P. lucens, P. nodosus, P. natans, P. gramineus, P. crispus, P. perfoliatus* and *P. acutus*. The majority of abundant species appearing in the Danube are submerged hydrophytes, and only a few of them develop natant leaves. The most abundant among amphiphytes is *Butomus umbellatus* that mainly appears as *f. aquatica*. We also detected few free-floating species, namely spermatophytes *Wolffia arrhyza, Lemna minor* and *L. gibba, Spirodela polyrhiza* and ferns *Salvinia natans* and *Azolla filiculoides*. In spite of a high number of recorded species, we did not find any species from the EU Red list.

The occurrence and relative abundance of these most abundant native species along the flow is presented in Figure 2. The majority of species, with the exception of *T. natans* and *Ceratophyllum submersum*, were found at different parts of the river course. *T. natans* was found mainly in the section from 1150 to 850 rkm, where also several other species occurred with the highest abundance.

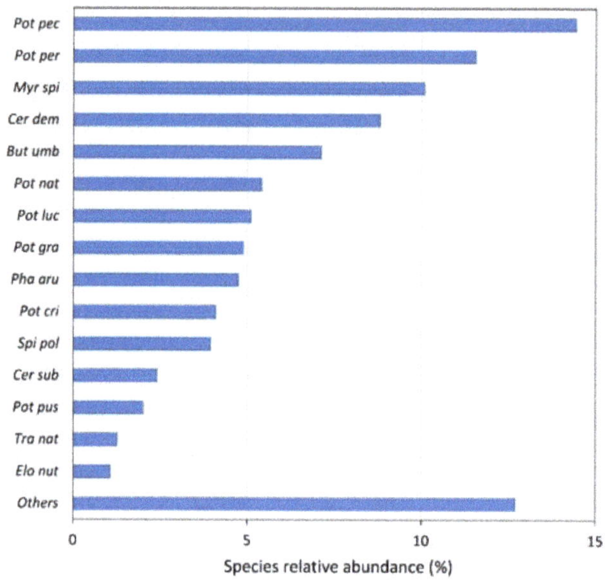

Figure 1. Relative values of total abundance of different species occurring in the Danube River. "Others" indicates the abundance of the remaining species. Species abbreviations: *Pot pec—Potamogeton pectinatus, Pot per—Potamogeton perfoliatus, Myr spi—Myriophyllum spicatum, Cer dem—Ceratophyllum demersum, But umb—Butomus umbellatus, Pot nat—Potamogeton natans, Pot luc—Potamogeton lucens, Pot gra—Potamogeton gramineus, Pha aru—Phalaris arundinacea, Pot cri—Potamogeton crispus, Spi pol—Spirodela polyrhiza, Cer sub—Ceratophyllum submersum, Pot pus—Potamogeton pusillus, Tra nat—Trapa natans, Elo nut—Elodea nuttallii.*

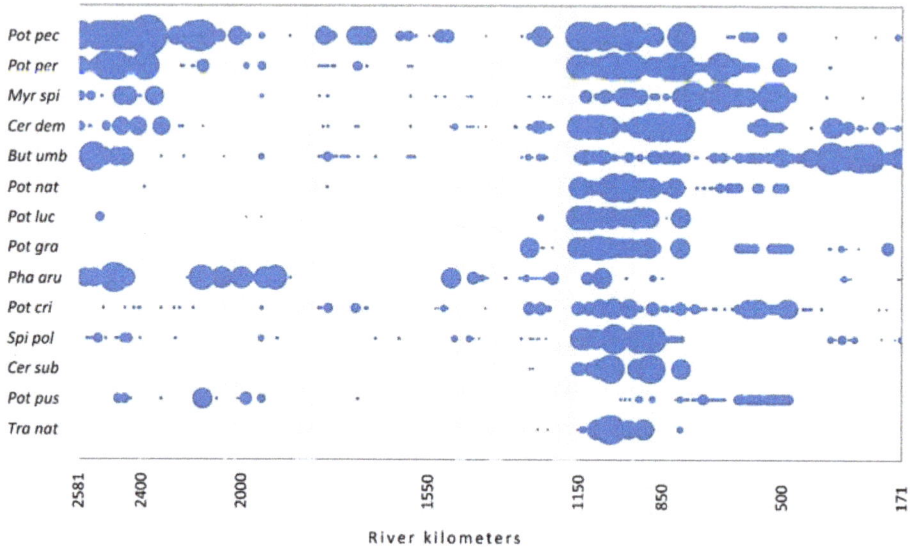

Figure 2. The occurrence of the most abundant native species (>1%) along the flow. The size of the circles shows the relative abundance of each species. Light gray vertical bars represent the sections with no data. For species abbreviations, see Figure 1.

The number of species recorded in a single stretch varied along the flow, and this holds true also for species abundance. The pattern of values of both parameters along the flow reveals a similar picture showing two peaks. The first appears at the beginning of the upper section, then it gradually decreases, reaching the second peak at the end of the middle section and at the beginning of the lower section of the river (Figure 3). Both parameters were the highest in the second part of the middle section, while the lowest values were detected at the first part of this section.

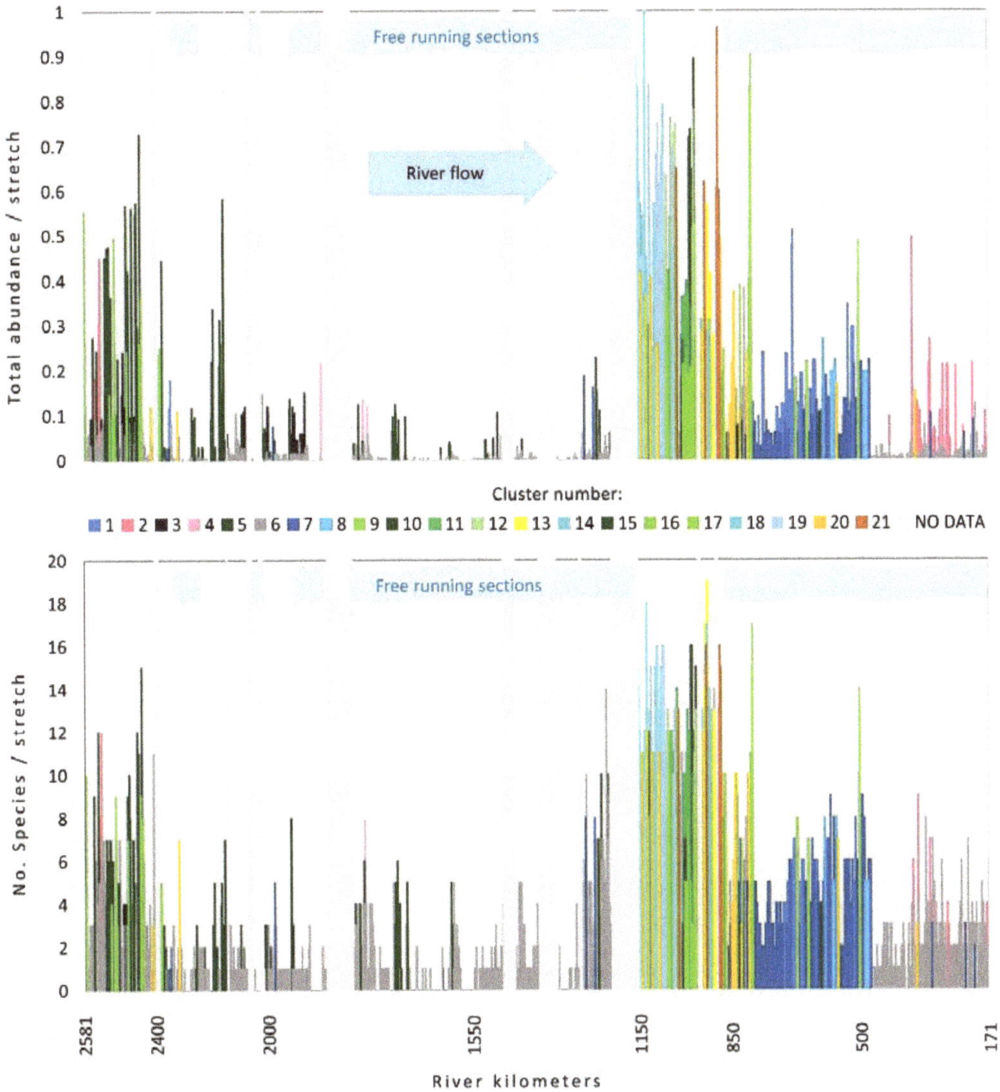

Figure 3. Changes in total species number (lower graph) and relative species abundance along the Danube River flow. Different colors indicate different clusters. Light grey vertical bars represent the sections with no data. Light blue stripes indicate sections where no influence of dams on the flow was detected. For cluster number information, see Table 1.

3.2. Neophyte Species Presence and Abundance

The presence and abundance of four species from the EU neophyte list were analyzed: *Elodea nuttallii, E. canadensis, Vallisneria spiralis* and *Azolla filiculoides*. Regarding the surveyed flow, *E. nuttallii* was only the 15th with 1.1% of total abundance, *V. spiralis* with 0.7% on the 23rd, *E. canadensis* on the 26th with 0.6%, and *A. filiculoides* with only 0.1% of total abundance the 40th. (Figure 1). *E. nuttallii* was found in 110 stretches and mainly occurred with low abundance together with more abundant species like *P. perfoliatum, C. demersum, B. umbellatus* and *S. polyrhiza*, and just in one case with *E. canadensis*. The latter species was found in 52 stretches and mainly occurred in species-rich plant communities in combination with the majority of species recorded. *V. spiralis* was found at 58 locations in a community with *P. natans, B. umbellatus, C. demersum, M. spicatum, P. perfoliatus* and *P. pectinatus*. However, regarding distribution along the flow, it is still limited to its native range. *A. filiculoides* was found at 51 stretches but with very low abundance in a company with a variety of species (Figure 4).

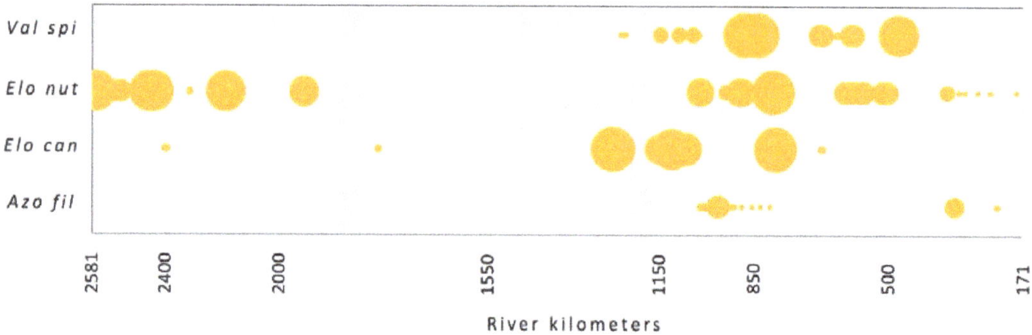

Figure 4. The occurrence of neophyte species (according to EU list) along the flow. The size of the circles shows the relative abundance of each species. Light gray vertical bars represent the sections with no data. Species abbreviations: *Val spi—Vallisneria spiralis, Elo nut—Elodea nuttallii, Elo can—E. canadensis, Azo fil—Azolla filiculoides*.

3.3. Plant Communities along the Course

Then we examined a similarity of macrophyte communities along the river course and obtained 2 main clusters subdivided into 9 and 12 clusters (Figure 5). The sample locations within the clusters did not show the river continuum but are scattered in different groups (Figure 6). Cluster 6, which comprises more than half of the stretches (1344), is marked by very scarce vegetation, mainly consisting of amphiphytes.

The description of different clusters regarding mean Secchi depth, mean number of species, mean total relative abundance, dominant species, dominant species mean relative abundance, species frequency and the type of association is presented in Table 1.

Based on the analyses of records and definitions of vegetation units, three orders and five alliances encompassing 15 associations were distinguished, which at the same time represented vegetation of three classes, namely vegetation of rooted hydrophytes Potamogetonetea, the vegetation of free-floating hydrophytes Lemnetea and vegetation of marshes of the class Phragmitetea.

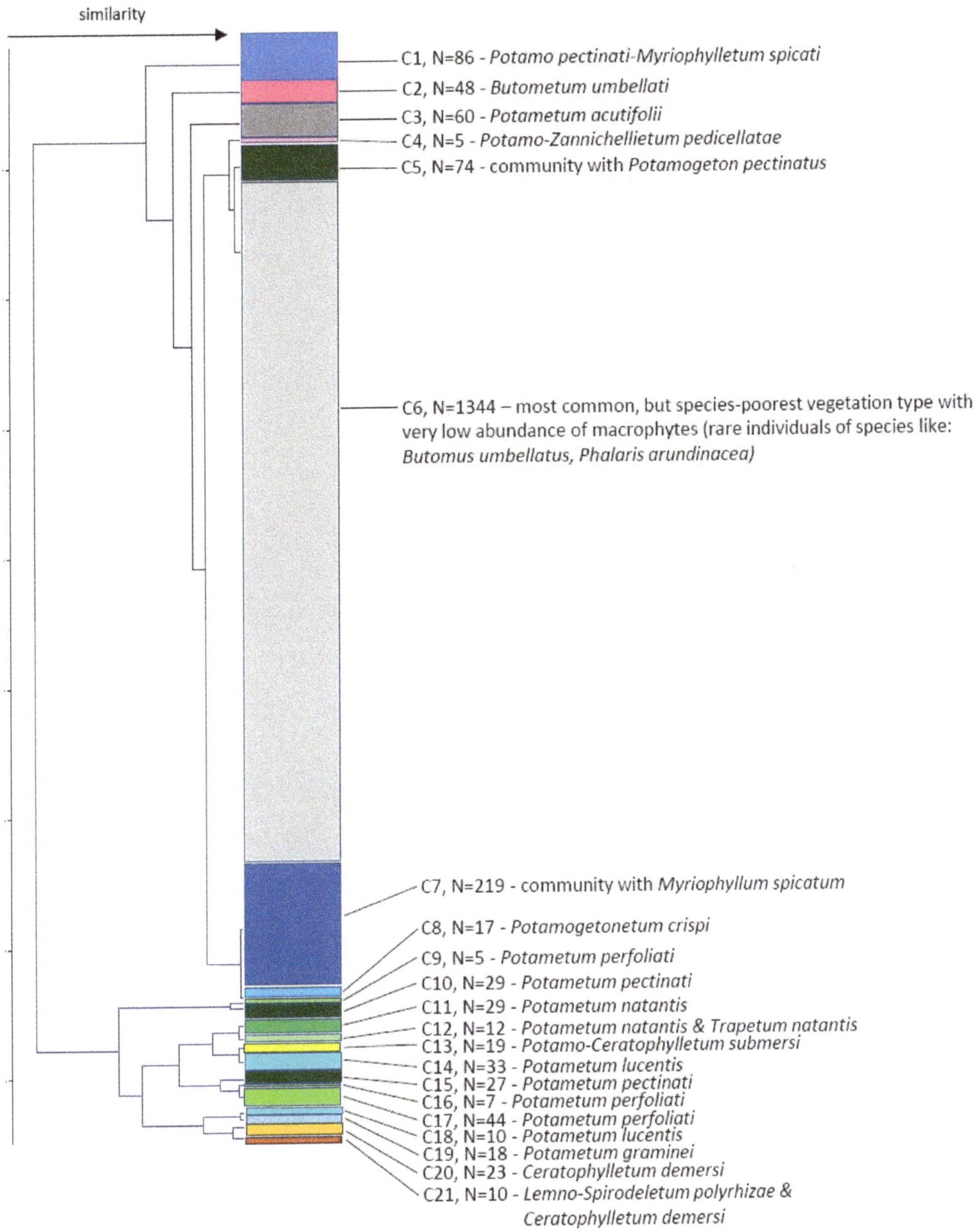

Figure 5. Similarity of the macrophyte communities comparing different stretches along the River Danube using Ward's linkage method and Euclidean distance similarity index. C—cluster, N—number of stretches in the cluster.

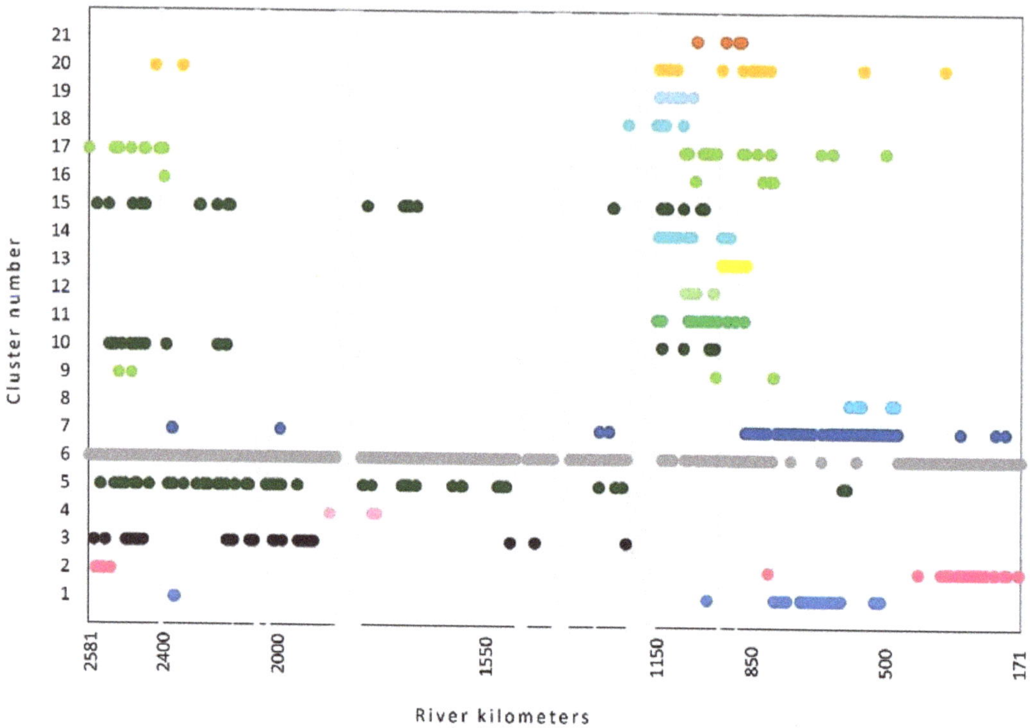

Figure 6. Occurrence of the different clusters along the Danube River flow derived from Figure 5. Different colors indicate different clusters and plant communities (see Table 1). Light gray vertical bars represent sections with no data.

3.3.1. Order Potamogetonetalia Koch 1926

The majority of the plants in different stretches were classified into plant communities of this order, which was also the most diverse and common order of the class *Potamogetonetea* Klika et Novak 1941, encompassing the rooted vegetation thriving in lentic or lotic ecosystems, which could be entirely submerged or with floating leaves.

Alliance *Potamogetonion* Libbert 1931

Association *Potamo pectinati-Myriophylletum spicati* Rivas Goday 1964

(cluster 1, 86 stretches)

The far most abundant (33%) and frequent (100%) macrophyte in this plant community (cluster 1) was *Myriophyllum spicatum*, which represented about two-thirds within total abundance. Consequently, this association is species-poor (on average 4.5), despite the high transparency measured in these stretches (118 ± 25 cm). It was found in the lower part of the Danube (Figure 6).

Plant community with *Myriophyllum spicatum*

(cluster 7, 219 stretches)

We classified this second largest group of stretches (cluster 7, Figure 5) into the suboptimal form of the above-described association since the characteristic species did not reach high abundance nor dominance. However, it was twice more frequent than the second most abundant species, *Ceratophyllum demersum*. This plant community also has the same pattern of distribution (Figure 6) along the course as the better-developed association *Potamo pectinati-Myriophylletum spicati* also occurs in the middle and upper part of the Danube.

Association *Potametum acutifolii* Segal ex Šumberová et Hrivnák in Chytrý 2011

(cluster 3, 60 stretches)

The most outstanding characteristics of this association are its lowest species richness (average 1.8) and total abundance. The characteristic and dominant species *Potamogeton acutifolius* is the most common (100%) and also the most abundant species. It is mainly found in turbid water of the upper part of the Danube (Figure 6), with some occurrence in the middle part.

Association *Parvo-Potamo-Zannichellietum pedicellatae Soó* 1947

(cluster 4, 5 stretches)

This is the rarest plant association in this study, found only in five stretches, where the characteristic and dominant species. *Zanichellia palustris* represents the vast majority of plant biomass, which tolerates low transparency (average 66 cm). Consequently, few other species thrive in these species-poor stands (average 3.8). These stretches are found in Slovakia within the Čunovo Reservoir.

Association *Potametum crispi Soo* 1927

(cluster 8, 17 stretches)

The species composition of this association is most similar to the plant community with *Myriophyllum spicatum* (Figure 5), but the dominant species here is *Potamogeton crispus*, which is present in all these stretches and is about three times more abundant than *Myriophyllum*. The abundance of this vegetation cover type also appears twice in the dendrogram, mostly on account of *Potamogeton crispus*. This association is found only in the lower part of the Danube.

Association *Potametum pectinati* Carstensen ex Hilbig 1971

(cluster 10, 29 stretches; cluster 15, 27 stretches)

This plant association occurs in two forms that are found in clusters 10 and 15 (Figure 5). The stretches found in cluster 10 are better defined by stretches of the dominant characteristic species *P. pectinatus* (*Potamogeton pectinatus*) (average 65%), so we considered it as the typical form of association. The species richness is moderate (average 8.1). The average water transparency is 78 cm. The majority of stretches are found in the upper part of the Danube, some in the middle (Figure 6).

The stretches in cluster 15 are poorer in species (average 5.9) as well as in their abundance (average 68%). The abundance of the dominant species *P. pectinatus* is two times lower than in typical form. One of the possible reasons is also lower transparency (69 cm) as compared with the typical form. These stretches are more widely distributed in the upper and middle parts of the Danube (Figure 6).

Plant community with *Potamogeton pectinatus*

(cluster 5, 74 stretches)

The stretches found in this cluster were classified as the plant community with *P. pectinatus*, which is in a suboptimal stage in the development of the above-listed association *Potametum pectinati*. Although the characteristic species *P. pectinatus* is present in almost any stretch (97%) with at least some individuals and is also the most abundant species, its abundance is too low to define the association (avg. 8.2%). This vegetation type is species-poor (avg. 3.3) and low in abundance (18%). The characteristic condition for this type is the lowest water transparency in the entire Danube (63 cm). This vegetation type has an even wider distribution range than the above association and could be found throughout the upper and middle part of the Danube, as well as in some of the stretches in the lower part (Figure 6).

Association *Potametum perfoliati* Miljan 1933

(cluster 9, 5 stretches; cluster 16, 7 stretches; cluster 17, 44 stretches)

This plant association occurs in three forms that are found in three separate clusters: 9, 16 and 17 (Figure 5). The stretches in cluster 16 are best defined by the most outstanding dominance of the characteristic species *P. perfoliatus* (average 62%), which represented half of the total abundance, so we considered it as the typical form of association. The second most abundant species was *Ceratophyllum demersum* (average abundance 21%, present in 86% of stretches). The species richness was moderate (average 9.6). The average water transparency was 123 cm.

A higher similarity was found with the stretches in the biggest cluster 17 (Figure 5), where *P. perfoliatus* retained its dominance despite a much lower abundance (26%). The second most abundant species in these stretches was *Potamogeton pectinatus*. These stretches were poorer in species (average 7.2). A possible reason is lower water transparency (avg.109 cm).

In the smallest cluster 9 (5 stretches), the characteristic species *P. perfoliatus* is codominant with *Potamogeton pectinatus*, which reached the same average abundances (63%). The third most abundant species was *Ceratophyllum demersum* (average 16%). These stretches are most species-rich (avg. 12) and very abundant with macrophytes. Despite the division of this association into three clusters, their distribution pattern is the same (Figure 5)—one group of stretches is in the uppermost part of the Danube in Bavaria, while the other is in the transition between the middle and lower part and also in the lower part of the Danube. The majority of stretches are found in the upper part of the Danube, some in the middle (Figure 6).

The stretches in cluster 15 are poorer in species (average 5.9). The abundance of the dominant species *Potamogeton pectinatus* is two times lower than in typical form. One of the possible reasons is also lower transparency (69 cm) than in the typical form. These stretches are more widely distributed in the upper and middle parts of the Danube (Figure 6).

Association *Potametum natantis* Hild 1959

(cluster 11, 29 stretches; cluster 12, 4 of 12 stretches)

This plant association occurs in two forms that are found in clusters 11 and 12 (Figure 5). The stretches in cluster 11 are more uniform than in the second cluster 12, but in both clusters, the characteristic species with floating leaves, *P. natans*, is most abundant and present in every stretch. In the first group, three species are subdominant (*P. perfoliatus*, *P. graminoides*, *P. pectinatus*), the species richness and total abundance are moderate. The second cluster is joined with the association *Trapetum natantis* in the same cluster and is richer in species (avg. 11.5) as well as in their abundance. The subdominant species there is *P. pectinatus*. This association is found in both Iron Gate Reservoirs.

Association *Potametum lucentis* Hueck 1931

(cluster 14, 33 stretches; cluster 18, 10 stretches)

This plant association occurs in two forms that are found in clusters 14 and 18 (Figure 5). The stretches in smaller cluster 18 are better defined due to the higher abundance of characteristic species *P. lucens* (avg. 47%). The codominant species here is *P. pectinatus* (avg. 42%), while *C. demersum* is subdominant (avg. 23%). These stretches are moderately species-rich (12.4) despite the highest total abundance within the studied stretches and low water transparency (68 cm). The larger cluster encompasses stretches with significantly lower total abundance (avg. 36%), while the number of species is almost the same (11.8). Although with lower abundances, the three most abundant species are the same as in the first cluster. The characteristic species *P. lucens* is most abundant here (avg. 24%). The distribution pattern of both clusters is the same, which is downstream the inflow of the river Sava and Iron Gate I Reservoir.

Association *Potametum graminei* Lang 1967

(cluster 19, 18 stretches)

The stretches in this cluster are classified into association *Potametum graminei* due to the high abundance of the characteristic species *P. gramineus* (avg. 29.6%), despite the fact that codominant species *P. pectinatus* and *Ceratophyllum demersum* have a bit higher abundance. However, for the definition of other associations, their abundances are not sufficient. The species-richness is moderate (12.9). The distribution pattern is very similar to the pattern of *Potametum lucentis*, which is recorded downstream the inflow of the river Sava and the Iron Gate Reservoir.

Species-poorest vegetation type in sections with the lowest abundance of macrophytes.

(cluster 6, 1345 stretches)

This was by far the largest group (cluster 6, Figure 5) and the most common vegetation type, which encompassed more than half of the recorded stretches (1345). At the same time, it was also the species-poorest group with only 1.5 macrophytes species on average (min–max 0–6) per stretch, which had a very low abundance value, which is 4% on average. The most common species in this type were actually marsh species (helophytes) *Butomus umbellatus* found within one-fifth of these stretches, *Phalaris arundinacea* within 15% of stretches. This type is scattered along the entire course of the Danube, from the source to the delta (Figure 6), with the exception of the stretches where communities with the dominant species *Myriophyllum spicatum* (clusters 1 and 7) are found.

Alliance Nymphaeion albae Oberdorfer 1957

Association *Trapetum natantis* Kárpáti 1963

(cluster 12, 8 of 12 stretches)

This plant association is defined within cluster 12 (Figure 5), which also contains 4 stretches with dominant species *P. natans*, which were classified into the association *Potametum natantis*. The species richness of the entire cluster is moderate (avg. 11.5) but with high total abundance. The dominant or codominant species in these stretches is *Trapa natans*; subdominant species there is *P. pectinatus*. This plant association is found only in the Iron Gate I Reservoir.

3.3.2. Order Lemnetalia minoris O. de Bolòs et Masclans 1955

Alliance Stratiotion Den Hartog et Segal 1964

Association *Potamo-Ceratophylletum submersi* Pop 1962

(cluster 13, 19 stretches)

This plant association is defined within cluster 13 (Figure 5) and is defined by the characteristic species *Ceratophyllum submersum*, which has the highest abundance (avg. 21%). Specific for this community is the high abundance and constancy of the pleustophyte *Spirodela polyrhiza* (avg. 18%) and also *Lemna minor*. The highest water transparency was measured in these stretches (130 cm). This association is found only in the Iron Gate II Reservoir, downstream the stretches with *Trapetum natantis*.

Association *Ceratophylletum demersi* Corillion 1957

(cluster 20, 23 stretches)

This pleustophyte association is defined with characteristic species *Ceratophyllum demersum*, which has the highest abundance (avg. 31%) and is accompanied by *Spirodela polyrhiza* (avg. 13%). The species-richness was lower (8.8) as in the previous association. The most significant difference was much lower water transparency (70 cm). The distribution of this association is wide since it is found in all three parts of the Danube (Figure 6).

Alliance *Lemnion minoris* O. de Bolòs et Masclans 1955

Association *Lemno-Spirodeletum polyrhizae* Koch 1954

(cluster 21, 8 of 10 stretches)

This most typical pleustophyte association is dominated by the characteristic species *Spirodela polyrhiza* (avg. 56%), which is accompanied by other pleustophytes like *Lemna minor*, *Salvinia natans* and *Wolffia arrhiza*. The most abundant submerged species were *Ceratophyllum demersum* (avg. 30%), which dominated in two assemblages (→ *Ceratophylletum demersi*) and *C. submersum* (avg. 20%). This association was the richest in species (avg. 13.2), while the average water transparency was the second-highest (125 cm). This association is found only in the Iron Gate Reservoirs (Figure 6).

3.3.3. Order Phragmitetalia

Association *Butometum umbellati* Philippi 1973

(cluster 2, 48 stretches)

This is the only association that represents the vegetation of marshes, which consists of different species of helophytes. In these stretches, the characteristic and dominant species *B. umbellatus* was found with the highest abundance (average 29%), mostly as *f. aquatica*. The second most abundant plant taxa were *Agrostis stolonifera* and *Sparganium erectum*, but with significantly lower abundances. This plant-community is species-poor (average 3.6) and has specific distribution since it was found in the section close to the source of the Danube and in the last section downstream to the rkm 400. Considering the most abundant plant taxa, this community is well adapted to the oscillation of the water level.

3.4. *The Influence of Environmental Factors on the Composition of Macrophyte Communities*

Using CCA and RDA, we tried to explain species presence and abundance with selected environmental parameters detected at species locations. When effects of single variables were examined (marginal effects), the specific variable at right (R) and left (L) bank explained similar levels of species variation. However, when all variables were taken together (conditional effects), in the majority of cases, only variables for one bank were revealed to be significant due to co-linearity. In the first run, the RDA was performed for the whole flow, but the variance explained by environmental factors was 8% in whole. When only species with relative abundance more than 1% were considered, the explained variance was somewhat higher (21%) (Table 2); water transparency is the most important by explaining 6% of species variability.

Table 2. Summary of multivariate analyses aimed to explain the species presence and abundance with environmental selected environmental parameters. RDA—Redundancy analysis, CCA—Canonical correspondence analysis, DCA—Detrended correspondence analysis.

Analysis *	Data Set	Total Explained Variance (%)	Significant Parameters/Explained Variance
RDA	Whole course (all species)	8	All, with the exception of land use
CCA	Whole course; only species with total abundance more than 1%	21	All parameters, with exception of flow velocity Secchi depth (R) (6%, $p = 0.001$) Distance from the source (4%, $p = 0.001$) Current velocity (R) (3%, $p = 0.001$) Other parameters (8%, $p \leq 0.014$)
CCA	Neophyte species only	45	Current velocity (L) (17%, $p = 0.001$) Secchi depth (L) (12%, $p = 0.001$) Sediment type (L) (7%, $p = 0.002$) No of species per stretch (5%, $p = 0.001$) Bank structure (R) (3%, $p = 0.001$) Current velocity (R) (1%, $p = 0.022$)

* Selected based on DCA regarding eigenvalue for the first axis and gradient length.

In the following run, we applied CCA to explore the relation between neophyte species and environmental parameters and explained as much as 41% of the variability of neophyte presence and abundance. The most influential parameters were a current velocity that explained 17% (L) and an additional 1% (R), followed by Secchi depth with 12%, number of species with 5% and bank structure with an additional 3% (Table 2).

CCA plot reveals different positions of four neophyte species with respect to the vectors of significant environmental variables (Figure 7). *E. canadensis* occurs at locations with a higher number of plant species, higher current velocity and finer sediment, while *E. nuttallii* occurs at the opposite side of these vectors, which reveals a much different habitat type characterized by lower current velocity, coarser sediment and lower number of species in the community. The presence and abundance of *V. spiralis* seem to be related to stretches with higher water transparency and banks rich in a finer substrate. *A. filiculoides* is a natant fern, which occurred close to the middle values of examined parameters.

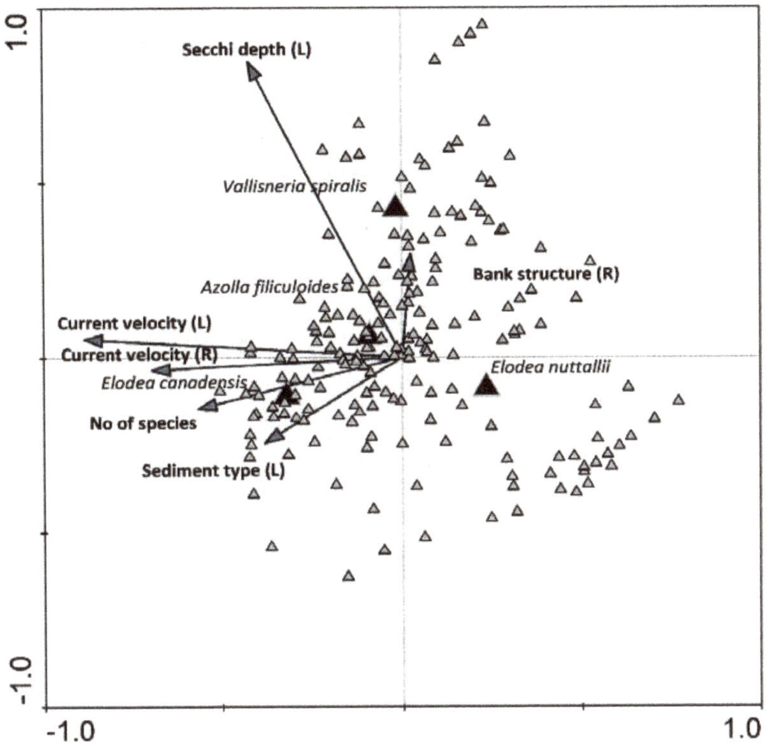

Figure 7. Canonical correspondence analysis (CCA) plot showing the relation between neophyte species presence and abundance and significant environmental parameters and a total number of species in the stretch detected in neophyte species habitats.

4. Discussion

4.1. Species Distribution and Abundance

In large rivers, helophytes and free-floating taxa are found in the shallow parts near the banks, while the deeper parts with possible macrophyte growth are characterized by a reduced number of submerged spermatophytes, which are resistant to high current velocity [23]. We recorded only a few helophyte species; the only species with a high total abundance (about 5%) was *P. arundinacea*, which appeared along the whole river course. This was the case for the majority of species that have a total relative abundance of more than 1%. The exception was floating leaved *T. natans*, which was present in abundance after the confluence of the Sava River, where also the total species number and abundance were the highest. Dense *T. natans* stands can offer valuable habitats to invertebrates and fish and increase their biodiversity and production, while they may also cause hypoxic or even anoxic conditions [70]. Skliar and Skliar [71] report that the decrease of water level negatively affected the growth of *T. natans* and increased the number of representatives of *Nymphaeaceae*. Among abundant species, we also detected two other species with leaves floating on the water surface, one with natant leaves, namely *P. natans* and one pleustophyte *S. polyrhiza*, that were also more abundant in the lower part of the river. The position of leaves/fronds on the water surface enables successful competition for space and light due to efficient shading of other species [72], which was also evident in our study. Within the group of more abundant species (total abundance was >2%) were 7 species of the genus *Potamogeton*, altogether 10 species. The representatives of this widespread genus may thrive in a variety of habitats, from running to stagnant waters [73] with different water regimes and nutrient enrichment [74]. The species of this genus are usually subjected to interspecific hybridization [75]; however, no hybrids were found in this survey. The most abundant species detected in our survey was *P. pectinatus* that occurred along the whole river course. The survey of Case and Madsen [76] revealed that the distribution of this species was negatively affected by water depth, wave action and water transparency and positively by the ratio of silt in the sediment. This species occurs nearly worldwide at locations with water velocity <1 m/s and water depth <2.5 m [77]. This holds true for other species found in great abundance. CCA runs with abundant species only explained a relatively low share in their presence and abundance, revealing the importance of water transparency and current velocity. Current velocity is an important factor affecting macrophyte production in rivers, and even a moderate increase reduces their abundance [78]. Most of these species have a broad ecological valence and are widely distributed around Europe, showing no specific regionality [38,39,54]. Even more, some of them, like *P. pectinatus* and *M. spicatum*, also became an invasive nuisance weed in different parts of the world. When we used the whole list of species for RDA, a very low proportion of the variability was explained, even though the majority of environmental parameters were significant.

4.2. Neophytes in the Danube Main Channel

E. canadensis is the most widespread invasive alien aquatic plant in Europe, *A. filiculoides* the second, followed by *V. spiralis* and *E. nuttallii* [79]. *E. canadensis* and *E. nuttallii* are native to most of North America [80]. *A. filiculoides* are native to the Americas and spread widely around the world by a variety of mechanisms, of which man has become the most significant [79]. *V. spiralis* is native to southern Europe, northern Africa, the Middle East and southwest Asia [79]. According to Hérault et al. [81], *E. canadensis* and *E. nuttallii* share similar ecological niches and biological traits; however, in the present study, they thrived in different environmental conditions. The presence and abundance of both species were best correlated to vectors that indicated species number, current velocity and sediment type, however in the CCA plot, the species are positioned in opposite sides of vectors, *E. nuttallii* being related to species-poorer community, coarser sediment and lower current velocity, while *E. canadensis* was found in species-richer assemblages. In Slovenian watercourses, *E. canadensis* usually thrived in relatively species-rich communities, frequently together with the species *P. natans*, *P. nodosus*, *P. perfoliatus*, *P. crispus* and *P. pectinatus* and in the company of *E. nuttallii* [82]. The coexistence of both *Elodea* species was not detected in the Danube River. *E. canadensis* can be outcompeted by other more invasive species, especially in the case of nutrient enrichment [83,84]. This was also the case in Great Britain, where *E. nuttallii* substituted *E. canadensis* [85], and in France, where it was dominated by *Egeria densa* and *E. nuttallii* [86]. Barrat-Segretain and Lemoine [87] also report that *E. nuttallii* was more competitive than *E. canadensis* and also more palatable to herbivores. For both species, it was shown that they have allelopathic potential [88]. In addition, *E. nuttallii* also showed higher survival rates under the influence of artificial disturbances due to its high regeneration potential from vegetative fragments [89], and it is likely to become a successful alternative of *E. canadensis* [86]. *A. filiculoides* was found in low abundance in the lower course of the Danube. It thrives in eutrophic waters where it forms dense mats up to 10 cm thick [90,91]. It was shown that in favorable habitats, the representatives of the genus *Azolla* grow quickly with a doubling time of only 2–5 days [92]. Their competitive advantage is the symbiosis with the nitrogen-fixing cyanobacteria of the genus *Anabaena* that enable the fern's high productivity [93]. In Poland, *A. filiculoides* occurred in different water habitats, where they substituted *Lemna minor* [91]. The same effect was confirmed by Paolacci et al. [94], where *A. filiculoides* exerted a negative influence on growth rates of *L. minor* and *L. minuta*. The fourth species in the EU neophyte list is *V. spiralis*. It is an obligate lowland macrophyte with an altitudinal range up to 300 m thriving in meso- to eutrophic aquatic habitats [95], as was the case in the Danube River. There it occurs mainly within the native range of distribution despite the report of Anačkov et al. [96], showing that in Serbia, *V. spiralis* occupied high-quality natural habitats along with some other alien species that show high potential for these habitats alternation.

4.3. The Distribution of Macrophyte Communities along the Course

Our study revealed that the distribution of macrophyte species and their communities reflect neither the river continuum concept described by Vannote et al. [9] nor a specific regionality as proposed by Illies [97] (see Figures 2, 3 and 6). Connell [98] showed that maximum species richness occurs under intermediate disturbance frequency and/or intensity (intermediate disturbance hypothesis) because only a few species tolerate either intense disturbance regime or habitats with little or no disturbances due to competitive exclusion [99]. Different parts of the Danube channel are subjected to different intensities of disturbance caused by large commercial navigation and also maintenance activities in the main channel [47], as well as higher current velocities at peak discharges. The most important effect is exerted by hydro-power plants [49]. According to the serial discontinuity concept, such barriers significantly alter habitat conditions and thus community structure, depending on dam position along the river continuum [11]. In addition, disturbances may also affect species presence and abundance directly by changing habitats or by disturbing longitudinal linkages within a river continuum preventing downstream transport of matter, including particulate organic carbon that is usually metabolized a considerable distance downstream from where it enters the stream [100]. According to Remmert [101] there is a significant relation between diversity of the environmental conditions and the number of species, while extreme conditions result in a lower number of species favoring those habitats with wider ecological range, as it is the case in our study.

However, we found a general pattern of low species richness and abundance of macrophytes (see Figure 3) as well as the presence of optimally developed plant-communities (see Figure 6) within the section between rkm 2400 at Kehlheim and rkm 1170 at Belgrade. The Danube River becomes navigable at rkm 2400 [47], which could be the reason for a general decrease in abundance of macrophyte abundance. Heavy traffic and frequent maintenance of the river channel cause habitat disturbances and decrease water transparency [49]. Within this over 1200 km long section of the Danube, only four plant-communities were found (see Table 1), namely *Potametum pectinati*, *Potametum acutifolii*, suboptimally, developed community with *Potamogeton pectinatus*. The far most frequent were the species-poorest and abundance lowest stretches where vegetation was too sparse to be classified into a certain vegetation type, habitats with different plant assemblages were found within the 628 km long Hungarian section [102]. All of these were species-poor and abundance-low, while average water transparency was also among the lowest (<70 cm). Other plant-communities occurred only sporadically within some of the stretches, which is negligible regarding the length of this section encompassing almost half of the Danube River. Specific for this section is the plant association with *Zannichelia palustris*, which was found only in the Čunovo Reservoir in Slovakia, although *Z. palustris* was recorded in numerous other stretches in this part of the Danube [102,103].

Upstream, within the headwater section reaching from the source to the beginning of the navigable river section, which is only 180 km long, we found 6 plant communities (Figure 6) despite its shortness and the fact that headwaters are mostly poorer in the number of macrophytes and their communities [9]. The possible reason could be the absence of disturbances influenced by ship traffic. It is also interesting that three of those plant communities were also found in the lowest section of the Danube, apart from the most poorly developed vegetation type.

On the downstream end of the longest section, we can notice a proliferation of macrophyte communities (Figure 6) as well as species richness and abundances (Figure 3). Conditions there change significantly due to various factors. The confluence with the river Sava brings water rich in calcium and other basic cations [104] since it drains the catchment, which predominantly consists of carbonate bedrock. Besides, the river Sava is also the tributary with the largest discharge [47]. Secondly, the metropolitan area of Belgrade city has no wastewater treatment plant and contributes a great number of nutrients to the Danube [18]. In addition, only 15 km downstream of Belgrade, the Danube reduces flow speed creating more "lentic" habitats since the huge impoundment of the Iron Gate I HPP already begins there. The mentioned factors create conditions with fewer disturbances and higher concentrations of calcium and other nutrients, which are much more favorable for the development of several macrophyte communities [105]. The dam of the Iron Gate II HPP prolongs these conditions further to rkm 843. Consequently, 12 different plant communities were recorded within this section, 5 of them exclusively here, e.g., *Potametum natantis*, *Trapetum natantis*, *Potametum graminei*, *Lemno-Spirodeletum polyrhizae* and *Potamo-Ceratophylletum submersi*. First two associations are characterized by floating-leaved dominant species, while the last two with pleustophytes, which explains their preference for lentic habitats.

The last section of the Danube is characterized by its lowest slope [14] but highest discharge [47]. In general, the conditions seem to also be favorable in this section since the plant species richness and abundances are considerably high (Figure 3); in some sections, they can reach the same values as in the previous section with impoundments. This is evident also in Figure 6, which displays the distribution of plant communities, and eight of them are found in this section of the Danube, the plant association *Potametum crispi* exclusively here. Several researchers connect the spreading of this association with the eutrophication of the water [67,106].

5. Conclusions

The Danube is far from being a natural river, but some free-running sections have reasonably low human interference. Still, the study on species distribution and abundance showed that the free-running sections of the river are poor in number and abundance of species, whereas impounded reaches mainly show the opposite result. According to the distribution of aquatic macrophytes and the communities, they are forming, four different sections of the Danube can be distinguished. However, we should be aware that the main reason driving this distribution pattern is human activities generating disturbances (navigation, hydropower-plants, pollution, etc.).

Author Contributions: Conceptualization, G.A.J., A.G., I.Z. and M.H.; methodology, G.A.J., N.E., A.G. and I.Z.; validation, G.A.J., N.E., G.A., V.B., Á.B., P.B., M.D., V.G., R.H., R.I., S.O., A.S., B.S., U.S.-M., W.S., K.S., E.S., J.T., S.T., M.V., V.V., D.V.; formal analysis, N.E., M.H., I.Z., and A.G.; investigation, G.A.J., N.E., G.A., V.B., Á.B., P.B., M.D., M.G., V.G., R.H., R.I., S.O., A.S., B.S., U.S.-M., W.S., K.S., E.S., J.T., S.T., M.V., V.V., D.V; data curation, N.E., M.H., A.G. and I.Z.; writing—original draft preparation, A.G., I.Z. and G.A.J.; writing—review and editing, A.G., I.Z., G.A.J., M.H. and R.H.; visualization, M.H., A.G. and I.Z.; supervision, G.A.J. and A.G.; project administration, G.A.J., N.E., A.G. and I.Z.; funding acquisition, G.A.J. All authors have read and agreed to the published version of the manuscript.

Funding: This research was part of the MIDCC Project supported by the Austrian Federal Ministry of Education, Sciences and Culture (Austria) and by the Austrian Committee of the International Association for the Danube Research (AC-IAD), Vienna. The financial supports of national foundations are also acknowledged: the Landesanstalt für Umweltschutz Baden-Württemberg and the Regierungspräsidium Tübingen, the Government of the Free State of Bavaria (Germany), Agency for Water Management, VERBUND Hydro Power AG (Austria), Grant Agency of the Slovak Academy of Sciences (Project No. 1/0004/11, Slovakia), Hungarian Scientific Research Grant—OTKA K106177

(Hungary), Ministry of Higher Education, Science and Technology—Research Program P1-0212, Biology of plants (Slovenia) and Ministry of Education, Science and Technological Development—project No III 43002 (Serbia).

Institutional Review Board Statement: Not applicable.

Informed Consent Statement: Not applicable.

Data Availability Statement: Data are stored within the documentation of outcomes of MIDCC project. Data are available on special request.

Conflicts of Interest: The authors declare no conflict of interest. The funders had no role in the design of the study, in the collection, analyses, or interpretation of data, in the writing of the manuscript, or in the decision to publish the results.

References

1. Sommerwerk, N.; Hein, T.; Schneider-Jacoby, M.; Baumgartner, C.; Ostojić, A.; Siber, R.; Bloesch, J.; Paunović, M.; Tockner, K. The Danube River Basin. In *Rivers of Europe*; Elsevier: Amsterdam, The Netherlands, 2009; pp. 59–112.
2. Lenhardt, M.; Smederevac-Lalić, M.; Hegediš, A.; Skorić, S.; Cvijanović, G.; Višnjić-Jeftić, Ž.; Djikanović, V.; Jovičić, K.; Jaćimović, M.; Jarić, I. Human Impacts on Fish Fauna in the Danube River in Serbia: Current Status and Ecological Implications. In *Geobotany Studies*; Springer: Berlin/Heidelberg, Germany, 2020; pp. 257–279.
3. Haslam, S.M. *The Historic River: Rovers and Cultures Down the Ages*; Cobden of Cambridge Press: Cambridge, UK, 1999; p. 324.
4. Winiwarter, V.; Schmid, M.; Hohensinner, S.; Haidvogl, G. The Environmental History of the Danube River Basin as an Issue of Long-Term Socio-ecological Research. In *Long Term Socio-Ecological Research*; Springer: Dordrecht, The Netherlands, 2013; pp. 103–122.
5. Hynes, H.B.N. *The Ecology of Running Waters*; University of Toronto Press: Toronto, ON, Canada, 1970; p. 555.
6. Odum, H.T. Trophic Structure and Productivity of Silver Springs, Florida. *Ecol. Monogr.* **1957**, *27*, 55–112. [CrossRef]
7. Illies, J.; Botosaneanu, L. Problèmes et méthodes de la classification et de la zonation écologique des eaux courantes, considerées surtout du point de vue faunistique. *Int. Ver. Theor. Angew. Limnol. Mitt.* **1963**, *12*, 1–57. [CrossRef]
8. Hawkes, H.A. River zonation and classification. In *River Ecology*; Whitton, B.A., Ed.; Blackwell Scientific Publishing: Oxford, UK, 1975; pp. 312–374.
9. Vannote, R.L.; Minshall, G.W.; Cummins, K.W.; Sedell, J.R.; Cushing, C.E. The River Continuum Concept. *Can. J. Fish. Aquat. Sci.* **1980**, *37*, 130–137. [CrossRef]
10. Ward, J.V.; Stanford, J. The serial discontinuity concept of lotic ecosystems. In *Dynamics of Lotic Ecosystems*; Fontane, T.D., Bartell, S.M., Eds.; Ann Arbor Scientific Publications: Ann Arbor, MI, USA, 1983; pp. 29–42.
11. Ward, J.V.; Stanford, J.A. The serial discontinuity concept: Extending the model to floodplain rivers. *Regul. Rivers Res. Manag.* **1995**, *10*, 159–168. [CrossRef]
12. Bornette, G.; Amoros, C.; Lamouroux, N. Aquatic plant diversity in riverine wetlands: The role of connectivity. *Freshw. Biol.* **1998**, *39*, 267–283. [CrossRef]
13. Moog, O.; Sommerhäuser, M.; Robert, S.; Battisti, T.; Birk, S.; Hering, D.; Ofenböck, T.; Schmedtje, U.; Schmidt-Kloiber, A.; Vogel, B. Typology of the Danube River based on "top-down" and "bottom-up" approaches. In Proceedings of the Proceedings 36th International Conference of IAD. Austrian Committee DanubeResearch/IAD, Vienna, Austria, 4–8 September 2006; pp. 260–264.
14. Laszloffy, W. Die Hydrographie der Donau. Der Fluss als Lebensraum. In *Limnologie der Donau—Eine Monographische Darstellung, Kapitel II*; Liepolt, R., Ed.; Schweizerbart: Stuttgart, Germany, 1965; pp. 16–57.
15. Illies, J. *Limnofauna Europaea: Eine Zusammenstellung aller die Europäischen Binnengewässer Bewohnenden Mehrzelligen Tierarten mit Angaben über ihre Verbreitung und Ökologie*; Fischer: Stuttgart, Germany, 1978.
16. Olson, K.R.; Krug, E. The Danube, an Empire Boundary River: Settlements, Invasions, Navigation, and Trade Pathway. *J. Water Resour. Prot.* **2020**, *12*, 884–897. [CrossRef]
17. Stoica, C.; Camejo, J.; Banciu, A.; Nita-Lazar, M.; Paun, I.; Cristofor, S.; Pacheco, O.R.; Guevara, M. Water quality of Danube Delta systems: Ecological status and prediction using machine-learning algorithms. *Water Sci. Technol.* **2016**, *73*, 2413–2421. [CrossRef]
18. Takić, L.; Mladenović-Ranisavljević, I.; Vasović, D.; Đorđević, L. The Assessment of the Danube River Water Pollution in Serbia. *Water Air Soil Pollut.* **2017**, *228*, 380. [CrossRef]
19. Hein, T.; Funk, A.; Pletterbauer, F.; Graf, W.; Zsuffa, I.; Haidvogl, G.; Schinegger, R.; Weigelhofer, G. Management challenges related to long-term ecological impacts, complex stressor interactions, and different assessment approaches in the Danube River Basin. *River Res. Appl.* **2019**, *35*, 500–509. [CrossRef]
20. Janauer, G.A.; Gaberščik, A.; Květ, J.; Germ, M.; Exler, N. *Macrophytes of the River Danube Basin*; Academia: Praha, Czech Republic, 2018; p. 408. ISBN 978-80-200-2743-6.
21. Wilzbach, M.A.; Cummins, K.W. Rivers and Streams: Physical Setting and Adapted Biota. In *Encyclopedia of Ecology*; Elsevier: Amsterdam, The Netherlands, 2019; pp. 594–606.
22. Neuhold, B.; Janauer, J.D.; Janauer, G.A. Within-weed bed architectural adaptation of branching pattern in *Myriophyllum spicatum* L. *Acta Biol. Slov.* **2018**, *61*, 55–63.

23. Breugnot, E.; Dutartre, A.; Laplace-Treyture, C.; Haury, J. Local distribution of macrophytes and consequences for sampling methods in large rivers. *Hydrobiologia* **2008**, *610*, 13–23. [CrossRef]
24. Baattrup-Pedersen, A.; Riis, T. Macrophyte diversity and composition in relation to substratum characteristics in regulated and unregulated Danish streams. *Freshw. Biol.* **1999**, *42*, 375–385. [CrossRef]
25. Clarke, S.J. Vegetation growth in rivers: Influences upon sediment and nutrient dynamics. *Prog. Phys. Geogr. Earth Environ.* **2002**, *26*, 159–172. [CrossRef]
26. Chambers, P.A.; Lacoul, P.; Murphy, K.J.; Thomaz, S.M. Global diversity of aquatic macrophytes in freshwater. *Hydrobiologia* **2008**, *595*, 9–26. [CrossRef]
27. Preiner, S.; Dai, Y.; Pucher, M.; Reitsema, R.E.; Schoelynck, J.; Meire, P.; Hein, T. Effects of macrophytes on ecosystem metabolism and net nutrient uptake in a groundwater fed lowland river. *Sci. Total Environ.* **2020**, *721*, 137620. [CrossRef]
28. Gibbs, H.; Gurnell, A.; Heppell, K.; Spencer, K. Macrophytes: Ecosystem engineers in UK urban rivers. *Geophys. Res. Abstr.* **2012**, *14*, 490.
29. Sand-Jensen, K. Macrophytes as biological engineers in the ecology of Danish streams. In *Freshwater Biology. Priorities and Development in Danish Research*; Gad: København, Denmark, 1997; pp. 74–101.
30. Rolland, D.C.; Haury, J.; Marmonier, P.; Lagadeuc, Y. Effect of Macrophytes on Flow Conditions and Deposition of Suspended Particles in Small Streams: An Experimental Study Using Artificial Vegetation. *Rev. Sci. l'eau/J. Water Sci.* **2015**, *28*, 231–245. [CrossRef]
31. Mechora, Š.; Germ, M.; Stibilj, V. Selenium compounds in selenium-enriched cabbage. *Pure Appl. Chem.* **2012**, *84*, 259–268. [CrossRef]
32. Rejmankova, E. The role of macrophytes in wetland ecosystems. *J. Ecol. Environ.* **2011**, *34*, 333–345. [CrossRef]
33. Bornette, G.; Amoros, C.; Chessel, D. Effect of allogenic processes on successional rates in former river channels. *J. Veg. Sci.* **1994**, *5*, 237–246. [CrossRef]
34. Šraj-Kržič, N.; Germ, M.; Urbanc-Berčič, O.; Kuhar, U.; Janauer, G.A.; Gaberščik, A. The quality of the aquatic environment and macrophytes of karstic watercourses. *Plant Ecol.* **2007**, *192*, 107–118. [CrossRef]
35. Jusik, S.; Szoszkiewicz, K.; Kupiec, J.M.; Lewin, I.; Samecka-Cymerman, A. Development of comprehensive river typology based on macrophytes in the mountain-lowland gradient of different Central European ecoregions. *Hydrobiologia* **2015**, *745*, 241–262. [CrossRef]
36. Graham, S.E.; O'Brien, J.M.; Burrell, T.K.; McIntosh, A.R. Aquatic macrophytes alter productivity-richness relationships in eutrophic stream food webs. *Ecosphere* **2015**, *6*, 1–18. [CrossRef]
37. Pip, E. Water temperature and freshwater macrophyte distribution. *Aquat. Bot.* **1989**, *34*, 367–373. [CrossRef]
38. Haslam, S.M. *River plants of Western Europe: The Macrophytic Vegetation of Watercourses of the European Economic Community*; Cambridge University Press: Cambridge, UK, 1987; p. 512.
39. Haslam, S.M. *River Plants: The Macrophytic Vegetation of Watercourses*; Forrest Text: Ceredigion, UK, 2006; p. 450.
40. Kuhar, U.; Germ, M.; Gaberščik, A.; Urbanič, G. Development of a River Macrophyte Index (RMI) for assessing river ecological status. *Limnologica* **2011**, *41*, 235–243. [CrossRef]
41. Mechora, Š.; Germ, M.; Stibilj, V. Monitoring of selenium in macrophytes—The case of Slovenia. *Chemosphere* **2014**, *111*, 464–470. [CrossRef]
42. Kohler, A.; Schneider, S. Macrophytes as bioindicators. *River Syst.* **2003**, *14*, 17–31. [CrossRef]
43. Kuhar, U.; Germ, M.; Gaberščik, A. Macrophytes of watercourses in the Slovenian Danube Basin. In *Macrophytes of Watercourses in the Slovenian Danube Basin*; Academia: Praha, Czech Republic, 2018; pp. 234–251.
44. Kuhar, U.; Gregorc, T.; Renčelj, M.; Šraj-Kržič, N.; Gaberščik, A. Distribution of macrophytes and condition of the physical environment of streams flowing through agricultural landscape in north-eastern Slovenia. *Limnologica* **2007**, *37*, 146–154. [CrossRef]
45. Demars, B.O.L.; Edwards, A.C. Distribution of aquatic macrophytes in contrasting river systems: A critique of compositional-based assessment of water quality. *Sci. Total Environ.* **2009**, *407*, 975–990. [CrossRef]
46. Sedell, J.R.; Richey, J.E.; Swanson, F.J. The river continuum concept: A basis for the expected ecosystem behavior of very large rivers? In Proceedings of the International Large River Symposium, Honey Harbour, ON, Canada, 14–19 September 1986.
47. Hasenbichler, H.-P. *Manual on Danube Navigation*; Thomas, H., Vera, H., Eds.; via donau–Österreichische Wasserstraßen-Gesellschaft mbH: Vienna, Austria, 2013; p. 218.
48. Vukov, D.; Ilic, M.; Cuk, M.; Igic, R.; Janauer, G. The relationship between habitat factors and aquatic macrophyte assemblages in the Danube river in Serbia. *Arch. Biol. Sci.* **2017**, *69*, 427–437. [CrossRef]
49. Fuller, M.R.; Doyle, M.W.; Strayer, D.L. Causes and consequences of habitat fragmentation in river networks. *Ann. N. Y. Acad. Sci.* **2015**, *1355*, 31–51. [CrossRef]
50. ISTER. Donau-Stromkilometer-Stromauf-Stromab.pdf. Available online: http://www.ister.at/Ister1/wp-content/uploads/2019/06 (accessed on 8 January 2021).
51. HPP-Bavaria. Available online: http://www.bew-augsburg.de/CBP/Wasserkraft/Daten/ODK.asp (accessed on 7 January 2021).
52. Danube Commission. List of Locks. Available online: https://de.wikipedia.org/wiki/Liste_der_Schleusen_in_der_Donau (accessed on 7 January 2021).

53. Kohler, A.; Janauer, G.A. Zur Methodik der Untersuchung von aquatischen Makrophyten in Fließgewässern. *Handb. Angew. Limnol. Ecomed-Verlag* **1995**, *VIII*, 1–22. [CrossRef]
54. Atlas Florae Europaeae (AFE)—Distribution of Vascular Plants in Europe. Available online: https://www.luomus.fi/en/atlas-florae-europaeae-afe-distribution-vascular-plants-europe (accessed on 23 December 2020).
55. *Corine Land Cover. Copernicus Land Monitoring Service—High Resolution Layer Water and Wetness*; European Environment Agency: Copenhagen, Denmark, 2018; Available online: http://land.copernicus.eu (accessed on 22 December 2020).
56. Schaumburg, J.; Schranz, C.; Foerster, J.; Gutowski, A.; Hofmann, G.; Meilinger, P.; Schneider, S.; Schmedtje, U. Ecological classification of macrophytes and phytobenthos for rivers in Germany according to the Water Framework Directive. *Limnologica* **2004**, *34*, 283–301. [CrossRef]
57. Braun-Blanquet, J. *Pflanzensoziologie, Grundzüge der Vegetationskunde*, 3rd ed.; Springer: Berlin/Heidelberg, Germany, 1964; p. 631.
58. Lukács, B.A.; Tóthmérész, B.; Borics, G.; Várbíró, G.; Juhász, P.; Kiss, B.; Müller, Z.; G-Tóth, L.; Eros, T. Macrophyte diversity of lakes in the Pannon Ecoregion (Hungary). *Limnologica* **2015**, *53*, 74–83. [CrossRef]
59. Pall, K.; Gecheva, G.; Soaru-Minea, A.; Lukacs, B.S.P. *Intercalibration of the National Classifications of Ecological Status for Eastern Continental Lakes*; Publications Office of the European Union: Luxembourg, 2018; p. 81.
60. Hammer, Ø.; Harper, D.A.T.; Ryan, P.D. PAST: Paleontological Statistics Software Package for Education and Data Analysis. *Palaeontol. Electron.* **2001**, *4*, 1–9.
61. Murtagh, F.; Legendre, P. Ward's Hierarchical Agglomerative Clustering Method: Which Algorithms Implement Ward's Criterion? *J. Classif.* **2014**, *31*, 274–295. [CrossRef]
62. Swaine, M.D.; Adomako, J.; Ameka, G.; de Graft-Johnston, K.A.A.; Cheek, M. Forest river plants and water quality in Ghana. *Aquat. Bot.* **2006**, *85*, 299–308. [CrossRef]
63. Wiegleb, G.; Bröring, U.; Filetti, M.; Brux, H.; Herr, W. Long-term dynamics of macrophyte dominance and growth-form types in two north-west German lowland streams. *Freshw. Biol.* **2014**, *59*, 1012–1025. [CrossRef]
64. Zelnik, I. Vegetation of the Meadows from the Order Molinietalia W. Koch 1926 and Contact Sites in Slovenia. Ph.D. Thesis, University of Ljubljana (Slovenia), Ljubljana, Slovenia, 2005; p. 196.
65. Hrivnák, R.; Bubíková, K.; Oťaheľová, H.; Šumberová, K. Formalised classification of aquatic vegetation in Slovakia. *Phytocoenologia* **2019**, *49*, 107–133. [CrossRef]
66. Landucci, F.; Tichý, L.; Šumberová, K.; Chytrý, M. Formalized classification of species-poor vegetation: A proposal of a consistent protocol for aquatic vegetation. *J. Veg. Sci.* **2015**, *26*, 791–803. [CrossRef]
67. Chytrý, M. *Vegetation of the Czech Republic 3. Aquatic and wetland vegetation*; Academia: Praha, Czech Republic, 2011; p. 828.
68. Ter Braak, C.J.F.; Šmilauer, P. *CANOCO Reference Manual and CanoDraw for Windows User's Guide: Software for Canonical Community Ordination (Version 4.5)*; www.canoco.com: Ithaca, NY, USA, 2002.
69. Hudon, C.; Gagnon, P.; Amyot, J.P.; Létourneau, G.; Jean, M.; Plante, C.; Rioux, D.; Deschênes, M. Historical changes in herbaceous wetland distribution induced by hydrological conditions in Lake Saint-Pierre (St. Lawrence River, Quebec, Canada). In *Proceedings of the Hydrobiologia*; Springer: Berlin/Heidelberg, Germany, 2005; Volume 539, pp. 205–224.
70. Kornijów, R.; Strayer, D.L.; Caraco, N.F. Macroinvertebrate communities of hypoxic habitats created by an invasive plant (Trapa natans) in the freshwater tidal Hudson River. *Fundam. Appl. Limnol.* **2010**, *176*, 199–207. [CrossRef]
71. Skliar, I.L.; Skliar, V.H. Trapa natans L. s.l. growth features in Desna River basin. *Ukr. J. Ecol.* **2017**, *7*, 239–245. [CrossRef]
72. Wetzel, R.G. Land–water interfaces: Larger plants. In *Limnology*; Elsevier: Amsterdam, The Netherlands, 2001; pp. 527–575.
73. Preston, C. *Pondweeds of Great Britain and Ireland*; Botanical Society of the British Isles: London, UK, 1995; p. 352.
74. Germ, M.; Kuhar, U.; Gaberščik, A. Abundance and Diversity of Taxa Within the Genus Potamogeton in Slovenian Watercourses. In *Natural and Constructed Wetlands*; Vymazal, J., Ed.; Springer International Publishing: Cham, Switzerland, 2016; pp. 283–291.
75. Kaplan, Z.; Jarolímová, V.; Fehrer, J. Revision of Chromosome Numbers of Potamogetonaceae: A New Basis for Taxonomic and Evolutionary Implications. *Preslia* **2013**, *85*, 421–482.
76. Case, M.L.; Madsen, J.D. Factors limiting the growth of Stuckenia pectinata (sago pondweed) in heron lake, minnesota. *J. Freshw. Ecol.* **2004**, *19*, 17–23. [CrossRef]
77. Kantrud, H.A. *Sago Pondweed (Potamogeton pectinatus L.): A Literature Review*; U.S. Fish and Wildlife Service: Washington, DC, USA, 1990; p. 89.
78. Chambers, P.A.; Prepas, E.E.; Hamilton, H.R.; Bothwell, M.L. Current Velocity and Its Effect on Aquatic Macrophytes in Flowing Waters. *Ecol. Appl.* **1991**, *1*, 249–257. [CrossRef] [PubMed]
79. Hussner, A. Alien aquatic plant species in European countries. *Weed Res.* **2012**, *52*, 297–306. [CrossRef]
80. Zehnsdorf, A.; Hussner, A.; Eismann, F.; Rönicke, H.; Melzer, A. Management options of invasive Elodea nuttallii and Elodea canadensis. *Limnologica* **2015**, *51*, 110–117. [CrossRef]
81. Hérault, B.; Bornet, A.; Trémolières, M. Redundancy and niche differentiation among the European invasive Elodea species. *Biol. Invasions* **2008**, *10*, 1099–1107. [CrossRef]
82. Kuhar, U.; Germ, M.; Gaberščik, A. Habitat characteristics of an alien species Elodea canadensis in Slovenian watercourses. *Hydrobiologia* **2010**, *656*, 205–212. [CrossRef]
83. Thiébaut, G. Does competition for phosphate supply explain the invasion pattern of Elodea species? *Water Res.* **2005**, *39*, 3385–3393. [CrossRef] [PubMed]

84. Grudnik, Z.M.; Jelenko, I.; Germ, M. Influence of abiotic factors on invasive behaviour of alien species Elodea nuttallii in the Drava River (Slovenia). *Ann. Limnol. Int. J. Limnol.* **2014**, *50*, 1–8. [CrossRef]
85. Simpson, D.A. Displacement of *Elodea canadensis* Michx by *Elodea nuttallii* (Planch.) H. St John in the British Isles. *Watsonia* **1990**, *18*, 173–177.
86. Thiébaut, G. Invasion success of non-indigenous aquatic and semi-aquatic plants in their native and introduced ranges. A comparison between their invasiveness in North America and in France. *Biol. Invasions* **2006**, *9*, 1–12. [CrossRef]
87. Barrat-Segretain, M.H.; Lemoine, D.G. Can snail herbivory influence the outcome of competition between Elodea species? *Aquat. Bot.* **2007**, *86*, 157–162. [CrossRef]
88. Erhard, D.; Gross, E.M. Allelopathic activity of Elodea canadensis and Elodea nuttallii against epiphytes and phytoplankton. *Aquat. Bot.* **2006**, *85*, 203–211. [CrossRef]
89. Barrat-Segretain, M.H.; Cellot, B. Response of invasive macrophyte species to drawdown: The case of *Elodea* sp. *Aquat. Bot.* **2007**, *87*, 255–261. [CrossRef]
90. Lumpkin, T.A.; Plucknett, D.L. Azolla: Botany, physiology, and use as a green manure. *Econ. Bot.* **1980**, *34*, 111–153. [CrossRef]
91. Szczęśniak, E.; Błachuta, J.; Krukowski, M.; Picińska-Fałtynowicz, J. Distribution of Azolla filiculoides Lam. (Azollaceae) in Poland. *Acta Soc. Bot. Pol.* **2011**, *78*, 241–246. [CrossRef]
92. Sabetraftar, K.; Zarkami, R.; Sadeghi, R.; Van Damme, P. A review of some ecological factors affecting the growth of *Azolla* spp. *Casp. J. Environ. Sci.* **2013**, *11*, 65–76.
93. Wagner, G.M. Azolla: A Review of Its Biology and Utilization. *Bot. Rev.* **1997**, *63*, 1–26. [CrossRef]
94. Paolacci, S.; Jansen, M.A.K.; Harrison, S. Competition Between Lemna minuta, Lemna minor, and Azolla filiculoides. Growing Fast or Being Steadfast? *Front. Chem.* **2018**, *6*, 207. [CrossRef] [PubMed]
95. Bolpagni, R.; Laini, A.; Soana, E.; Tomaselli, M.; Nascimbene, J. Growth performance of Vallisneria spiralis under oligotrophic conditions supports its potential invasiveness in mid-elevation freshwaters. *Weed Res.* **2015**, *55*, 185–194. [CrossRef]
96. Anačkov, G.T.; Rat, M.M.; Radak, B.D.; Igić, R.S.; Vukov, D.M.; Rućando, M.M.; Krstivojević, M.M.; Radulović, S.B.; Cvijanović, D.L.; Milić, D.M.; et al. Alien invasive neophytes of the Southeastern part of the Pannonian Plain. *Cent. Eur. J. Biol.* **2013**, *8*, 1032–1043. [CrossRef]
97. Illies, J. Versuch einer allgemeinen biozönotischen Gliederung der Fließgewässer. *Int. Rev. Gesamten Hydrobiol. Hydrogr.* **1961**, *46*, 205–213. [CrossRef]
98. Connell, J.H. Diversity in Tropical Rain Forests and Coral Reefs. *Science* **1978**, *199*, 1302–1310. [CrossRef] [PubMed]
99. Grime, J.P. Competitive Exclusion in Herbaceous Vegetation. *Nature* **1973**, *242*, 344–347. [CrossRef]
100. Webster, J.R. Spiraling down the river continuum: Stream ecology and the U-shaped curve. *J. N. Am. Benthol. Soc.* **2007**, *26*, 375–389. [CrossRef]
101. *The Mosaic-Cycle Concept of Ecosystems*; Remmert, H. (Ed.) Ecological Studies; Springer: Berlin/Heidelberg, Germany, 1991; Volume 85, p. 168.
102. Engloner, A.; Szalma, E.; Sipos, K.; Dinka, M. Occurrence and habitat preference of aquatic macrophytes in a large river channel. *Community Ecol.* **2013**, *14*, 243–248. [CrossRef]
103. Oťaheľová, H.; Valachovic˘, M.; Hrivnák, R. The impact of environmental factors on the distribution pattern of aquatic plants along the Danube River corridor (Slovakia). *Limnologica* **2007**, *37*, 290–302. [CrossRef]
104. Markovics, R.; Kanduč, T.; Szramek, K.; Globočanin, D.; Milačič, R.; Ogrinc, N. Chemical dynamics of the Sava riverine system—A stable isotopic approach. *J. Environ. Monit.* **2010**, *12*, 2165–2176. [CrossRef] [PubMed]
105. Chmara, R.; Banaś, K.; Szmeja, J. Changes in the structural and functional diversity of macrophyte communities along an acidity gradient in softwater lakes. *Flora-Morphol. Distrib. Funct. Ecol. Plants* **2015**, *216*, 57–64. [CrossRef]
106. Riis, T.; Sand-Jensen, K. Historical changes in species composition and richness accompanying perturbation and eutrophication of Danish lowland streams over 100 years. *Freshw. Biol.* **2001**, *46*, 269–280. [CrossRef]

water

MDPI

Article

Dynamic of Dominant Plant Communities in Kettle Holes (Northeast Germany) during a Five-Year Period of Extreme Weather Conditions

Marlene Pätzig [1,*] and Eveline Düker [1,2]

[1] Leibniz Centre for Agricultural Landscape Research (ZALF) e.V., 15374 Müncheberg, Germany; edueker@gmx.de
[2] City University of Applied Sciences (HSB), 28199 Bremen, Germany
* Correspondence: marlene.paetzig@zalf.de

Abstract: Understanding the ecosystem functions and services of central European kettle holes (small wetlands) requires knowledge about their spatiotemporal dynamics. A lot of existing research has been conducted on the wet–dry cycles of North American potholes, but much less is known about kettle holes. Based on the extreme weather conditions between 2015 and 2020, we aimed to quantify differences among dominant plant communities of kettle holes using unmanned aerial systems. Different dominant plant communities were differently affected by dry and wet intervals with a major increase in terrestrial plants. Multivariate analysis showed strong variability in plant community composition for permanent and semi-permanent kettle holes, where hydrophytes decreased and nitrophilous perennials increased. Although we cannot provide quantitative results in succession over a five-year observation period, we found indications of progressive succession towards irreversible alternative stable states with woody plants for some kettle holes, especially of the "storage type". Therefore, we assume stronger changes in proportion of wetland types in kettle holes compared to potholes and we expect the proportion of wood-dominated kettle holes to increase in the central European landscape in the future, leading to enhanced homogenization of the landscape accompanied by a loss of ecosystem functions and services.

Keywords: ponds; small wetlands; macrophytes; changes in the water surface; hydrological variability; succession; climate change; unmanned aerial system

Citation: Pätzig, M.; Düker, E. Dynamic of Dominant Plant Communities in Kettle Holes (Northeast Germany) during a Five-Year Period of Extreme Weather Conditions. *Water* **2021**, *13*, 688. https://doi.org/10.3390/w13050688

Academic Editor: Alenka Gaberščik

Received: 28 January 2021
Accepted: 26 February 2021
Published: 3 March 2021

1. Introduction

The understanding of ecosystem functioning in small, highly variable wetland ecosystems such as potholes or kettle holes requires sufficient knowledge about their spatiotemporal dynamics [1–3]. This is even more important under climate change conditions where changing precipitation patterns cause extreme weather conditions [4]. The functioning of small wetlands strongly depends on water level fluctuations through enhanced sensitivity to changes in the frequency and duration of drought and deluge periods [5]. Cyclic changes in precipitation are normal climatic phenomena and result in oscillating water levels in small wetlands around a long-term mean [6–8]. Plant communities respond to the resulting wet–dry cycles [9] and are key drivers of the functioning of small wetland ecosystems, as low water levels and desiccation enables them to colonize the entire pond bottom and/or to occupy the whole water column. This, in turn, affects biodiversity, productivity and biogeochemical processes of small wetlands [10–13]. Thus, differences in oscillating water levels—due to natural meteorological variability or climate change—directly affect the vegetation dynamics of small wetlands and may shift them to alternative stable states [14–16]. Hence, a prediction of changes in the functioning of small wetland ecosystems requires a good understanding of oscillating water levels and their effect on wet–dry cycles of vegetation and their interactions.

Many studies have analyzed the dependence of vegetation in small wetlands on water levels and moisture [17–19]. In particular, intensive research was done on potholes in North America's Prairie Pothole Region (PPR) in the last century. Van der Valk [7] summarized the research about the spatiotemporal dynamics of potholes and emphasized that earliest studies observed cyclic changes in the vegetation resulting from water level fluctuation due to changes in amounts of precipitation between years. He concluded that joint research efforts in the 20th century enabled a comprehensive understanding of the driving factors of vegetation dynamics, including particularly seasonal wetlands (defined by an extended period of water during the growing season). Important knowledge and a conceptual basis for vegetation zonation and change have also been found in previous research [9,20–22]. For example, plant communities during all stages of a wet–dry cycle have been described and three types of vegetation change defined, i.e., fluctuation (change in species abundance with water level variations <0.5 m), "gleasonian" succession (change in community composition with water level variations >1.5 m), and maturation (change in biomass). Key life history features form the basis to understand changing vegetation stages due to altered water-level oscillations [7,9,22]. Based on this knowledge, Seabloom et al. [23] were able to *"predict the composition and distribution of the vegetation during all stages of a wet–dry cycle"* [7], which is important under climate change conditions.

In addition to these findings, the "wetland continuum concept" developed by Euliss et al. (2004) [24] and modified by Mushet et al. [25], explains how changes in water levels and salinity from different water sources, i.e., groundwater and rainwater, potentially influence wetland vegetation. Placing vegetation at any time point at the right position of two gradients ("groundwater axis" and "rainwater axis" as per Mushet et al. [25]) facilitates the interpretation of wetland vegetation and prediction of communities and potential state shifts by knowing groundwater and rainwater components [8,24,25].

The "wetland continuum concept" with the included basic types of vegetation change suggested by van der Valk [9] does not directly incorporate irreversible succession towards late states or even a final climax state. Progressive succession is a natural, slow phenomenon of small wetlands that has rapidly increased in recent decades, specifically in Europe due to discontinued wetland management and increased nutrient enrichment through intensive agricultural practices. The resulting terrestrialization leads to sediment deposition and bush encroachment, for example with *Salix cinerea*, until overgrowth and anoxic conditions occur, accompanied by declining biodiversity [26]. Further increases in wetlands overgrown with woody vegetation will dramatically decrease the ecosystem services provided by small wetlands in agricultural landscapes [27,28]. Climate change could accelerate succession and terrestrialization processes by reduced water availability, which hampers the reset of succession by flooding disturbance until tipping points towards irreversible state shifts may be reached [14,29]. Nevertheless, climate change may conversely lead to increased water availability creating wetland lake marshes, such as in the PPR in recent decades [8,30]. Depending on the underlying climate scenario, van der Valk et al. [31] predicted a potential loss or gain of wetland vegetation zones due to increased or decreased precipitation and subsequent water level changes. For the PPR, this could result in an altered proportion of wetland classes (Stewart and Kantrud [20]) [32]. However, van de Valk and Mushet [32] also state that although proportions of wetland classes may change, wet–dry cycles and related plant communities of prairie potholes will persist. The question arises if this also applies to kettle holes with their potentially different ecosystem constraints.

Although vegetation changes due to oscillating water levels of potholes in North America are well researched, these wet–dry cycles are less well studied in central European kettle holes [33]. Lehsten et al. [34] simulated water level changes over 50 years, showing oscillating water levels that were correlated with climatic indices. Moreover, they suggested that predicted climate change, inter alia, will strongly affect the hydrology and vegetation of kettle holes, and noted an increase in *Salix* species during the last 30 years of their studies. The need for longer time series that account for the effects of altered hydrological conditions and potential state shifts of macrophytes in kettle holes was already emphasized by

Pätzig et al. [3]. Although the authors of this study showed that variability in macrophyte communities of the flooded area among individual kettle holes explains more variance than among years, they also found non-consistent interannual patterns within communities of single kettle holes. Thus, the observed interannual changes in the vegetation of the studied kettle holes during the hydrologically stable period of 1990–1999 (Northeast Germany) suggested a higher importance of interannual effects during periods of higher hydrological variability [3]. Since the 1990s, the climatic water balance has been persistently negative and comparably extreme weather conditions between 2015 and 2020 gave us the possibility to study the interannual vegetation dynamics of central European kettle holes based on rapidly collected data using unmanned aerial systems (UAS). The considered time period will not allow us to answer questions about future persistence of all kettle holes types in central Europe but will enhance our knowledge of their interannual vegetation dynamics. This is a prerequisite for understanding and quantifying the functioning and ecosystem services of such kettle holes under changing climatic conditions.

On the basis of these temporal constraints, we aimed at quantifying differences among dominant plant communities of kettle holes during five years with extreme weather events based on manual digitization using maps derived by UAS. We first (1) hypothesized that different groups of dominant plant communities will be significantly affected to different degrees by dry and wet periods, with an increase in terrestrial plants and helophytes during drought and in moisture indicators during wet periods. Second (2), we calculated turnover, loss, and gain rates between time intervals to determine the effects of different hydrological conditions on changes in dominant plant communities. Third (3), we expected larger interannual variability in plant community composition than in Pätzig et al. (2020) [3], because the present study also includes kettle holes with usually long dry periods starting in early summer (episodic) or latest in autumn (periodic), unlike the permanent and semi-permanent kettle holes mainly examined by Pätzig et al. [3]. Fourth (4), we predicted to find a strong relation between vegetation changes and water availability (area), resulting from extreme weather events but considered morphological "fixed" factors because these are well known to influence vegetation in small wetlands [17,35,36].

The results will be embedded in existing knowledge about small wetland vegetation dynamics and discussed regarding potential future trends of central European kettle holes. In particular, we focus on the extent to which a drought in the last three studied years can lead towards potentially irreversible effects with increases in shrubs and trees and whether we may face a future loss of kettle hole types.

2. Materials and Methods

2.1. Study Sites

We monitored 14 out of more than 1500 kettle holes located in the "AgroScapeLab Quillow" (ASLQ, Leibniz Centre for Agricultural Landscape Research (ZALF)) that covers the catchment of the river Quillow with an area of about 170 km^2. This landscape laboratory is located in the lowlands of northeast Germany and part of a hummocky young moraine landscape. The Quillow catchment is a fertile region with about three-quarters coverage mainly characterized by intensive agriculture, primarily used for cultivating crops. The climate is sub-humid with a mean annual temperature of 8.8 °C, corrected mean annual precipitation of 573 mm and mean annual evapotranspiration of 658 mm for the period from 1992 to 2019 (ZALF weather stations at the Dedelow Research Station) resulting in a mean annual negative climatic water balance of −84 mm.

At the beginning of the study period in summer 2016, four of the 14 selected kettle holes belonged to the edge type with an open water area, and 10 to the full reed type with no or scant open water (Table 1). Regarding the dominant vegetation, three kettle holes belonged to reed type (*Phragmites australis* (Cav.) Trin. ex Steud., *Typha* spp.), five to sedge type (*Carex* spp.), and six to reed canary grass (*Phalaris arundinacea* L.). Five of them were additionally characterized by dominant shore woods at the edge (Table 1).

Table 1. Hydrogeomorphological characteristics (Kalettka and Rudat [37]) and dominant vegetation types for the year 2016 (after Kalettka and Rudat, unpublished) of the selected 14 kettle holes. **KH_No** = kettle hole number according to numbering within the SWBTrans project (ZALF), **Vegtype2016** = dominant vegetation type in 2016. **HGM** = hydrogeomorphological type: BS-S = big and shallow storage type, BS-SO = big and shallow shore overflow type, SS-S = small and shallow storage type, SS-SO= small and shallow shore overflow type. **KA** = mean area of the kettle hole between 2016 and 2020, measured up to the tilled area of the adjacent field in square meter. **KA_C** = size class of the mean area of the kettle hole in hectare: 1 = very small (≤0.03 ha), 2 = small (≤0.10 ha), 3 = medium (≤0.32 ha), 4 = large (≤1.0 ha). **SS_C** = class of shore slope in percent: 1 = flat (≤10%), 2 = oblique (≤20%), 3 = very oblique (≤30%), 4 = steep (≤40%). **LP_2011-2017** = long-term pond permanence class for the period 2011 to 2017, **LSOT_2011-2017** = long-term shore overflow tendency class for the period 2011 to 2017.

KH_No	VegType 2016	HGM	KA	KA_C	SS_C	LP_2011–2017	LSOT_2011–2017
1	Edge type with shore woods and sedges	SS-S	2271 ± 89	≤0.32	≤40	5_permanent	1_no overflow
4	Edge type with sedges	BS-SO	3224 ± 143	≤1.0	≤20	4_semi-permanent	3_fully overflowing
7	Full reed type with sedges	SS-S	1878 ±32	≤0.32	≤30	3_periodic	1_no overflow
9	Full reed type with reeds and shore woods	SS-S	2910 ± 70	≤0.32	≤30	4_semi-permanent	1_no overflow
10	Full reed type with sedges	SS-SO	789 ± 25	≤0.10	≤20	2_episodic	2_partly overflowing
11	Full reed type with reed canary grass	SS-SO	1286 ± 8	≤0.32	≤20	3_periodic	3_fully overflowing
13	Edge type with reed canary grass	SS-S	1017 ± 27	≤0.10	≤40	3_periodic	1_no overflow
14	Full reed type with reed canary grass	SS-S	379 ± 23	≤0.10	≤40	2_episodic	1_no overflow
15	Full reed type with shore woods and reed canary grass	SS-S	1560 ± 55	≤0.32	≤40	2_episodic	1_no overflow
16	Edge type with reed canary grass	SS-SO	1014 ± 52	≤0.10	≤30	4_semi-permanent	3_fully overflowing
17	Full reed type with reed	SS-SO	1660 ± 93	≤0.32	≤20	4_semi-permanent	2_partly overflowing
18	Full reed type with shore woods and reed canary grass	SS-S	3852 ± 17	≤1.0	≤30	4_semi-permanent	1_no overflow
20	Full reed type with reed	BS-SO	4644 ± 356	≤1.0	≤20	3_periodic	2_partly overflowing
24	Full reed type with shore woods and sedges	BS-S	4834 ± 369	≤1.0	≤30	3_periodic	1_no overflow

Almost 60% of the 14 kettle holes showed a long-term hydroperiod, subsequently referred to as long-term pond permanence according to the definition by Hayashi et al. [8], with episodic/ephemeral or periodic/seasonal pond permanence, while the remaining kettle holes are semi-permanent or permanent (Table 1). With respect to long-term shore overflow tendencies, eight kettle holes were classified as storage types (no shore overflow) and six kettle holes as partly or fully able to overflow the top shore. With regard to morphological characteristics, they covered a range of areal sizes from small (min = 0.037 ha) to large (max = 0.464 ha) with oblique to steep, predominantly narrow shores and a range of depths from wadeable (>1.6 m) to shallow ($\leq$4 m) (Table 1). In summary, the 14 kettle holes were assigned to the following hydrogeomorphological (HGM) types according to the classification by Kalettka and Rudat [37], i.e., seven small shallow storage types, four small shallow overflow types, two big shallow-overflow types and one big shallow storage type (Table 1).

Water quality of the kettle holes was usually measured in spring and summer when water was available within the ZALF monitoring program supported by third-party funded projects. Since none of the kettle holes had enough water for sampling in the summers of 2019 and 2020 and the number of kettle holes that contained water was generally low in summer, we only considered spring data (except for 2018). For 2018, we included summer values because no spring sampling was carried out due to inaccessibility of the kettle holes during this extremely wet season. An overview of the water quality data is provided in Table 2.

2.2. UAS Data Acquisition and Image Processing

UAS flight campaigns were conducted in 2016, 2018 and 2020 within 14 days between mid-July and the beginning of August at the climax of the vegetation growing season. In 2016 and 2018, we used a Goliath 4 quadrocopter (Cadmic GmbH) carrying a GoPro Hero4 RGB camera with 4 Megapixel and a fixed 16 mm lens. In 2020, we used a Phantom 4 RTK (DJI) carrying a 20 Megapixel CMOS sensor with a fixed 8.8 mm lens. The cameras were attached to a gimbal to compensate for pitch and roll movements of the aircraft. Since our knowledge of drone technique, image capturing and processing improved over the years, the entire workflow from image capturing and processing to the final image/orthomosaic differed from year to year. Generally, in 2016 a single image was taken from the center above each kettle hole. In 2018 we manually flew over the kettle holes taking images at a constant interval of ~2 s. In 2020 we used the integrated mission planning software DJI GS RTK app of the P4 RTK to plan missions with parallel flight tracks between 25 and 35 m altitude, connecting to the D-RTK 2 mobile station for high relative position accuracy. Additionally, 3 to 6 ground control points (GCPs) were distributed around the periphery of the kettle hole prior to the flights in 2018 and 2020 using a real-time kinematic (RTK) GNSS (Reach RS, Emlid Ltd., Saint Petersburg, Russia). Camera orientation was nadir during all flight campaigns in the three years.

All images taken in 2016 were corrected for distortion that was caused by the fish-eye effect of the GoPro Hero4. Correction was carried out with Adobe Lightroom Classic Version 9.3 software, which automatically detects the camera model GoPro Hero4 Black Edition. The distortion-corrected images were finally exported as tiff-images with 600 dpi resolution and georeferenced by distinctive, stationary structures from 2018 and 2020 orthomosaics of respective kettle holes in QGIS using the extension GDAL Georeference 3.1.9.

All images from 2018 and 2020 were quality-checked and subsequently processed in a SfM photogrammetric workflow using Agisoft Metashape Professional 1.6.2 build 6205 (Agisoft LLC). Finally, for each kettle hole mapped in 2018 and 2020 an orthomosaic was computed on the basis of the DPC and exported in the desired image format (Tiff) and reference system (WGS 84).

Table 2. Summary of water quality parameters for spring 2016, 2017, 2019 and 2020 as well as summer 2018 with ± 95% confidence intervals and number of values in brackets. KH_No = kettle hole number according to numbering within the SWBTrans project (ZALF), EC = conductivity in µS/cm, Eh = redox potential in mV, O2 = oxygen in mg L^{-1}, SRP = soluble reactive phosphorous in mg L^{-1}, NH4-N = ammonium nitrogen in mg L^{-1}, TN = total nitrogen mg L^{-1}, TP = total phosphorous mg L^{-1}, CH = carbonate hardness in dH.

KH_No	pH	EC	Eh	O2	SRP	NH4-N	TN	TP	CH
1	7.6 ± 0.6 (4)	286 ± 36 (4)	426 ± 56 (3)	10.7 ± 3.3 (4)	0.026 ± 0.030 (4)	0.212 ± 0.277 (4)	1.51 ± 0.40 (4)	0.09 ± 0.03 (4)	9.7 ± 1.7 (3)
4	8.0 ± 0.4 (4)	452 ± 50 (4)	431 ± 24 (3)	10.9 ± 2.4 (4)	0.076 ± 0.123 (4)	0.054 ± 0.029 (4)	2.73 ± 0.62 (4)	0.18 ± 0.18 (4)	13.3 ± 8.5 (3)
7	7.4 ± 0.3 (5)	611 ± 253 (5)	347 ± 195 (4)	14.3 ± 9.1 (5)	0.075 ± 0.115 (5)	0.072 ± 0.070 (5)	4.15 ± 1.97 (5)	0.23 ± 0.18 (5)	9.0 ± 2.9 (4)
9	6.5 ± 1.2 (4)	744 ± 564 (4)	340 ± 160 (3)	13.2 ± 7.6 (4)	0.521 ± 0.504 (4)	1.671 ± 1.557 (4)	4.55 ± 2.90 (4)	0.77 ± 0.46 (4)	5.5 ± 1.7
10	7.0 ± 0.5 (4)	433 ± 199 (4)	414 ± 19 (3)	12.2 ± 2.7 (4)	0.078 ± 0.046 (4)	0.122 ± 0.095 (4)	2.50 ± 1.35 (4)	0.17 ± 0.10 (4)	7.5 ± 4.4 (4)
11	7.2 (1)	459 (1)	203 (1)	9.2 (1)	Na	Na	Na	Na	10.0 (1)
13	7.5 (1)	871 (1)	259 (1)	3.7 (1)	0.068 (1)	0.126 (1)	2.65 (1)	0.30 (1)	21.0 (1)
14	6.7 (1)	155 (1)	379 (1)	13.7 (1)	0.264 (1)	0.204 (1)	2.76 (1)	0.42 (1)	4.0 (1)
15	6.7 (1)	175 (1)	456 (1)	13.2 (1)	0.455 (1)	0.128 (1)	1.99 (1)	0.60 (1)	5.0 (1)
16	8.5 ± 0.44 (5)	653 ± 166 (5)	395 ± 50 (4)	14.0 ± 5.3 (5)	0.050 ± 0.049 (5)	0.055 ± 0.042 (5)	3.33 ± 1.86 (5)	0.11 ± 0.08 (5)	10 ± 2.8 (4)
17	7.6 ± 0.6 (5)	686 ± 124 (5)	450 ± 31 (4)	13.6 ± 3.1 (5)	0.027 ± 0.030 (5)	0.097 ± 0.065 (5)	4.28 ± 2.66 (5)	0.15 ± 0.10 (5)	11.5 ± 3.3 (4)
18	7.5 ± 0.7 (5)	676 ± 248 (5)	388 ± 72 (4)	14.3 ± 8.4 (5)	0.407 ± 0.356 (5)	1.219 ± 1.617 (5)	14.11 ± 10.82 (5)	0.89 ± 0.61 (5)	11.8 ± 7.7 (4)
20	7.0 ± 0.5 (4)	740 ± 124 (4)	396 ± 14 (3)	6.8 ± 3.9 (4)	0.235 ± 0.331 (4)	0.830 ± 1.416 (4)	5.48 ± 3.15 (4)	0.40 ± 0.51 (4)	19.0 ± 5.2 (3)
24	7.1 ± 0.8 (2)	296 ± 162 (2)	480 (1)	7.1 ± 4.4 (2)	0.155 ± 0.187 (2)	0.027 ± 0.189 (2)	2.63 ± 0.86 (2)	0.26 ± 0.25 (2)	7.0 ± 3.9 (2)

2.3. Vegetation Data, Kettle Hole and Water Area

UAS are increasingly used in ecological studies due to their numerous advantages, such as cost-efficiency, flexibility and accessibility over satellite data, and extension of ground surveys. Workflows for automated vegetation classification were successfully tested and the increased time-efficiency of digital plant mapping [38,39]. Chabot and Bird [40] previously recommended the use of UAS for complex wetland vegetation. However, with increased vegetation complexity, the classification accuracy has been shown to decrease [41]. Since kettle holes can feature multiple spatial and temporal plant species transitions and especially due to the different UAS image acquisition approaches, we manually mapped the dominant plant communities in this study. Based on the UAS images of all three years (2016, 2018, 2020) and different sources of ground surveys, i.e., presence–absence data in 2016, abundance data in 2017, sketches of the dominant vegetation in 2018 and 2020, we trained our image interpretation skills and finally created 42 vegetation maps (14 kettle holes $\times$ 3 years).

Homogenous plant stands were visually separated through color, texture and/or shape from surrounding vegetation patches, water and soil [41]. The naming of homogenous plant stands during the digitization procedure was done according to the dominant plant species/taxon with more than 75% visually estimated coverage. In case of one dominant species with undefinable or low coverage of other plant species, we mentioned the dominant species/taxon and added "mix", for example, "*Phalaris*-mix". If two or three species with lower coverage than 75% were recognized as mixed stands, we created two or three taxon classes, and finally, divided them into the respective dominant plant community. Finally, we created 14 dominant plant communities that were considered in the analysis of this study (Table 3).

We defined the kettle hole boundary as the sharp edge between the natural vegetation of the kettle holes and adjacent managed fields. Accordingly, we mapped the total area of the kettle holes for each year, as differing meteorological conditions can change the management intensity around kettle holes. Total water area was determined by mapping the shore line visible in the UAS derived orthomosaics. In case of dense vegetation, we estimated the shore line based on vegetation characteristics and expert knowledge.

The prepared data for dominant plant communities, water area and hydromorphological information are open accessible under DOI: 10.4228/ZALF.DK.165.

2.4. Meteorological Data

We used meteorological data collected every 10 min between August 2015 and July 2020 by two replacing weather stations at the ZALF research station Dedelow (Germany) located in the east of the AgroScapeLab Quillow [42–47]. Prior to further calculation, precipitation was corrected—as per Richter [48]—on a daily basis. Potential evapotranspiration for the entire observation period was calculated as per Wendling et al. [49], using daily sums of global radiation and averaged daily air temperatures. From August 2015 until April 2016, global radiation was complemented from further local and regional stations in order to also calculate potential evapotranspiration for the operation period of the first weather station with a non-functioning global radiation sensor. Based on the daily corrected precipitation and evapotranspiration values, we calculated monthly and annual precipitation and evapotranspiration heights, and climatic water balances for the observation period.

Table 3. Assignment of the digitally mapped dominant species/taxa into 14 final dominant plant communities.

	Abbreviation	Dominant Plant Community	Dominant Species/Taxa Included
1	Amp_veg	Amphibious vegetation	*Alisma plantago-aquatica* L., *Oenanthe aquatica* (L.) Poiret, *Rorippa amphibia* (L.) Besser, *Rorippa palustris* (L.) Besser, *Rumex maritimus* L., amphibious seedlings (not identifiable)
2	Car_mix	Carex mix	*Carex riparia* Curtis, *Carex* spec.
3	Edge_mix	Edge mix	Poaceae (not further identified), but including *Elymus repens* (L.) Gould s. str., *Arrhenatherum elatius* (L.) P. Beauv. ex J. Presl & C. Presl, *Bromus sterilis* L., non-dominant stands of *Phalaris arundinacea* L., non-dominant stands of *Nitrophilous perennials* (see group 5), unidentified segetal species
4	Hyd	Hydrophytes	*Lemna minor* L., *Spirodela polyrhiza* (L.) Schleid., *Potamogeton natans* L., *Batrachium* spec. (characterized by floating leaves); *Ceratophyllum submersum* L. (potential area for submersed hydrophytes, characterized as open water area)
5	Nit_per	Nitrophilous perennials	*Arctium lappa* L., *Cirsium arvense* (L.) Scop., *Crisium vulgaris* (Savi) Ten., *Galeopsis speciosa* Mill., *Urtica dioica* L. s. l., *Rumex obtusifolius* L., *Rumex* spec. (not *R. maritimus*)
6	Flood	Flood-meadow	*Alopecurus aequalis* Sobol., *Glyceria fluitans* (L.) R. Br.
7	Others	Others	*Anthriscus sylvestris* (L.) Hoffm., *Iris pseudacorus* L., *Juncus effusus* L., *Juncus* spec.
8	Pha_mix	Phalaris mix	*Phalaris arundinacea* L.
9	Poa_mix	Poa mix	*Poa palustris* L., *Poa trivialis* L.
10	Phr_mix	Phragmites mix	*Phragmites australis* (Cav.) Trin. ex Steud.
11	Woody	Woody plants	*Humulus lupulus* L., *Fraxinus excelsior* L., *Rosa* spec., *Salix alba* L., *Salix cinerea* L., *Salix* spec., *Sambucus nigra* L., *Sorbus aucuparia* L., unidentified woody species
12	Spa_mix	Sparganium mix	*Sparganium emersum* Rehmann
13	Typ_mix	Typha latifolia	*Typha latifolia* L.
14	Unveg	Unvegetated	No vegetation

2.5. Statistical Analysis

We used response ratios to quantify the effect of the three observed periods (2016–2018, 2018–2020, 2016–2020) with different hydrological conditions on the dominant vegetation. The response ratio L as an effect size is expressed as the natural logarithm of the response ratio R [50]:

$$R = XE/XC, \ln(R) = L = \ln(XE) - \ln(XC), \quad (1)$$

In our case XE (the experimental group) was described by the second year of each considered interval, whereas XC (the control group) referred to the first year of each interval. We calculated the response ratio L individually for each kettle hole pair and separately for each dominant plant community, if the community occurred at least once in the three years. Subsequently, response ratios L were averaged and the 95% confidence intervals (CI) were calculated per period and dominant plant community, where non-overlapping CIs were interpreted as significantly different between the periods.

To determine whether changes between different groups of vegetation were related to each other, we calculated Spearman rank correlations between the proportional coverage of each dominant plant community using the R package psych v. 2.0.12 [51]. To quantify the change in vegetation composition and the speed of change between the observation periods, we applied a formula typically used in relation to species change:

$$T = \frac{groups\ gained + groups\ lost}{total\ number\ of\ groups\ observed \in both\ years} \quad (2)$$

Proportional appearance (gain rates; G) and disappearance (loss rates; L), as well as change of dominant groups (turnover rates; T) were calculated relative to the total number of dominance groups counted across the respective sampling years. Calculations were based on presence–absence data for the full set of sites, using the R codyn package v. 2.0.5 [52] and following Anderson [53].

Spearman rank correlation was used to determine relations between turnover rates and total area, water area, percentage change in water area and HGM types.

A two-way permutational multivariate analysis of variance (PERMANOVA) was conducted with PERMANOVA+ for Primer v6 [54] to estimate compositional variability of dominant plant communities between the factors "kettle hole" and "year". The same analysis was performed for three data-subsets. First, excluding shrubs, edge vegetation and nitrophilous perennials to be better comparable with Pätzig et al. [3], second, the reduced vegetation dataset only for permanent and semi-permanent kettle holes, and third, the reduced vegetation dataset only for episodic and periodic kettles holes. The underlying resemblance matrices were based on Bray–Curtis similarities. We did not transform the data prior to statistical analysis to down-weight dominant taxa, because the focus of this study was on the dominant plant communities. Permutations of residuals were calculated under a reduced model with 9999 permutations and type III of sum of squares was chosen [55]. In addition, for the entire dataset we tested the differences in multivariate dispersion using the primer routine PERMDISP [54] to allow for a clear interpretation of the results, with regard to differences in group means (location) and variance (multivariate dispersions) among groups. Spearman rank correlation was calculated between the obtained multivariate dispersion indices and total amounts of percentage changes in water area over three years of data collection. Finally, the composition of dominant plant communities was ordinated using non-metric multidimensional scaling (nMDS) and contour lines were overlaid at the 40% similarity level of a dendrogram created by a previously performed cluster analysis in Primer v6 [54]. Together with the depicted dominant plant communities, the contour lines aided us in placing individual vegetation characteristics of the kettle holes over the observation period in relation to each other.

A distance-based linear model (DISTLM) was used to examine the strength of relationships between the dominant plant communities and the environmental variable water area, total kettle hole area, as well as the hydromorphological type and shore slope class. The latter two variables were included as dummy variables. Total kettle hole area and water area were log-transformed prior to the analysis due to a positively skewed distribution. Although water area was not normally distributed after log-transformation, mainly caused by the multiple zero values, we still included this variable in the analysis, due to the high flexibility of the method. The DISTLM applied was based on a stepwise selection procedure and the "An Information Criterion" (AIC) [56]. Inter-correlation was reviewed in PRIMER v6 with a correlation matrix. The usual cut-off of inter-correlated variables is 0.95 [55]. None of the variables exceeded this limit. The scores of the corresponding distance-based redundancy analysis (dbRDA) were used to test for differences between years by running an analysis of variance (ANOVA) and Tukey's HSD test using the package "psych" in R [51].

3. Results

3.1. Meteorological and Hydrological Conditions

Average annual air temperature was lowest in the first two observation intervals (9.7 and 9.4 °C) and about one degree higher in the last two of five observation intervals (10.7 and 10.6 °C) (Figure 1). Total annual precipitation was highest in the interval 2016–2017 (698 mm) and lowest in the interval 2018–2019 (399 mm). Total annual potential evaporation was higher in the last three intervals between August 2017 to July 2020 (795 to 804 mm) compared to the first two intervals of the observation period (733 and 703 mm). Consequently, annual climatic water balances (CWB) were generally negative, but nearly balanced in the extreme wet interval 2016–2017 (−4.7 mm) (Figure 1). Despite high precipitation in summer 2017 (May–June: 338 mm), further high rain events in autumn 2017 and winter 2018 led to the second highest precipitation value with 537 mm in the interval 2017–2018, although the precipitation amount in summer 2018 was the lowest (May–June: 85 mm). The most negative value of the CWB was reached in 2018–2019 (−406 mm) (Figure 1). In summary, the mean annual climatic water balance during the study period (August 2015–July 2020) was −260 mm, much more negative compared to the long-term mean (1992–2019: −84 mm).

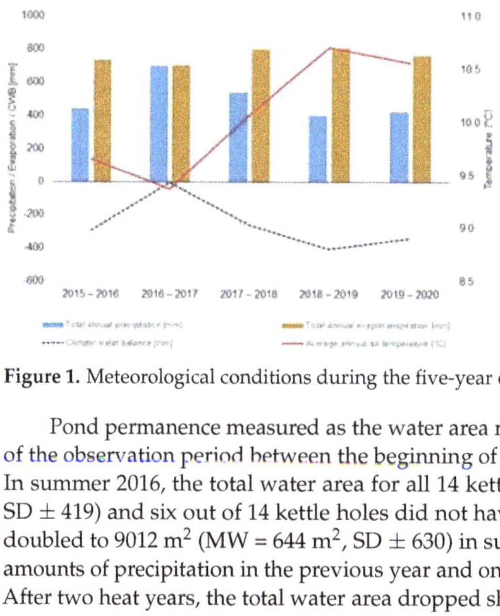

Figure 1. Meteorological conditions during the five-year observation period.

Pond permanence measured as the water area reflected the meteorological conditions of the observation period between the beginning of August 2015 and the end of July 2020. In summer 2016, the total water area for all 14 kettle holes was 4984 m^2 (MW = 356 m^2, SD ± 419) and six out of 14 kettle holes did not have water. The total water area almost doubled to 9012 m^2 (MW = 644 m^2, SD ± 630) in summer 2018 as a consequence of high amounts of precipitation in the previous year and only four kettle holes had no water at all. After two heat years, the total water area dropped sharply to 64 m^2 (MW = 5 m^2, SD ± 17) in summer 2020, which was measured at one out of the 14 kettle holes.

3.2. Effect of Dry and Wet Periods on Dominant Plant Communities

From the 14 identified dominant plant communities, *Phalaris* and *Carex*-mix were most abundant with a total share of 17.5 and 17.3%, respectively, over all kettle holes and years. These were followed by woody plants (14.7%), *Phragmites*-mix (10.4%) and nitrophilous perennials (12.3%). These groups corresponded to the classified dominant vegetation types of the kettle holes (see Table 1). The area of potential occurrence of hydrophytes accounted for 9.7%, while important indicators of water level fluctuations, such as amphibious vegetation, flood-meadow or *Typha latifolia* only shared 1, 1.3 or 2.1%, respectively, of the total area. The unvegetated area covered 3.8% of the digitized kettle hole area. Edge mix (8.1%), *Sparganium*-mix, *Poa*-mix and "Others" accounted for the remaining investigated area.

All three observed intervals with contrasting hydrological conditions differently affected the dominant plant communities. The response ratios showed the largest variation in cover changes for potential hydrophytes with a strong negative effect over the entire period (2016–2020) (Figure 2). In 2016 potential hydrophytes covered 8.8% of the investigated kettle hole area, while this decreased to 0.1% in 2020, after being 20.1% in 2018. Amphibious vegetation also showed a strong negative effect over the entire period, with 1.8% less coverage in 2020 compared to 2016. Potential hydrophytes increased from 2016 to 2018, but amphibious plants and all other helophytes showed negative or no response (Figure 2). The opposite was true between 2018 and 2020. We revealed positive effects, and thus, an increase in vegetation cover over the entire period from 2016 to 2020 for terrestrial plants (nitrophilous perennials and woody plants, 4.2% more coverage), as well as unvegetated areas. Moreover, flood-meadow vegetation increased over the entire observation period due to a strong increase in the water sparse period from 2018 to 2020 (Figure 2). Flood-meadow covered 1.4% of the area in 2016, was not observed in 2018, but covered 2.5% in 2020.

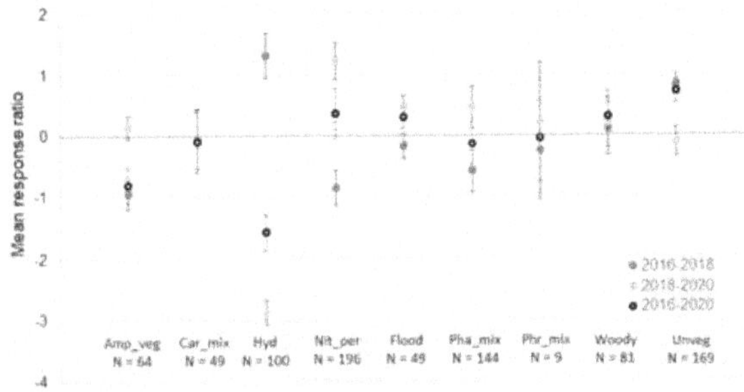

Figure 2. Effect sizes of the three different periods on the dominant plant communities. Effect sizes were calculated as the natural logarithm of the response ratio R and presented with 95% confidence intervals. A positive response ratio above the zero line indicates an increase in the compared abundances and a negative response ratio below the zero line indicates a decrease. The periods were considered significantly different when 95% CIs did not overlap.

The opposing effects between some dominant plant communities shown by the response ratios are supported by the Spearman's correlation coefficient between dominant plant communities, where negative relations were found between potential hydrophytes and nitrophilous perennials ($r_s = -0.37$, $p = 0.02$), amphibious vegetation and woody plants ($r_s = -0.31$, $p = 0.046$), as well as mixed *Phalaris* stands and edge vegetation ($r_s = -0.33$, $p = 0.03$) (Figure 3). Among helophytes, adverse correlations were found between mixed *Phragmites* and *Carex*, as well as *Phalaris* stands ($r_s = -0.48$, $p = 0.001$ and $r_s = -0.42$, $p = 0.005$). On the other hand, terrestrial plants (woody plants in relation to nitrophilous perennials and mixed *Poa* stands) showed positive correlations ($r_s = 0.42$, $p = 0.01$ and $r_s = 0.39$, $p = 0.01$). The same was found between potential hydrophytes and mixed *Sparganium* stands ($r_s = 0.43$, $p = 0.004$) (Figure 3).

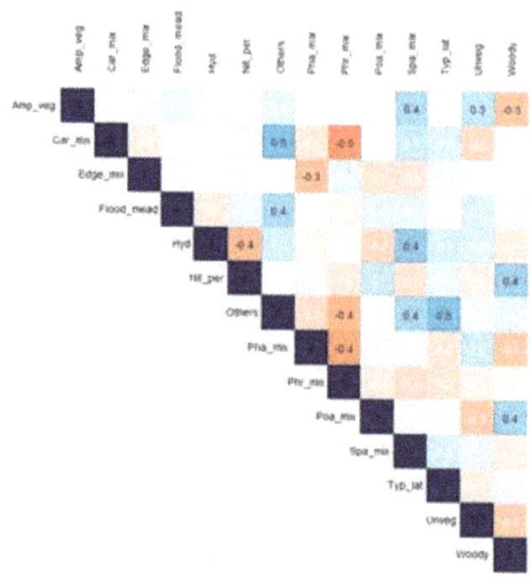

Figure 3. Spearman's Correlation Coefficient between coverages of dominant plant communities of 14 kettle holes over five years; significance level = 0.05; significant values = black, insignificant values = white; color scale represents strength (between 0 and ±1) and direction (+, −) of Spearman rs.

When calculating turnover rates for the dominant plant communities, we found a 34% change in the dominant plant communities; specifically, a 16% gain and 18% loss of groups averaged across years and kettle holes. While the biannual temporal turnover and gain rate in dominant groups was higher between 2018 and 2020, the loss rate across sites was higher between 2016 and 2018 (Table S1). Percentage differences in water area between 2016 and 2018 were positively correlated with concurrent gain rates and turnover rates from both intervals (G: rs = 0.68, p = 0.01; T16-18: rs = 0.73, p = 0.003; T18-20: rs = 0.79, p = 0.001). Water area differences between 2018 and 2020 correlated positively only with concurrent turnover rates (rs = 0.66, p = 0.01). Further positive correlations were found between turnover and gain rates averaged across both intervals and summed changes in water area over five years (T: rs = 0.69, p = 0.01; G: rs = 0.70, p =0.01). Correlations between turnover rates and HGM types were not found.

3.3. *Interannual Variability in Dominant Plant Community Composition*

We obtained significant differences between the 14 kettle holes (Df = 13, $p < 0.05$) and the three sampled years (Df = 2, $p < 0.05$) for the composition of the dominant plant communities running the PERMANOVA routine. As we had expected the variance explained by differences between "kettle holes" (45.7% variance explained) exceeded the variance explained by differences between "years" (11.8% variance explained). However, the percentage gap was clearly reduced compared to Pätzig et al. (2020), where 60% of the variation in macrophyte composition was explained by differences between "kettle holes" and only 7% by interannual differences. When we excluded woody plants, edge vegetation and nitrophilous perennials from the resemblance matrix the PERMANOVA routine yielded 15.4% explained variability for the factor "year" and 50.4% for "kettle holes". The explained variance for the factor "year" increased to 27.5% when we only considered semi-permanent and permanent kettle holes (N = 6), while it decreased to 8.6% for episodic and periodic kettle holes (N = 8) based on the resemblance matrix without woody plants, edge vegetation and nitrophilous perennials.

A test for multivariate homogeneity of dispersion for the entire data set revealed nearly significant differences in the dispersion of macrophyte composition among the analyzed kettle holes (PERMDISP pseudo-F = 3.9, p = 0.05), but differences in location were much stronger (PERMANOVA pseudo-F = 16.1). For the factor "year" the dispersion of dominant plant community composition was not significant (PERMDISP pseudo-F = 0.03), which underlines the significant differences in location (PERMDISP pseudo-F = 5.7). Likewise, the multivariate dispersion index showed similar values among the years, the lowest value of 0.97 for the wet year 2018 and slightly higher values of 1.01 for 2016 and 2020, respectively. Kettle holes showed a comparably larger range in the multivariate dispersion index, with a minimum index of 0.26 for kettle hole 20 (periodic, partly regulated, big and shallow shore overflow type (BS-SO), full reed type with *Phragmites australis*), and a maximum index of 1.66 for kettle hole 16 (semi-permanent, small and shallow shore overflow type (SS-SO), mainly edge type with *Phalaris arundinacea*) (Table S1). The multivariate dispersion indices of the kettle holes were strongly correlated with the total amounts of percentage changes in water area over the five-year observation period ($r_s = 0.69$, $p = 0.006$).

The corresponding nMDS plot showed groupings of the individual kettle holes, but did not always show a clear separation among them (Figure 4). The overlaid cluster analysis with 40% similarity also emphasizes the gradual changes in the kettle hole vegetation and concomitant difficulties to group them into vegetation types. Thus, the identified groups were not considered in further analysis. Nonetheless, the empirically known dominant vegetation types, i.e., *Phalaris*, *Carex*, and *Phragmites*-mix as well as woody plants, were depicted in four opposing directions of the two-dimensional realization of the multivariate vegetation data (Figure 4).

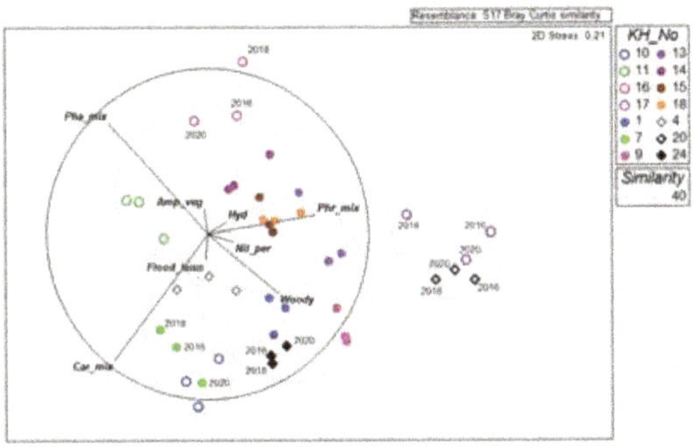

Figure 4. nMDS ordination representing differences in dominant plant communities among kettle holes and the three studied years (labeled for some examples) overlaid with dominant plant communities and groups of cluster analysis with 40% similarity.

3.4. *Relation between Dominant Plant Communities, Interannual Scale and Environmental Variables*

The marginal test of the DISTLM analysis showed that all variables apart from the shore slope class "very oblique" had significant relationships to the dominant plant communities (Table 4). Seven out of nine variables were chosen as the best set of variables explaining the variability in dominant plant community composition, i.e., shore slope classes "oblique" and "very oblique", water area, total kettle hole area and the two shore overflow types (AIC = 317, $R^2 = 0.45$). The corresponding dbRDA revealed a strong positive relation between shore slope classes "oblique" and the dbRDA axis 1 followed by the shore overflow types and the total area of the kettle holes (Table 4, Figure 5). Water area was highly correlated with dbRDA axis 2 (Table 4, Figure 5).

Table 4. Marginal test for environmental variables and relationships between dbRDA coordinate axes and best set of environmental variables identified by a distance-based linear model (DISTLM) stepwise procedure. Area_tot = total kettle hole area, Water_Area = water surface, BS-S = big and shallow storage type, BS-SO = big and shallow shore overflow typ, SS-S = small and shallow storage type, SS-SO = small and shallow shore overflow type, o = oblique, vo = very oblique, s = steep. Pseudo-F = Pseudo-F statistic, P = probability value, Prob. = probability in percent/explained variability. Significant values = bold.

Marginal Test				Relationship		
Variable	**Pseudo-F**	**P**	**Prop.**	**dbRDA1**	**dbRDA2**	**dbRDA3**
Area_tot	3.401	**0.004**	7.8	0.409	0.046	**0.726**
Water_Area	4.962	**0.000**	11.0	−0.306	**0.793**	0.358
BS-S	2.397	**0.027**	5.7			
BS-SO	2.295	**0.039**	5.4	−0.393	0.051	−0.145
SS-S	4.077	**0.002**	9.3			
SS-SO	3.451	**0.006**	7.9	−0.414	0.255	−0.192
o	6.345	**0.000**	13.7	**−0.641**	**−0.538**	**0.515**
vo	2.842	0.222	6.6	−0.043	−0.111	−0.146
s	1.413	**0.017**	3.4			

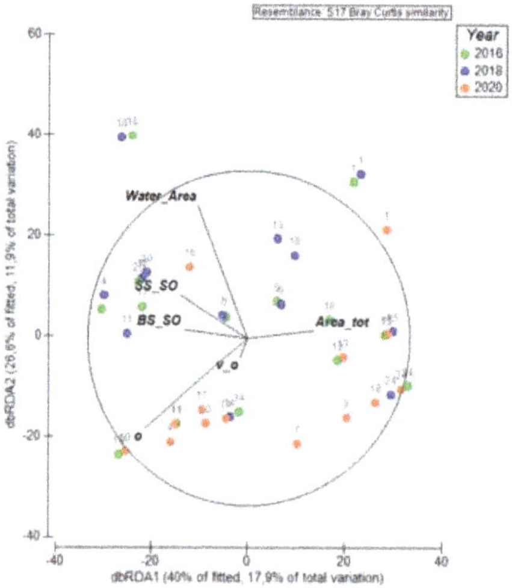

Figure 5. Distance-based redundancy analysis (dbRDA) ordination representing differences in dominant plant communities among kettle holes over the three years. Abbreviation of environmental variables see Table 4.

For the log-transformed scores of the second axis (dbRDA2) that were strongly related to water area, a global analysis of variance (ANOVA) revealed significant differences between years (p = 0.045). TukeyHSD post hoc test confirmed significant differences between 2018 and 2020 (p = 0.016), but not between the other two pairs of intervals.

The four dominating plant communities (*Phalaris*, *Carex*, *Phragmites*-mix and woody plants) showed rather indifferent or stronger relations to the morphological variables shore slope and hydromorphological type (Figure 6). In contrast, the hydrophytes and nitrophilous perennials were strongly positive or negative correlated with the second dbRDA axis, and thus, with water area in the different years according to the previous analysis in combination with the bubble plots (Figure 6).

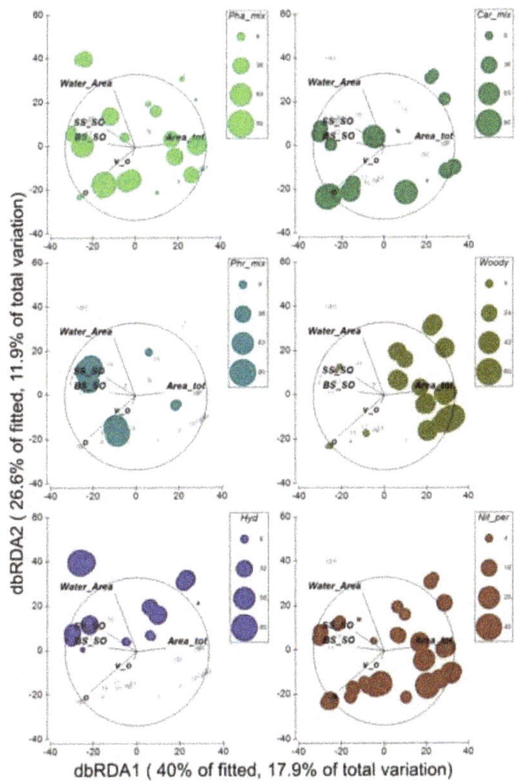

Figure 6. Bubble plots for each of the four dominant groups *Phalaris* (Pha_mix), *Carex* (Car_mix), *Phragmites* (Phr_mix)-mix and woody plants (woody) and the two most contrasting groups hydrophytes (Hyd) and nitrophilous perennials (Nit_per) with regard to water area on the basis of the dbRDA ordination (see Figure 5). For abbreviation of environmental variables see Table 4.

4. Discussion

4.1. Different Groups of Dominant Plant Communities Are Differently Affected by Dry and Wet Periods

Here, we provide evidence that the variability in pond permanence in the studied kettle holes can be attributed to extreme changes in climatic water balance (CWB) over five investigated years. This was particularly apparent in 2020 and the preceding two years, when increased mean evapotranspiration caused by increased air temperatures, together with decreased precipitation levels, corresponded with nearly absent water availability. Lehsten et al. [34] found a delay of one to two years between increased evaporation rates and concomitant changes in the water levels of kettle holes. Such a time lag became evident in our study in 2018, where high precipitation and low evapotranspiration levels in the 2016–2017 interval seems to explain the high number of kettle holes with pond permanence in summer 2018, despite a clear negative CWB in this time interval (2017–2018). For our study, kettle hole hydrology can, overall, be seen as a product of the meteorological conditions [57], and a causal link seems to be plausible to explain the high proportions of pond permanence in 2018 and the low in 2020.

A high variability in water area between the three years of vegetation mapping (2016, 2018, and 2020) caused a shift in the occurrence and dominance of plant communities, reflected by distinct variability between the response ratios. In particular, the potential area for hydrophytes and nitrophilous perennials showed significant large effects and contrasting correlations, but amphibious and flood-meadow vegetation also showed a large level of variability. Although the total coverage of the latter was not high, they are important indicators of water level variations, together with the presence or absence of hydrophytes. Van der Valk [7,9] and Seabloom et al. [23] among others, have drawn attention to the ability of amphibious species, such as *Rorippa amphibia* and mudflat annuals, to react instantly to oscillating water levels by recruiting seeds from underlying seed banks and subsequent reestablishment on the open ground surface, following pronounced decreases in water levels. Earlier research from kettle holes proved the existence of positive relationships between amphibious plants and the duration of annual inundation in kettle holes [35]. The separation of flood-meadow species, such as *Alopecurus aequalis* and *Glyceria fluitians*, emphasized a different reaction to decreases in water levels and resettlement to herbaceous amphiphytes. Flood-meadow vegetation mainly occurred in kettle holes with generally low or no pond permanence, where the advanced stage of succession was reset in 2018, and they then recovered in 2020, before further succession of helophytes or terrestrial plant species will probably occur again. Therefore, they were the only moisture indicating group with an increase over the five-year period.

Except for the flood-meadow vegetation, an overall positive effect and high correlation with decreased pond permanence was observed only for nitrophilous perennials and woody plants. The increase in the cover of woody plants was not significantly different but was steady over the years, while nitrophilous perennials were strongly reduced in 2018 but reproduced rapidly under the general conditions of high nutrient availability [58,59] and the absence of waterlogged soils in 2020. Hence, although we observed a reset of advanced succession stages in 2018, the general negative climatic water balance of the last two decades and the extreme drought conditions in the years after 2017 fostered further succession and terrestrialization of the studied kettle holes. Changes in plant composition along moisture gradients leading—in dry phases—towards stages without moisture indicators, but dominated by terrestrial (common distribution outside the ecosystem) species, were described in several studies of small wetlands [17,60,61].

Besides woody plants, the other three groups with dominant taxa *Carex* spp., *Phragmites australis* and *Phalaris arundinacea* showed no significant effects over the entire study period, suggesting different responses to changes in pond permanence. Obviously, these dominant groups occur in many kettle holes with different abiotic characteristics, which makes it difficult to capture the reaction to changes in pond permanence for individual kettle holes. This becomes apparent in the bubble plots (Figure 6), where they are arranged along the morphological gradient—especially the shore slope and the shore overflow types—and do not group with the years like the hydrophytes and nitrophilous perennials do. The importance of morphology for helophytes, such as a wide and gentle shore, has been established in other studies [35,36,62]. In accordance with these results, we interpret the negative correlation coefficient between *Phragmites* and *Carex* as well as *Phalaris* stands as non-causal in our study, because these types often do not occur simultaneously at one kettle hole (Figure 4). Yet, the negative correlation between proportions of *Phragmites* and *Phalaris* as well as *Carex* stands underlines the competitive ability of *Phalaris* stands on sites experiencing intensive intermittent fluctuations in water levels, as shown in other studies [63], while sedges prefer drier locations compared to *Phragmites* stands [58]. Finally, variations in hydrology, geomorphology and subsequent on-site microclimate form the basis for existing high heterogeneity among kettle holes in the study region (ASQL, northeast Germany), mirrored by observed high variances between different ponds in the nMDS. The results also emphasize the gradual change of vegetation among kettle holes, which stresses the need to move away from kettle hole types and implement gradual concepts such as the "wetland continuum concept" [24] (see below).

Rising turnover rates over time, that mainly resulted from significantly elevated gain rates in 2018–2020, suggest early stages of succession, as gain rates usually decline with progressing succession [53]. Higher loss rates, together with significantly lower gain rates across sites in the first interval, indicate limited space for additionally appearing vegetation and a loss of groups due to elevated flooding in the wet interval following 2016, respectively. On the other hand, dry conditions in 2018–2020 accelerated the spread of the dominant groups due to the provision of additional space in desiccated kettle holes. Gain in species groups (primarily amphibious plants and hydrophytes) was, therefore, reduced in the wetter interval of 2016–2018 compared to the appearance of dominant plant communities (e.g., graminoids) in the dryer interval of 2018–2020. This is in line with the findings of Mulhouse et al. [64], who reported the establishment and expansion of additional species (helophytes, wetland grasses and terrestrial plants), as a result of exposed substrates and a more open vegetation structure in herbaceous ponds following drought and recruitment from soil seed banks or via dispersal. Repulsing effects of high water levels on formerly established species due to deluge were likewise reported in the case of potholes by van der Valk [7]. Positive correlations between gain rates and water area differences suggest promoting effects of water level changes on the appearance of dominant plant communities, caused by increased availability of varying habitats for vegetation with different preferences [65,66]. The highest turnover rates were associated with kettle holes exhibiting shore overflow characteristics (No. 10, 11, 17; SS-SO-types/oblique shore-slopes). Plant communities, here, seem to be affected more strongly from gradual and more frequent flooding compared to more stable storage types with less variation in the area and volume of water. This result is indirectly supported by the significant relation between variability in vegetation composition and hydromorphological type, shore slope and water area found in the DISTLM analysis and additionally by corresponding dbRDA-result with its strong positive relation between shore slope classes "oblique" and axis 1 (Table 4, Figure 5) and between the water area and axis 2. Van Leeuwen et al. [67] revealed that drawdown water regimes in experimental ponds increased recruitment and the species richness of plant communities and the number of seedlings on gradual sloping shores, but not on steep shores. In the dbRDA bubble plots (Figure 6), hydrophytes and nitrophilous perennials were positively or negatively correlated with years. The same groups accounted for the temporal changes in the kettle holes with highest turnover rates (No. 10, 11, 17). These groups, thus, seem to mirror the differences in hydroperiodicity over time, through changes in abundances and occurrence most clearly.

4.2. Interannual Variability of Kettle Hole Vegetation Increases during Extreme Weather Conditions

The variability in the composition of plant communities explained by the factor "year" was larger with 11.8% compared to the 7% reported for the macrophyte composition of the inundated kettle hole area in the study by Pätzig et al. [3], because the observed five-year period (2016–2020) with contrasting extreme weather conditions in this study showed larger hydrological variability compared to the rather wet period with lower hydrological variability during the 1990s. Although the datasets are not directly comparable, because this study contains dominant plant communities instead of species, the entire kettle hole area (and not just the flooded area), and kettle holes including episodic and periodic, compared to the mainly permanent and semi-permanent kettle holes examined by Pätzig et al. [3], we are still able to draw important conclusions about the role of interannual hydrological variability on the vegetation dynamics of kettle holes.

The increased explanatory power for the factor "year" that was caused by more systematic differences in hydrological conditions between the years, confirms the expectation of Pätzig et al. [3] that the importance of multiple-year observations will increase in periods of higher hydrological variability. When we adapted our dataset to the one analyzed by Pätzig et al. [3], by excluding the woody plants, edge vegetation and nitrophilous perennials that were not included in the former study, the variance explained by "year" increased to 15.4%, and even to 27.5% when we included only permanent and semi-permanent kettle holes. For the permanent and semi-permanent kettle holes, even this 27.5% is a rather conservative estimate, because we merely considered dominant plant communities and not species in our analysis. In contrast, the explained variability by "year" for the dominant plant communities of episodic kettle holes was only 8.6%, which again, is a result of lower hydrological variability, but this time as consequence of no or less water availability and not because of high water levels with low fluctuations as in the study by Pätzig et al. [3].

From this, we can conclude that water level fluctuations and accompanied vegetation changes are stronger in episodic to periodic kettle holes during wet periods with rather balanced precipitation, while this sets in for permanent and semi-permanent kettle holes during dry periods. Thus, kettle holes are highly variable not only in space but also at an interannual scale, but the types of kettle holes respond differently in different hydrological periods, which has important consequences for the assessment of those small wetland ecosystems. This requires long-term monitoring for all types of kettle holes to sufficiently monitor their ecological state, their functioning, and their ecosystem services.

4.3. Future Perspectives for Kettle Hole Occurrence in Central European Pleistocene Landscapes

The continuous increase in woody plants and even greater increase in nitrophilous perennials within our five-year observation period, despite an extreme wet year in between, suggest that we are facing a serious bush encroachment of kettle holes, which fosters irreversible vegetation stages under persistent meteorological conditions and predicted climate scenarios. For the Pleistocene landscape of Germany, the climate scenarios involve a decrease in precipitation—although winter rain will increase—and rising temperature [68]. As a consequence, the inundation timing of temporary wetlands will change and the resulting effects on wetland communities may be additionally amplified by higher temperatures, generally leading to increased productivity [69]. Bourgeois et al. [70] found a substantial impact of trees on shifts between low-growth species structuring plant communities within intermittently inundated floodplains. Whenever canopies exceeded thresholds, abrupt species turnover or shifts between rank orders led towards taxa with high shading tolerance. Increased bush encroachment was already observed in the second half of the 20th century in central European kettle holes [34] and is an important topic in England, where science-based restoration efforts take place at overgrown sites [26].

Since future climate scenarios predict less precipitation but no increase in heavy precipitation events for the Pleistocene landscapes of Germany [68], even in extreme wet periods flooding disturbances might be too short to completely reset succession stages. Already, van der Valk [7] stated that a change in vegetation stage may only be permanent after a certain threshold of hydrological conditions have been reached. One year of extreme wet conditions was not sufficient to reset succession completely in our study, especially when woody plants were already established. The succession process towards wood types seems to be accelerated due to the competitive advantage of terrestrial plants under drier conditions and the absence of wetland management [71]. This increases sedimentation, evapotranspiration and terrestrialization, and leads to higher numbers of wood types in the landscape. Continuous high nutrient input from surrounding agricultural fields into kettle holes, especially during runoff events in wet phases, will further accelerate this development [7,72]. There is evidence that increased nutrient load addition in aquatic phases can have negative effects on the diversity, coverage and composition of terrestrial plants in dry phases, which in turn affects the following aquatic phase [73]. Although we cannot quantitatively prove that wetland classes will be lost, the available data from kettle holes and experiences from England [26] suggest that the proportion of wood dominated kettle holes will increase in the future. This corresponds to increased homogenization of the landscape and will be accompanied by a loss of ecosystem functions and services of central European kettles holes and similar small wetlands.

The realization that kettle holes and comparable small wetland types develop towards irreversible wood dominated climax stages, shows that the "wetland continuum concept" developed on the basis of pothole research [24,25] is not fully transferable to all kinds of small wetland ecosystems. As Mushet et al. [15] pointed out, the "wetland continuum concept" incorporates the prairie–marsh cycle [9] at the vertical rainwater axis, with lake marsh during deluge and dry-marsh during drought, as two end-points that could be considered as stable systems and with the regenerating-marsh and degenerating-marsh as alternative stable or transitory states [15]. Alternatively, Mushet et al. [15] postulate that all predictable phases described by van der Valk and Davis [9] are the "stable state" of the highly dynamic potholes and need to be viewed over longer time scales. This idea becomes particularly clear when highlighting seedbanks in the concept, which are a component of the diversity described as "dark diversity" or "sleeping" diversity within one stable state (see 15 and references therein). Instead of the rainwater axis, shifts along the hydrological axis of the "wetland continuum concept", which describes the influence of groundwater on wetland hydrology, are considered to be alternative stable states [15]. Within these alternative stable states, all phases of the prairie–marsh cycle occur and are assumed to be reversible. However, for kettle holes and other small wetlands, the dry-marsh phase is not the end-point during drought conditions, but a more terrestrial phase dominated by shrubs and trees. A similar scenario was modeled for potholes that could shift to permanent states with reduced productivity and loss of ecosystem functions under climate change [14,29].

We suggest that terrestrial phases of kettle holes could also be regarded as alternative stable states, because if we assume irreversible succession the "dark diversity" will not reappear, resulting in irreversible changes to community composition. Hence, we agree with Mushet et al. [15] that alternative stable states exist for highly dynamic systems, not only for changes in the horizontal hydrological axis, but also in the vertical rainwater axis. We hypothesize that many central European kettle holes, especially from the storage type, will shift towards more advanced succession stages after "appropriate ecological time scales" of drought conditions with less extreme wet years. These drought conditions act as external drivers and push the kettle holes beyond a tipping-point of irreversible alternative stable states at the drought end-point of the rainwater axis of the "wetland continuum concept".

Moreover, we propose that compared to the potholes in North America, the proportion of wetland classes will change more dramatically at kettle holes in the future. Although the genesis of potholes and kettle holes is similar, the environmental conditions in North America, primarily the climate conditions, usually prevent any establishment of shrubs and trees at the edge of the potholes, which has important effects on interspecific competition and interactions with hydrology. Further research is needed to disentangle the commonalities and differences between North American potholes and central European kettle holes to be able to relate the existing knowledge about potholes to kettle holes.

5. Conclusions

Our study contributes to a better understanding about effects of water level variations on wet–dry cycles of vegetation in central European kettle holes. In summary, the different dominant plant communities were differently affected by dry and wet intervals, with an increase in terrestrial plants and a decrease in the potential hydrophyte area over the entire five-year observation period. The increase in terrestrial plants was more pronounced in the storage type kettle holes, characterized by steeper shores and potentially stronger changes in water levels and volume.

The hydrological history and current water area affected the occurrence of moisture indicators, resulting in an increase in flood-meadow taxa in the center of mainly episodic to periodic kettle holes after the extreme wet year of 2018. Meanwhile, herbaceous amphibious vegetation increased in this time interval, but decreased over the entire observation period due to enhanced dryness. As a consequence, temporal turnover differed between wet and dry phases, with higher gain rates in the dry intervals between 2018 and 2020. This result can be regarded as generally being in line with the existing knowledge about local habitat filtering and plant species sorting along hydrological gradients creating different wetland vegetation types. In addition, we conclude that water level fluctuations and accompanied changes in plant communities is stronger in episodic to periodic kettle holes during wet periods with rather balanced precipitation, while permanent and semi-permanent kettle holes are more greatly affected in dry periods.

The observed vegetation changes in our study can be assigned to cyclic alterations within one stable state, where seedling recruitment from the seed bank ("dark diversity") can be assumed to lead to visible changes in abundance and composition of the dominant plant communities, but this does not result in an alternative stable state. However, since we also observed a slight continuous increase in woody plants over the entire period, we interpret this together with the existing literature as evidence of progressive succession towards irreversible alternative stable states with woody plants for some kettle holes. We assume that this will cause stronger changes of wetland types in kettle holes compared to potholes. Consequently, the proportion of wood dominated kettle holes will increase in the central European landscape in the future, which corresponds to greater homogenization of the landscape and a loss of ecosystem functions and services.

In order to obtain more reliable data to capture the interannual variability and future development of kettle holes and other small wetlands under climate change and further degrading forces, long-term monitoring is needed, as already forcefully emphasized in many studies. New technologies, especially remote sensing at different spatial and temporal scales, artificial intelligence, and eDNA will be key elements in future research and monitoring of kettle holes and other small wetlands.

Supplementary Materials: The following are available online at 'https://www.mdpi.com/2073-4441/13/5/688/s1. Table S1: Turnover, gain and loss rate per interval and multivariate dispersion index (MPI) for each kettle hole.

Author Contributions: Conceptualization, M.P.; methodology, MP (lead) and E.D.; software, M.P. and E.D.; validation, M.P.; formal analysis, M.P. and E.D.; investigation, M.P. and E.D. (supporting); resources, M.P.; data curation, M.P. and E.D. (supporting); writing—original draft preparation, M.P. (lead) and E.D.; writing—review and editing, M.P. and E.D. (supporting); visualization, M.P. and E.D.; supervision, M.P.; project administration, M.P. All authors have read and agreed to the published version of the manuscript.

Funding: This research received no external funding.

Institutional Review Board Statement: Not applicable for studies not involving humans or animals.

Informed Consent Statement: Not applicable for studies not involving humans.

Data Availability Statement: The data that support the findings of this study are available from the corresponding author upon reasonable request.

Acknowledgments: This study would not have been possible without the contribution and support of several researchers, technical staff, and scientific assistants, not all of whom are explicitly mentioned. In particular, we thank Thomas Kalettka for his fundamental work and for allowing the research of kettle holes in the ASQL, in addition to collecting information about kettle holes in this region, in particular, related to their basic hydromorphological characteristics. Furthermore, we would like to thank Carlos Acame for flying the quadrocopter in 2016 and the BioMove Research Training Group (DFG-GRK 2118/1) for providing the species list of kettle holes for 2016, recorded by Benito Schöpke. We thank Jörg Haase for providing help with the meteorological data. Finally, we would like to thank Dorith Henning for technical assistance and her great support in the digital mapping of vegetation.

Conflicts of Interest: The authors declare no conflict of interest. The funders had no role in the design of the study, in the collection, analyses, or interpretation of data, in the writing of the manuscript, or in the decision to publish the results.

References

1. Jeffries, M.J. Flood, drought and the inter-annual variation to the number and size of ponds and small wetlands in an English lowland landscape over three years of weather extremes. *Hydrobiologia* **2016**, *768*, 255–272. [CrossRef]
2. Thornhill, I.A.; Biggs, J.; Hill, M.J.; Briers, R.; Gledhill, D.; Wood, P.J.; Gee, J.H.R.; Ledger, M.; Hassall, C. The functional response and resilience in small waterbodies along land-use and environmental gradients. *Glob. Chang. Biol.* **2018**, *24*, 3079–3092. [CrossRef]

3. Marlene, P.; Kalettka, T.; Onandia, G.; Balla, D.; Lischeid, G.; Pätzig, M.; Thomas, K.; Gabriela, O.; Dagmar, B.; Gunnar, L. How much information do we gain from multiple-year sampling in natural pond research? *Limnologica* **2020**, *80*, 125728. [CrossRef]
4. IPCC. *Climate Change 2014: Synthesis Report. Contribution of Working Groups I, II and III to the Fifth Assessment Report of the Intergovernmental Panel on Climate Change*; IPCC: Geneva, Switzerland, 2014.
5. Riley, W.D.; Potter, E.C.; Biggs, J.; Collins, A.L.; Jarvie, H.P.; Jones, J.I.; Kelly-Quinn, M.; Ormerod, S.J.; Sear, D.A.; Wilby, R.L.; et al. Small Water Bodies in Great Britain and Ireland: Ecosystem function, human-generated degradation, and options for restorative action. *Sci. Total. Environ.* **2018**, *645*, 1598–1616. [CrossRef]
6. Winter, T.C.; Rosenberry, D.O. Hydrology of Prairie Pothole Wetlands during Drought and Deluge: A 17-Year Study of the Cottonwood Lake Wetland Complex in North Dakota in the Perspective of Longer Term Measured and Proxy Hydrological Records. *Clim. Chang.* **1998**, *40*, 189–209. [CrossRef]
7. Valk, A.G.V.D. Water-level fluctuations in North American prairie wetlands. *Hydrobiologia* **2005**, *539*, 171–188. [CrossRef]
8. Hayashi, M.; Van Der Kamp, G.; Rosenberry, D.O. Hydrology of Prairie Wetlands: Understanding the Integrated Surface-Water and Groundwater Processes. *Wetlands* **2016**, *36*, 237–254. [CrossRef]
9. Van Der Valk, A.G.; Davis, C.B. The Role of Seed Banks in the Vegetation Dynamics of Prairie Glacial Marshes. *Ecology* **1978**, *59*, 322–335. [CrossRef]
10. Davies, B.; Biggs, J.; Williams, P.; Whitfield, M.; Nicolet, P.; Sear, D.; Bray, S.; Maund, S. Comparative biodiversity of aquatic habitats in the European agricultural landscape. *Agric. Ecosyst. Environ.* **2008**, *125*, 1–8. [CrossRef]
11. Onandia, G.; Lischeid, G.; Kalettka, T.; Kleeberg, A.; Omari, M.; Premke, K.; Arhonditsis, G.B. Biogeochemistry of natural ponds in agricultural landscape: Lessons learned from modeling a kettle hole in Northeast Germany. *Sci. Total. Environ.* **2018**, *634*, 1615–1630. [CrossRef]
12. Kazanjian, G.; Flury, S.; Attermeyer, K.; Kalettka, T.; Kleeberg, A.; Premke, K.; Köhler, J.; Hilt, S. Primary production in nutrient-rich kettle holes and consequences for nutrient and carbon cycling. *Hydrobiologia* **2017**, *806*, 77–93. [CrossRef]
13. Johnson, W.C.; Millett, B.V.; Gilmanov, T.; Voldseth, R.A.; Guntenspergen, G.R.; Naugle, D.E. Vulnerability of north-ern prairie wetlands to climate change. *Bioscience* **2005**, *55*, 863–872. [CrossRef]
14. Johnson, W.C.; Werner, B.; Guntenspergen, G.R. Non-linear responses of glaciated prairie wetlands to climate warming. *Clim. Chang.* **2015**, *134*, 209–223. [CrossRef]
15. Mushet, D.M.; McKenna, O.P.; McLean, K.I. Alternative stable states in inherently unstable systems. *Ecol. Evol.* **2019**, *10*, 843–850. [CrossRef]
16. Johnson, W.C.; Boettcher, S.E.; Poiani, K.A.; Guntenspergen, G.R. Influence of weather extremes on the water levels of glaciated prairie wetlands. *Wetlands* **2004**, *24*, 385–398. [CrossRef]
17. Bauder, E.T. Inundation effects on small-scale plant distributions in San Diego, California vernal pools. *Aquat. Ecol.* **2000**, *34*, 43–61. [CrossRef]
18. Altenfelder, S.; Raabe, U.; Albrecht, H. Effects of water regime and agricultural land use on diversity and species com-position of vascular plants inhabiting temporary ponds in northeastern Germany. *Tuexenia* **2014**, *34*, 145–162. [CrossRef]
19. Rocarpin, P.; Gachet, S.; Metzner, K.; Saatkamp, A. Moisture and soil parameters drive plant community assembly in Mediterranean temporary pools. *Hydrobiol.* **2015**, *781*, 55–66. [CrossRef]
20. Stewart, R.E.; Kantrud, H.A. *Classification of Natural Ponds and Lakes in the Glaciated Prairie Region*; Bureau of Sport Fisheries and Wildlife, U.S. Fish and Wildlife Service: Washington, DC, USA, 1971.
21. Kantrud, H.A.; Krapu, G.L.; Swanson, G.A. Prairie basin wetlands of the Dakotas—A community profile. *Biol. Rep.* **1989**, *85*, 1–111.
22. Van der Valk, A. Succession in Wetlands: A Gleasonian Approach. *Ecology* **1981**, *62*, 688–696. [CrossRef]
23. Seabloom, E.W.; Moloney, K.A.; van der Valk, A.G. Constraints on the establishment of plants along a fluctuating water-depth gradient. *Ecology* **2001**, *82*, 2216–2232. [CrossRef]
24. Euliss, N.H.; LaBaugh, J.W.; Fredrickson, L.H.; Mushet, D.M.; Laubhan, M.K.; Swanson, G.A.; Winter, T.C.; Rosenberry, D.O.; Nelson, R.D. The wetland continuum: A conceptual framework for interpreting biological studies. *Wetlands* **2004**, *24*, 448–458. [CrossRef]
25. Mushet, D.M.; McKenna, O.P.; La Baugh, J.W.; Euliss, N.H.; Rosenberry, D.O. Accommodating State Shifts within the Conceptual Framework of the Wetland Continuum. *Wetlands* **2018**, *38*, 647–651. [CrossRef]
26. Sayer, C.D.; Andrews, K.; Shilland, E.; Edmonds, N.; Edmonds-Brown, V.; Patmore, I.R.; Emson, D.; Axmacher, J.C. The role of pond management for biodiversity conservation in an agricultural landscape. *Aquat. Conserv. Mar. Freshw. Ecosyst.* **2012**, *22*, 626–638. [CrossRef]
27. Lewis-Phillips, J.; Brooks, S.J.; Sayer, C.D.; Patmore, I.R.; Hilton, G.M.; Harrison, A.; Robson, H.; Axmacher, J.C. Ponds as insect chimneys: Restoring overgrown farmland ponds benefits birds through elevated productivity of emerging aquatic insects. *Biol. Conserv.* **2020**, *241*, 108253. [CrossRef]
28. Lewis-Phillips, J.; Brooks, S.; Sayer, C.D.; McCrea, R.; Siriwardena, G.; Axmacher, J.C. Pond management enhances the local abundance and species richness of farmland bird communities. *Agric. Ecosyst. Environ.* **2019**, *273*, 130–140. [CrossRef]
29. Johnson, W.C.; Poiani, K.A. Climate Change Effects on Prairie Pothole Wetlands: Findings from a Twenty-five Year Numerical Modeling Project. *Wetlands* **2016**, *36*, 273–285. [CrossRef]

30. McKenna, O.P.; Mushet, D.M.; Rosenberry, D.O.; LaBaugh, J.W. Evidence for a climate-induced ecohydrological state shift in wetland ecosystems of the southern Prairie Pothole Region. *Clim. Chang.* **2017**, *145*, 273–287. [CrossRef]
31. Van Der Valk, A.G.; Volin, J.C.; Wetzel, P.R. Predicted Changes in Interannual Water-Level Fluctuations Due to Climate Change and Its Implications for the Vegetation of the Florida Everglades. *Environ. Manag.* **2015**, *55*, 799–806. [CrossRef]
32. Van Der Valk, A.; Mushet, D.M. Interannual Water-level Fluctuations and the Vegetation of Prairie Potholes: Potential Impacts of Climate Change. *Wetlands* **2016**, *36*, 397–406. [CrossRef]
33. Vasić, F.; Paul, C.; Strauss, V.; Helming, K. Ecosystem Services of Kettle Holes in Agricultural Landscapes. *Agronomy* **2020**, *10*, 1326. [CrossRef]
34. Lehsten, D.; Von Asmuth, J.R.; Kleyer, M. Simulation of Water Level Fluctuations in Kettle Holes Using a Time Series Model. *Wetlands* **2011**, *31*, 511–520. [CrossRef]
35. Pätzig, M.; Kalettka, T.; Glemnitz, M.; Berger, G. What governs macrophyte species richness in kettle hole types? A case study from Northeast Germany. *Limnologica* **2012**, *42*, 340–354. [CrossRef]
36. Raulings, E.J.; Morris, K.; Roache, M.C.; Boon, P.I. The importance of water regimes operating at small spatial scales for the diversity and structure of wetland vegetation. *Freshw. Biol.* **2010**, *55*, 701–715. [CrossRef]
37. Kalettka, T.; Rudat, C. Hydrogeomorphic types of glacially created kettle holes in North-East Germany. *Limnologica* **2006**, *36*, 54–64. [CrossRef]
38. Wich, S.A.; Koh, L.P. *Conservation Drones: Mapping and Monitoring Biodiversity*; Oxford University Press: New York, NY, USA, 2018.
39. Díaz-Delgado, R.; Mücher, S. Editorial of Special Issue "Drones for Biodiversity Conservation and Ecological Monitoring". *Drones* **2019**, *3*, 47. [CrossRef]
40. Chabot, D.; Bird, D.M. Small unmanned aircraft: Precise and convenient new tools for surveying wetlands. *J. Unmanned Veh. Syst.* **2013**, *1*, 15–24. [CrossRef]
41. Husson, E.; Hagner, O.; Ecke, F. Unmanned aircraft systems help to map aquatic vegetation. *Appl. Veg. Sci.* **2014**, *17*, 567–577. [CrossRef]
42. Verch, G. *Weather Data 2015, Dedelow, Germany*; Leibniz Centre for Agricultural Landscape Research (ZALF): Müncheberg, Germany, 2018. [CrossRef]
43. Verch, G. *Weather Data 2016, Dedelow, Germany*; Leibniz Centre for Agricultural Landscape Research (ZALF): Müncheberg, Germany, 2019. [CrossRef]
44. Verch, G. *Weather Data 2017, Dedelow, Germany*; Leibniz Centre for Agricultural Landscape Research (ZALF): Müncheberg, Germany, 2020. [CrossRef]
45. Verch, G. *Weather Data 2018, Dedelow, Germany*; Leibniz Centre for Agricultural Landscape Research (ZALF): Müncheberg, Germany, 2021. [CrossRef]
46. Verch, G. *Weather Data 2019, Dedelow, Germany*; Leibniz Centre for Agricultural Landscape Research (ZALF): Müncheberg, Germany, 2021. [CrossRef]
47. Verch, G. *Weather Data 2020, Dedelow, Germany*; Leibniz Centre for Agricultural Landscape Research (ZALF): Müncheberg, Germany, 2021. [CrossRef]
48. Richter, D. *Ergebnisse Methodischer Untersuchungen zu Korrektur des Systematischen Meßfehlers des Hellmann-Niederschlagsmessers (Berichte des Deutschen Wetterdienstes)*; Selbstverlag des Deutschen Wetterdienstes: Offenbach, Germany, 1995.
49. Wendling, U.; Müller, J.; Schwede, K. Ergebnisse von Verdunstungsmessungen über Gras mit einem offline-Datenerfassungssystem. *Z. Meteorol.* **1984**, *34*, 190–202.
50. Hedges, L.V.; Gurevitch, J.; Curtis, P.S. The Meta-Analisys of Response Ratios in Experimental Ecology. *Ecology* **1999**, *80*, 1150–1156. [CrossRef]
51. Revelle, W. *psych: Procedures for Psychological, Psychometric, and Personality Research*; R Package Version 2.0.12; Northwestern University: Evanston, IL, USA, 2020. Available online: https://CRAN.R-project.org/package=psych (accessed on 20 December 2020).
52. Hallett, L.; Avolio, M.; Carroll, I.T.; Jones, S.K.; MacDonald, A.A.; Flynn, D.F.B.; Slaughter, P.; Ripplinger, J.; Collins, S.L.; Gries, C.; et al. codyn: Community Dynamics Metrics. R Package Version 2.0.5. 2020. Available online: https://github.com/NCEAS/codyn (accessed on 18 December 2020). [CrossRef]
53. Anderson, K.J. Temporal Patterns in Rates of Community Change during Succession. *Am. Nat.* **2007**, *169*, 780–793. [CrossRef]
54. Clarke, K.R.; Gorley, R.N. *Primer v6: User Manual/Tutorial*; PRIMER-E: Plymouth, UK, 2006.
55. Anderson, M.J.; Gorley, R.N.; Clarke, K.R. *PERMANOVA+ for PRIMER: Guide to Software and Statisitcal Methods*; PRIMER-E: Plymouth, UK, 2008.
56. Akaike, H. Information theory as an extension of the maximum likelihood principle. In Proceedings of the Second International Symposium on Information Theory, Tsahkadsor, Armenia, 2–8 September 1973; pp. 267–281.
57. Erwin, K.L. Wetlands and global climate change: The role of wetland restoration in a changing world. *Wetl. Ecol. Manag.* **2009**, *17*, 71–84. [CrossRef]
58. Frey, W.; Lösch, R. *Lehrbuch der Geobotanik—Pflanze und Vegetation in Raum und Zeit*, 2nd ed.; Spektrum Akademischer Verlag: München, Germany, 2004.

59. Chapin, F.S.; Matson, P.A.; Vitousek, P.M. *Principles of Terrestrial Ecosystem Ecology*, 2nd ed.; Springer Science + Business Media: Berlin/Heidelberg, Germnay, 2011.
60. Jeffries, M. The spatial and temporal heterogeneity of macrophyte communities in thirty small, temporary ponds over a period of ten years. *Ecography* **2008**, *31*, 765–775. [CrossRef]
61. Rhazi, L.; Grillas, P.; Rhazi, M.; Aznar, J.-C. Ten-year dynamics of vegetation in a Mediterranean temporary pool in western Morocco. *Hydrobiologia* **2009**, *634*, 185–194. [CrossRef]
62. Francová, K.; Šumberová, K.; Kučerová, A.; Šorf, M.; Grill, S.; Exler, N.; Vrba, J. Drivers of plant species composition of ecotonal vegetation in two fishpond management types. *Wetl. Ecol. Manag.* **2021**, *29*, 93–110. [CrossRef]
63. Galatowitsch, S.M.; Anderson, N.O.; Ascher, P.D. Invasiveness in wetland plants in temperate North America. *Wetlands* **1999**, *19*, 733–755. [CrossRef]
64. Mulhouse, J.M.; De Steven, D.; Lide, R.F.; Sharitz, R.R. Effects of dominant species on vegetation change in Carolina bay wetlands following a multi-year drought1. *J. Torrey Bot. Soc.* **2005**, *132*, 411–420. [CrossRef]
65. Wilcox, D.A. Implications of hydrologic variability on the succession of plants in Great Lakes wetlands. *Aquat. Ecosyst. Health Manag.* **2004**, *7*, 223–231. [CrossRef]
66. Oertli, B.; Joye, D.A.; Castella, E.; Juge, R.; Cambin, D.G.; Lachavanne, J.-B. Does size matter? The relationship between pond area and biodiversity. *Biol. Conserv.* **2002**, *104*, 59–70. [CrossRef]
67. Van Leeuwen, C.H.A.; Sarneel, J.M.; Van Paassen, J.; Rip, W.J.; Bakker, E.S. Hydrology, shore morphology and species traits affect seed dispersal, germination and community assembly in shoreline plant communities. *J. Ecol.* **2014**, *102*, 998–1007. [CrossRef]
68. Buth, M.; Kahlenborn, W.; Savelsberg, J.; Becker, N.; Bubeck, P.; Kabisch, S.; Kofler, C. Germany's vulnerability to Climate Change—Summary. *Dessau-Roßlau* **2015**, *11*, 2015.
69. Shin, H.R.; Kneitel, J.M. Warming interacts with inundation timing to influence the species composition of California vernal pool communities. *Hydrobiologica* **2019**, *843*, 93–105. [CrossRef]
70. Bourgeois, B.; Vanasse, A.; González, E.; Andersen, R.; Poulin, M. Threshold dynamics in plant succession after tree planting in agricultural riparian zones. *J. Appl. Ecol.* **2016**, *53*, 1704–1713. [CrossRef]
71. Rheinhardt, R.D.; Faser, K. Relationship between hydrology and zonation of freshwater swale wetlands on lower hatteras island, North Carolina, USA. *Wetlands* **2001**, *21*, 265–273. [CrossRef]
72. Frielinghaus, M.; Vahrson, W.-G. Soil translocation by water erosion from agricultural cropland into wet depressions (morainic kettle holes). *Soil Tillage Res.* **1998**, *46*, 23–30. [CrossRef]
73. Kneitel, J.M.; Lessin, C.L. Ecosystem-phase interactions: Aquatic eutrophication decreases terrestrial plant diversity in California vernal pools. *Oecologia* **2009**, *163*, 461–469. [CrossRef]

water

MDPI

Article

Carbon Stocks of Hardwood Floodplain Forests along the Middle Elbe: The Influence of Forest Age, Structure, Species, and Hydrological Conditions

Heather A. Shupe [1,*], Timo Hartmann [2], Mathias Scholz [2], Kai Jensen [1] and Kristin Ludewig [1]

[1] Applied Plant Ecology, Institute of Plant Sciences and Microbiology, Universität Hamburg, Ohnhorststraße 18, 22609 Hamburg, Germany; kai.jensen@uni-hamburg.de (K.J.); Kristin.Ludewig@uni-hamburg.de (K.L.)

[2] Helmholtz-Centre for Environmental Research-UFZ, Department of Conservation Biology, Permoserstraße 15, 04318 Leipzig, Germany; timo.hartmann@ufz.de (T.H.); mathias.scholz@ufz.de (M.S.)

* Correspondence: Heather.Alyson.Shupe@uni-hamburg.de; Tel.: +49-40-42816-388

Abstract: Hardwood floodplain (HF) forests can store a considerable amount of carbon (C), and floodplains may be good candidates for reforestation to provide natural C sinks. In this study, we use nondestructive inventory methods to estimate the C stocks of different tree species and C pools within HF forests of varying age and structure and located at sites differing in hydrological conditions (low and high active floodplain, seepage water zone, tributaries). The study was carried out along the Elbe river (Germany). Average C stocks for young plantations in the active floodplain were significantly lower (50.2 ± 10.8 SE Mg ha^{-1}) than those of old dense (140.6 ± 11.6 SE Mg ha^{-1}) and old sparse forests (180.4 ± 26.6 SE Mg ha^{-1}) with comparable hydrological conditions. C stocks of old dense forests did not significantly vary from old sparse forests. Additionally, C stocks of old forests did not significantly vary according to hydrological conditions. The highest amount of C was stored in *Quercus robur* for all hydrological conditions. *Ulmus laevis* stored the second-highest amount of C on the active floodplain. We conclude that sparse and dense forests as well as forests under different hydrological conditions provide the same C storage function.

Keywords: carbon stocks; hardwood floodplain forest; hydrological conditions; floodplain ecology

Citation: Shupe, H.A.; Hartmann, T.; Scholz, M.; Jensen, K.; Ludewig, K. Carbon Stocks of Hardwood Floodplain Forests along the Middle Elbe: The Influence of Forest Age, Structure, Species, and Hydrological Conditions. *Water* **2021**, *13*, 670. https://doi.org/10.3390/w13050670

Academic Editor: Alenka Gaberščik

Received: 30 January 2021
Accepted: 25 February 2021
Published: 1 March 2021

Publisher's Note: MDPI stays neutral with regard to jurisdictional claims in published maps and institutional affiliations.

1. Introduction

Hardwood floodplain (HF) forests can store a considerable amount of carbon (C) in woody biomass and provide many ecological services such as climate regulation through long-term C storage [1–3]. Globally, HF forests have declined substantially during the last centuries, and in Germany, natural HF forests have been reduced to less than 1% of the active floodplain area [4–6]. The destruction of European HF forests began in the Middle Ages and continued with the expansion of agricultural land and the construction of dikes [7]. Only 10–20% of the former floodplains of the major river catchments in Germany are left, and these active floodplain areas are dominated by managed grasslands [5,6]. HF forests also grow on the seepage water zone behind the dike and on tributary floodplains, but they have different hydrological site conditions which may influence their function. Many of today's remnants of HF forests are patchy and sparse, and dense HF forests with successful natural regeneration are rare.

Land management which increases the C storage of ecosystems is known to be a natural climate solution, and reforestation has garnered global attention as a climate change mitigation measure [8]. Global and regional initiatives such as the Bonn Challenge and ECCA30 have set targets to restore millions of hectares of degraded and deforested lands by 2030 [9]. To meet these targets, suitable locations for reforestation must be identified. Highly productive floodplains are good candidates for reforestation [3], where other ecosystem services such as habitat provisioning to increase biodiversity and flood risk reduction of agricultural lands could also be maximized by reforestation [10]. Many studies have identified positive effects of floodplain vegetation on flood risk reduction, concluding that floodplain vegetation reduces flood risks by increasing hydraulic resistance, reducing flow velocity, and reducing peak magnitude at the catchment outflow [11–13]. However, vegetation with high roughness can also potentially increase flood risks in upstream areas, and to what degree the roughness of vegetation affects flooding in different locations is still under debate [13]. Flow resistance may be influenced by forest age and density [14,15], and the ecosystem services and functions (i.e., C storage) of different forest ages and structures must therefore be quantified and assessed in parallel with other ecosystem services such as flood risk reduction to determine suitable floodplain management advice [16]. Reforestation of HF forests is also possible on the seepage water zone behind dikes and on the tributary floodplains, but it is not well known whether the C storage function of these forests with different hydrological conditions is equal to the active floodplain of the main river channel.

The few available studies on C stocks in HF forests report a wide range of C stocks of the standing biomass (7.5–281 Mg ha^{-1}) [2]. On the Danube floodplain in Austria, C stock in aboveground biomass of HF forests is highest (281 $\pm$ 59 Mg ha^{-1}) compared to softwood (163 $\pm$ 26 Mg ha^{-1}), cottonwood (199 $\pm$ 29 Mg ha^{-1}), and reforestations (35 $\pm$ 17 Mg ha^{-1}) [17]. Interestingly, in another study for the same area of the Danube floodplain, much lower aboveground C stocks (123 $\pm$ 26 Mg ha^{-1}) were reported for mixed hardwood and softwood riparian forests [18]. While differences between inventory methods may have contributed to the wide range of C stock estimates, other site-specific conditions such as forest age and structure as well as hydrological and other abiotic conditions and species composition could also play a role here.

Stand structure and forest age affects the growth and yield of trees [19] and the C storage function of forests [20]. Management of forest stands, i.e., timber extraction, understory livestock grazing, and the clearance of deadwood and thinning of trees to enhance the growth and dimensions of a few harvestable trees, is a long-standing practice in Europe [21–23]. In Germany, timber extraction occurs in a majority of floodplain forests, and only a few near-natural stands remain [24]. The few remaining patches of dense forests which are multilayered and exhibit a well-developed shrub layer and overstory are contrasted with sparse forests which lack a well-developed shrub layer. Sparse and dense forests most likely represent differences in former and current management. Both sparse and dense forest are present on the active floodplain, but it is not known how C stocks differ according to these forest structures.

Trees in HF forests must withstand large hydrological fluctuations resulting in hypoxic or anoxic soils during flooding events and drought in dry periods. Annual flooding duration of HF forests on the active floodplain is related to elevation, with relatively low-lying HF forests subject to longer periods of hypoxia and those at higher elevations more prone to drought. Soil texture affects the water holding capacity and plant available water of the soil, with loamy soil able to hold more water than sandy soil during dry periods [25]. In lowland floodplains, soils are commonly loamy in low-elevated and sandy in high-elevated sites [26], where plant available water may therefore be additionally reduced during dry periods. HF forests are typically dominated by *Fraxinus*, *Ulmus*, and *Quercus* taxa [27]. *Quercus robur* and *Ulmus laevis* are two characteristic tree species for European HF forests which are adapted to cope with flooding and reduced soil oxygen availability through the development of adventitious roots and hypertrophied lenticels [28,29]. Additionally, the capacity to regulate stomatal conductance and long tap roots allow *Q. robur* to survive moderate drought stress. However, *Q. robur* is also prone to hydraulic failure due to vessel cavitation resulting in increased mortality under prolonged drought conditions [30]. *U. laevis* thrives in damp soils and is highly vulnerable to vessel cavitation and mortality as a result of drought stress [30,31]. It is not well known how much these adaptations affect the growth of trees in different hydrological conditions, and how this in turn influences C stocks of HF forests on the active floodplain, the seepage water zone and tributaries.

In this study, we aim to answer (a) how C storage of HF forests develops with age, (b) whether sparse forests fulfill the same C storage function as dense forests, (c) how C storage of HF forests differs between typical hydrological conditions and (d) how C storage differs by taxon under different hydrological conditions.

2. Materials and Methods

2.1. Study Area and Forest Types

The study area is a part of the UNESCO Biosphere Reserve River Landscape Elbe and spans approximately 100 km along the lower Middle Elbe river within the German states Saxony-Anhalt, Brandenburg, and Lower Saxony (Figure 1). The study area is located within the central European temperate climate region. The Lenzen weather station (53.08° N, 11.48° E) records a mean annual precipitation of 615 mm and a mean annual temperature of 9.3 °C for 1981–2010 [32]. Characteristic soils of the Elbe floodplain include Gleysols, Fluvisols, and Cambisols [26].

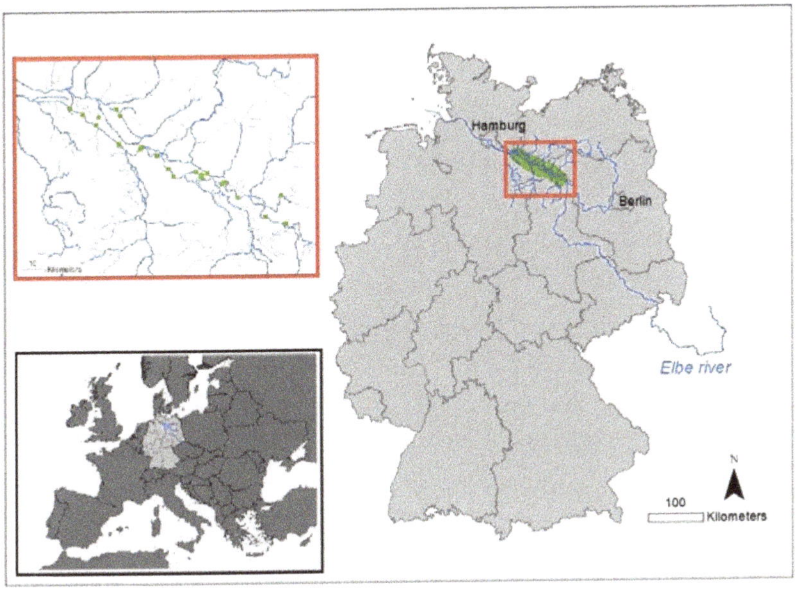

Figure 1. Map of the study area along the Middle Elbe river, Germany. The red box outlines the area which spans approximately 100 km along the Elbe river and the green dots represent the locations of each 2500 km^2 plot in the different studied forest types.

The UNESCO Biosphere Reserve River Landscape Elbe is used as a model system for anthropogenically altered European floodplains. With a history of diking, deforestation and agriculture on the active floodplain, todays' HF forests represent only small remnants of the former contiguous ecosystem type. On the active floodplain confined by dikes, HF forests are more frequently flooded on the lower sites and less frequently flooded on the higher sites. Flooding events mainly occur on the active floodplain [33] after snow melt during winter and spring and after intense rain events during summer. The duration of flooding is related to elevation on the active floodplain.

Typical species in HF forests, also referred to as mixed riparian forests (NATURA 2000 Code 91F0), include pedunculate oak (*Quercus robur*), European white, field, and wych elm (*Ulmus laevis*, *Ulmus minor*, and *Ulmus glabra*), European hornbeam (*Carpinus betula*), and European ash (*Fraxinus excelsior*). Typical understory vegetation includes *Crataegus monogyna*, *Sambucus nigra*, and *Cornus sanguinea*.

2.2. Study Sites and Sampling

This study investigates six different forest types (n = 5 per type). In total, 30 plots (2500 m^2 each) were studied. A sketch of the study design is shown in Figure 2 and site characteristics for each forest type are presented in Table 1.

The effect of forest age and structure on the C storage of HF forests was studied for 15 plots located on the low active floodplain only. Here, five replicate plots of young plantations, old sparse, and old dense HF forests were sampled. The five young plantations are composed of woody species (mainly *Q. robur* and *U. laevis*) which were actively planted on a mix of former grasslands and forests. Dense forests are characterized as multilayered forests with a well-developed overstory and shrub layer, while sparse forests lack the well-developed shrub layer. The age of the young plantations ranged from 18–27 years while the old forests ranged from 80–200 years.

HF forests are found behind the dikes in the seepage water zone of the fossil floodplain and at floodplains of the tributaries. In this study, the possible effects of hydrological conditions on C storage of old dense HF forests were analyzed by sampling five replicate plots of these four forest types (low active floodplain, high active floodplain, seepage water zone, tributary; Figure 2). Hydrological conditions for high and low plots on the active floodplain were selected based on the average number of days the sites were flooded per year: a categorical mean of 0–5 days of flooding for high plots and greater than five days of flooding for the low plots. Flooding duration was estimated using a 35 year mean from 1990–2016 with a 1-dimensional model that integrates data from various databases [34].

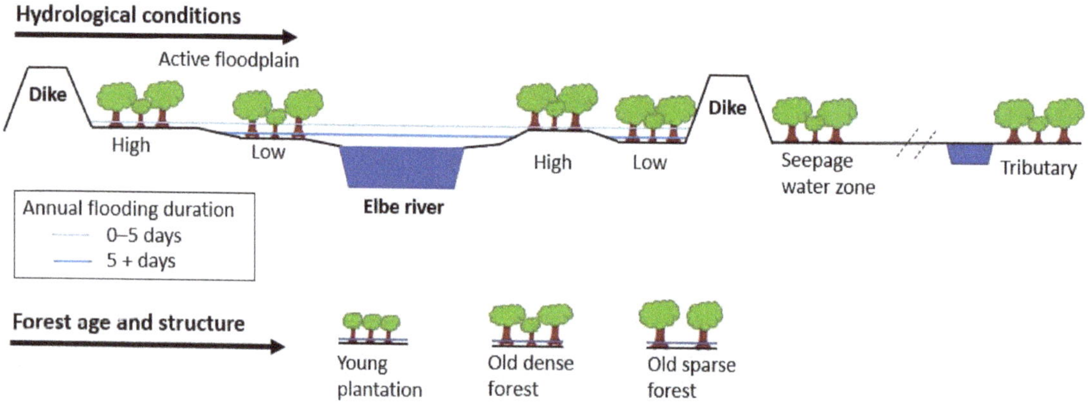

Figure 2. Study design showing old dense HF forests with different hydrological conditions (high active floodplain, low active floodplain, seepage water zone, and tributary) and HF forests with different ages and structure (young plantation, old sparse, and old dense) on the low active floodplain. Softwood floodplain forests and other land cover and ecosystem types are not represented in this sketch.

Table 1. Characteristics of each studied forest type showing minimum (min), maximum (max), and mean values with standard deviation (SD). Each forest type has a sample size of five replicate plots. DBH = diameter at breast height. x indicates a lack of data related to the spatial limitations of the flooding duration model.

	Low Active Floodplain, Young		Low Active Floodplain, Old Dense		Low Active Floodplain, Old Sparse		High Active Floodplain		Seepage Water Zone		Tributary	
	Min Max	Mean ± SD	Min Max	Mean ± SD	Min Max	Mean ± SD	Min Max	Mean ± SD	Min Max	Mean ± SD	Min Max	Mean ± SD
Forest age (years)	18 27	23 ± 4	104 200	141 ± 36	129 167	144 ± 14	108 186	134 ± 33	81 185	128 ± 39	82 170	130 ± 37
Basal area (m^2 ha^{-1})	12 30	22 ± 7	26 36	32 ± 4	29 52	38 ± 10	27 45	36 ± 8	29 43	35 ± 6	27 44	36 ± 7
Tree count (trees ha^{-1})	728 1576	1245 ± 346	181 496	291 ± 121	75 160	123 ± 40	192 325	248 ± 55	235 464	357 ± 100	331 763	531 ± 186
Mean tree height (m)	8 14	11 ± 3	11 20	15 ± 3	13 26	22 ± 5	13 21	16 ± 4	12 26	18 ± 6	14 20	17 ± 3
Mean tree DBH (cm)	11 16	14 ± 2	20 39	32 ± 7	43 85	60 ± 16	29 39	36 ± 5	20 45	31 ± 11	20 30	25 ± 4
Tree species richness	1 5	3 ± 2	2 3	2 ± 1	1 3	2 ± 1	2 5	3 ± 1	3 8	5 ± 2	3 10	5 ± 3
Flooding duration (days $year^{-1}$)	6 86	34 ± 28	11 59	36 ± 18	9 33	22 ± 9	0 9	4 ± 3	x	x	x	x

2.3. Carbon Stock Estimations

Individual C stocks of trees, shrubs, deadwood, and litter were analyzed in the winter months between January and April of 2018 and 2019. The total C stocks per plot were estimated by averaging the summed values for large trees, shrubs, standing dead trees, downed woody debris, and leaf litter. These values were then scaled to Mg ha^{-1}.

2.3.1. Trees

For quantifying C stocks of trees, four 625 m^2 square nested plots (quadrants) were delineated within each of the 30 plots. Within three quadrants for old forests and two quadrants for young plantations, the diameter at breast height (DBH; 1.3 m above ground level) of all trees $\geq$ 5 cm was measured using a standard diameter tape. Within the same quadrants, the height (H) of all trees with a DBH $\geq$ 5 cm were measured with a Vertex Laser Geo (Haglöf, Sweden). The species identity of each measured tree was recorded.

Based on the measured variables H and DBH, individual tree stem volumes were calculated with species-specific allometric equations (Table A1 in Appendix A) [35]. Aboveground tree biomass was calculated by multiplying estimated tree stem volume by species-specific average wood density (Table A2 in Appendix A) taken from the Global Wood Density Database [36]. Finally, a C content (CC) fraction of 0.47 was applied to estimate aboveground tree C stock [37]. To estimate the C stocks of tree roots, a root: shoot ratio of 0.3 was applied to the aboveground tree C stock [38].

2.3.2. Shrubs

All shrubs with a DBH $\geq$ 5 cm were inventoried using the line intersect transect method. Each quadrant chosen for the tree inventory was transected diagonally, and the DBH and H of any shrub crossing 1 m from each side of the transect was measured. The allometric volume equation, biomass factor and C factor for *Corylus* (Table A1) was used for all shrubs and the values were scaled to Mg ha^{-1}. The roots of the shrubs were estimated using a root: shoot factor of 0.4 [37].

2.3.3. Deadwood

The C stock of deadwood was measured following the methodologies and density reduction factors proposed by the United Nations to measure C stocks [37]. Two deadwood pools were measured: large standing dead trees (SDT) and downed woody debris (DWD).

To measure SDT's, the same allometric equations were used as for the estimates of C stocks of large trees, which were then multiplied by density reduction factors depending on the state of decay (sound = 1; intermediate = 0.8; rotten = 0.45). Unlike in the United Nations guidelines [37], if a tree was leaning or newly fallen and lay completely within the study plot, it was included in the SDT pool. The roots of SDTs were measured the same as the live trees, with a root: shoot ratio of 0.3.

Lying downed woody debris (DWD) was measured using transect lines diagonally crossing three quadrants for every plot (with a total length of 106 m per plot). All deadwood with a diameter $\geq$ 5 cm crossing the transect lines were measured horizontally and vertically at the point of intersect and the state of decay was recorded. Trees already accounted for in the SDT pool were omitted. Equation (1) was used to estimate the volume of DWD [39]. DWD volume estimates were then multiplied by 0.5 to obtain DWD biomass, density reduction factors depending on the state of decay, and finally by 0.5 to estimate C content.

$$\hat{X}_j = \frac{\pi^2}{8L_j}\sum_{i=1}^{N}\left(\frac{d_{1i}+d_{2i}}{2}\right)^2 \quad (1)$$

Equation (1) measures the volume of DWD ($\hat{X}$) in m^3 ha^{-1} for the individual sample plots (j). L_j is the horizontal length of the transect lines, while d_{1i} and d_{2i} are the horizontal and vertical diameter measurements (in cm) of individual pieces of dead wood intersected along the transect.

2.3.4. Litter

The winter stock of leaf litter was estimated from February to March in 2019. The litter was measured in winter, because the plots are mainly dominated by oak trees, which do not abscise their leaves until late winter to early spring. Within each of the three studied quadrants, a 1 m^2 quadrat frame was randomly placed along the diagonal transect and the dry weight of leaf litter was measured. Subsamples of fresh litter were brought to the lab and air dried until constant weight. The dry: wet weight ratio was applied to the field values, and the average quadrant values were taken as plot values. The biomass values were multiplied by 0.37 to estimate C stock [37].

2.4. Forest Ages

Forest ages were estimated using annual tree ring measurements [40]. Tree cores were taken from four dominant *Q. robur* trees per plot using a 5 mm Mora increment borer. Singular relic trees with an outlying DBH from the other dominant trees from the same plot were not sampled. The surface of each core was carefully scraped with a razor blade to increase the visibility of the tree ring vessel structure. A microscope connected to a LINTAB^{TM} 5 measuring table and the TSAP-Win^{TM} software program (RINNTECH, Heidelberg, Germany) were used to measure tree rings and establish tree ages. When the pith was not present, concentric circles were used to estimate missing rings [41].

2.5. Statistical Analyses

The C stock data were tested for normality by Shapiro-Wilk test ($p > 0.05$) and a visual inspection of Q-Q plots, box plots, and histograms. Nonparametric independent-samples Kruskall-Wallis Tests with pairwise comparisons were conducted to examine the differences in C stock for each C pool according to forest age and structure and for the forests with different hydrological conditions. Different curve estimation models with forest type as the independent variable and total C stocks as the dependent variable were assessed for best fit. Additionally, a univariate general linear model (GLM) was used to compare C stocks of old, dense forests with different hydrological conditions. The dependent variable was C stocks while the covariate was estimated forest age. Regression curve estimation models were explored to evaluate the best fit relationship between forest age and C stock. All tests were performed using SPSS version 26 (IBM Corp. 2019, Armonk, New York, NY, USA).

3. Results

3.1. Carbon Stocks Related to Forest Age and Structure

On the low active floodplain, the total C stock of young plantations was 50.2 ± 10.8 SE Mg ha^{-1} and thus significantly lower (H = 10.5, $p = 0.005$, df = 2) than that of old sparse (180.4 ± 26.6 SE Mg ha^{-1}) and of old dense forests (140.6 ± 11.6 SE Mg ha^{-1}). Pairwise comparisons found no significant difference between old sparse and old dense forests. Young plantations had significantly less C stock in the tree pool than old dense or sparse forests (H = 10.5, $p = 0.005$, df = 2). Young plantations also had significantly less DWD than old dense forests (H = 7.4, $p = 0.009$, df = 1). No other significant differences comparing C pools between different forest types on the low active floodplain were found. Overall, the most C was stored in the tree C pool than in any other pool (Figure 3, Table A3), and the SE of the tree pool was commonly larger than the stock estimated for other C pools. A positive logarithmic relationship ($r^2 = 0.741$, $p < 0.001$) was found between forest age and C stock (Figure 4).

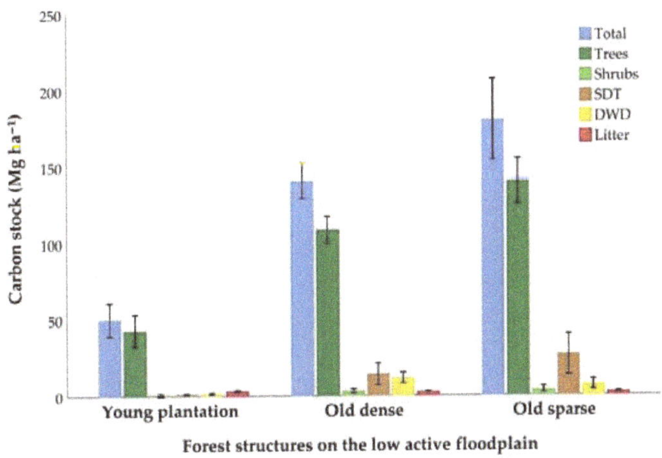

Figure 3. Carbon stocks in Mg ha^{-1} of hardwood forests with different ages and structures on the low active floodplain (mean ± SE, $n = 5$). Carbon pools include trees, shrubs, standing dead trees (SDT), downed woody debris (DWD), and leaf litter.

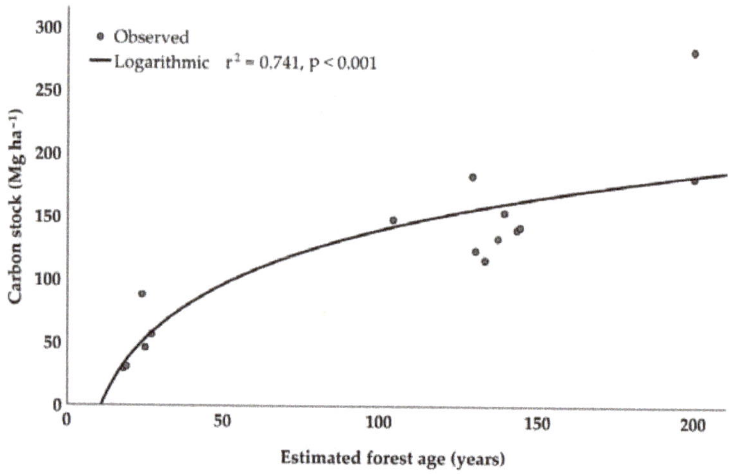

Figure 4. Carbon stocks in Mg ha^{-1} of young plantations, old dense, and old sparse hardwood forests on the low active floodplain are plotted against estimated forest age in years. A logarithmic fit curve with the output of the regression is included. Carbon stocks include trees, shrubs, standing dead trees (SDT), downed woody debris (DWD), and leaf litter.

3.2. C Stocks of Old HF Forests under Different Hydrological Conditions

C stocks of old dense forests under different hydrological conditions ranged from 140.5 ± 11.6 (low active floodplain) to 163.5 ± 8.3 SE Mg ha^{-1} (high active floodplain) (Figure 5, Table A4). Kruskall-Wallis tests revealed that there were no significant differences between the total C stock or any other C pool of old dense forests with different hydrological conditions. The GLM revealed that the covariate, forest age, was not significantly related to C stock, $F_{1,15} = 0.72$, $p > 0.05$, r = 0.41. There was also no significant effect of hydrological conditions on C stocks after controlling for forest age, $F_{3,26} = 0.54$, $p > 0.05$, partial $\eta^2 = 0.10$.

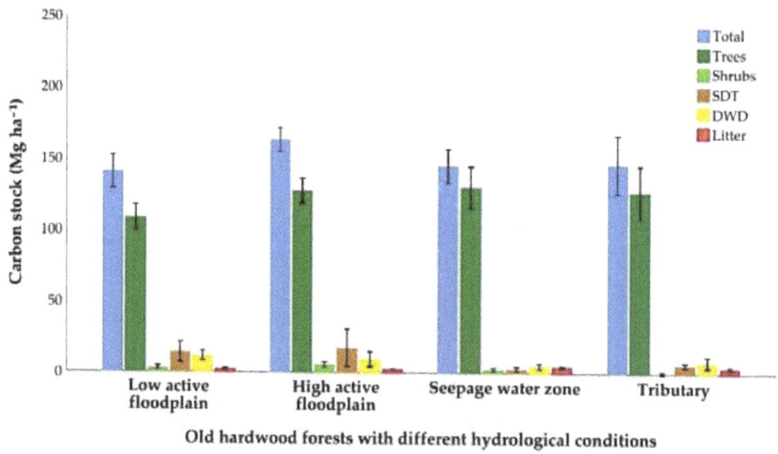

Figure 5. Carbon stocks in Mg ha^{-1} of dense hardwood forests with different hydrological conditions (mean ± SE, $n = 5$). Carbon pools include trees, shrubs, standing dead trees (SDT), downed woody debris (DWD), and leaf litter.

3.3. C Stocks by Species

Q. robur stored more C than any other species in all hydrological conditions (Figure 6). On the active floodplain, *Ulmus* spp. stored the second highest amount of C, whereas in the seepage water zone and on tributary floodplains, *Ulmus* spp. were rare and *C. betulus* stored the second highest amount of C.

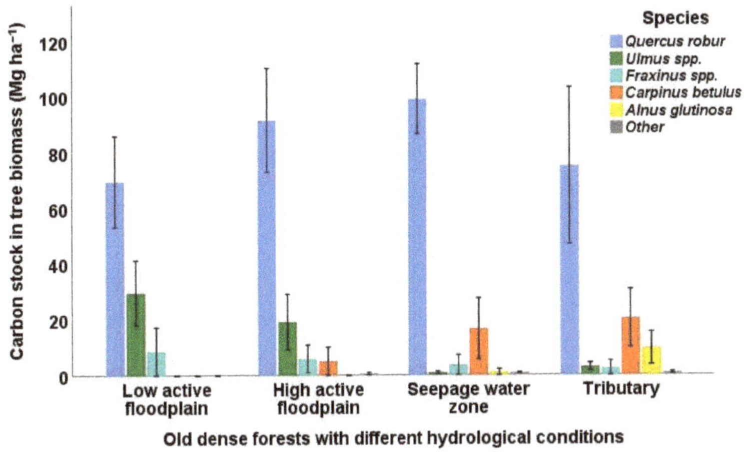

Figure 6. Carbon stock of large trees in Mg ha^{-1} including above- and below-ground biomass by tree species (mean ± SE, n = 5). Other tree taxa include *Acer*, *Corylus*, *Fagus*, *Picea*, *Pinus*, *Populus*, *Prunus*, *Salix*, *Sorbus*, and *Tilia* spp.

4. Discussion

C stocks positively developed with age, with young plantations storing less C than old forests. This finding supports research in the floodplains of the Danube, where young reforestations also showed significantly lower C stocks than mature HF forests [17]. Many years are required for young plantations to mature and provide the same ecosystem function as old forests, but from the projected path of the logarithmic age curve, the increase in C stock is greatest in the first fifty years, before the stock slowly begins to taper as the forest matures. This implies that the annual rate at which the young plantations store C, or the C sequestration rate, is larger than that of old forests. The age curve had one major outlier, where the total estimated C stock was 282.4 Mg ha^{-1}. This outlier forest was characterized as sparse, had the highest live tree stock (193.7 Mg ha^{-1}), and had the highest proportion of *U. laevis* trees with large dimensions compared to other studied plots. This forest also had a significantly higher deadwood stock (80.7 Mg ha^{-1}) than any other plot, with large fallen trees and SDTs left to naturally decompose. In most of the other old forest plots, deadwood may have been removed either by management or flooding disturbance. This rare outlier of a forest with ample deadwood suggests that the removal of deadwood decreases the C storage function of HF forests. However, because this is one outlier, more studies should be conducted specifically looking at the potential reduction in C stocks as a consequence of deadwood removal on the floodplains.

On the low active floodplain, sparse forests stored equally as much C as dense forests. While some studies find that thinning of floodplain forests can increase C stocks [42], this study suggests that the overall C stock of naturally dense forests are equal to sparse forests. If the only purpose of reforestation is to maximize C storage, then either forest structure would be an appropriate land management target. However, suitable land management decisions rely on the assessment of multiple ecosystem services [16] and must consider potential risks and the preference of local stakeholders. For example, how the different structures contribute to or alleviate flooding should also be quantified and used in the assessment to determine proper floodplain management. The potentially higher roughness of dense forests may increase or decrease flood protection, depending on the location along the river. While flood risk can be reduced downstream from a forest with high roughness through the reduction of flow velocity and peak magnitude at the catchment outflow [11–13], the flood risk upstream from the forest could be increased by the backwater effect [13]. Suitable locations for reforestation of either dense or sparse forests is therefore also dependent on the surrounding land use, and considerations should be made to maximize the benefits of reforestation while minimizing potential risks.

The total C stocks and all C pools of old dense HF forests with different hydrological conditions did not significantly differ between each other, which indicates that the C storage function of the HF forests is equal. This finding supports the results of Rieger et al. [18], who observed no significant difference between C stocks of HF forests on the active floodplain and HF forests behind dikes in the seepage water zone. Trees contributed the greatest to the total C stock, and the equal C stocks implies that the trees are well adapted to the different hydrological conditions. Although the dike severs the connection between the forests on the seepage water zone from the flood pulse and the nutrients that come with it [43], the trees still grow at a seemingly equal rate. To verify this, quantification of tree growth at an annual scale is needed. Additional research is also needed to assess the effects of climate change on tree vitality and productivity on different elevations of the active floodplain, the seepage water zone, and tributaries. Climate change models project increased temperatures, precipitation, and river discharge at the Elbe [44,45], with increasing drought conditions in spring and summer and increased precipitation in autumn and winter [46]. There is some evidence that increased flood frequency may reduce drought effects on the active floodplain [47], but these benefits may not be as pronounced in the seepage water zone behind dikes. If the flooding events occur in winter, drought conditions under high spring and summer temperatures may lead to tree mortality, which will greatly alter the distribution of C within pools, from C-fixing live trees to C-releasing dead trees. The finding that the C storage function of old hardwood forests is nearly equal on the low and high active floodplain, the seepage water zone, and tributaries reveals that all of these sites are potentially suitable for reforestation considering the past climatic conditions, but this may not be the case considering future climate change. Additionally, the other ecosystem services must now be quantified for these forests and stakeholder preferences taken into consideration to determine suitable land management decisions.

Our C stock estimates of 50.2 $\pm$ 10.8 SE Mg ha^{-1} for young plantations, 140.6 $\pm$ 11.6 SE Mg ha^{-1} for old dense forests, and 180.4 $\pm$ 26.6 SE Mg ha^{-1} for old sparse forests are within the 7.5–281 Mg ha^{-1} range of previously reviewed HF forests [2] and are similar to upland forests in Germany. Generally, the C storage in German forests is reported to be 120–190 Mg ha^{-1}, depending on age class and tree species [48]. *Quercus petraea* forests in Northern Germany have an estimated C stock of 107.82 $\pm$ 7.27 Mg ha^{-1} in aboveground live tree biomass and 9.35 $\pm$ 6.51 Mg ha^{-1} in deadwood [49]. We estimated a C stock in aboveground tree biomass of 33 $\pm$ 17.9 SE Mg ha^{-1} for young plantations, 83.6 $\pm$ 15.1 SE Mg ha^{-1} for old dense forests, and 107.6 $\pm$ 25.5 SE Mg ha^{-1} for old sparse forests. Our C stock estimate for aboveground trees not including roots in old sparse forests is therefore almost identical to the estimate for naturally developed upland forests dominated by *Q. petraea*. Our deadwood estimates for old sparse forests; however, are much higher (26 $\pm$ 13.3 Mg ha^{-1} for SDTs with an additional 7.2 $\pm$ 3.3 Mg ha^{-1} for DWD). In another study of different forest types in Germany, the average C stocks in aboveground and belowground biomass, deadwood and soils are reported to be 224 Mg ha^{-1} [50]. It is reported that 46% of C is stored in the aboveground and belowground biomass and 1% of C is stored in deadwood, which would mean a C stock of 105 Mg ha^{-1} in trees and deadwood. This estimate is one third less than in our studied HF forests. The tree pool C stocks calculated in our study are lower than the estimated 281 $\pm$ 59 Mg ha^{-1} for HF forests by the Danube river [17]. The large difference between C stocks estimated along the Danube and those estimated here may be attributed to abiotic and climatic differences, management, forest structures such as number of tree stems, or methodological differences in estimating C stocks. The HF forests along the Danube had a mean tree count of 590 ± 80, while the old forests studied here had a mean tree count ranging from 123 ± 40 to 531 ± 186. However, tree count is not a good indicator for determining C stocks, as shown by young plantations which have a much larger tree count than old forests, but a lower overall C stock. Additionally, a major setback in making accurate comparisons between studies reporting on C stocks of forests is the absence of a universally applied field inventory and C calculation method. There are proposed guidelines, such as the UNFCCC methods [37]; however, there are various national inventories [51] as well as research papers [17,48–50,52,53], which use different field measurements and computational methods to estimate C stocks. Various allometric equations are available, and the choice of the equations can greatly impact the estimated C stock values. Additionally, many old trees such as oaks and elms become hollow as they mature, and the allometric equations do not take into account this reduction in biomass [54], which may result in an over-estimation of C stock. Although allometric equations provide a non-destructive way of measuring C stocks, there can be large variations depending on the selected equations, which adds great uncertainties to the estimated C distribution in different forest ecosystems worldwide [55].

Q. robur had a dominating presence in all hydrological conditions compared to other taxa, which may be a consequence of the species' drought and flood tolerance [28] or the fact that forest managers in the past mainly preferred to plant and foster oaks for quality timber harvesting and to provide animal fodder and tanning agents [56]. Today, planting campaigns that include a diversity of species are recommended to increase resilience against biotic stressors and variability in abiotic conditions [57], as well as to enhance productivity and C storage [58,59]. Monoculture planting campaigns should be avoided to minimize pathogens and insect attacks [60]. This is especially true in the middle Elbe region, where outbreaks of oak processionary moth (*Thaumetopoea processionea*) are especially prevalent in plantations with high oak densities. Therefore, although *Q. robur* is a suitable tree species for reforestation under all hydrological conditions, other species should be interspersed. Many elms (mainly *U. laevis*) were found on the active floodplain, while very few elms were growing in the seepage water zone or along the tributaries. Although the results may be interpreted in a way that the hydrological conditions of the seepage water zone are not suitable for elms, the lack of elms behind the dike could also be a consequence of management and the preference of foresters to foster oaks. Compared to oaks and elms, very few ash trees (*F. excelsior*) were observed, except for one plot that had mostly non-native green ash (*F. pennsylvanica*). Although ash dieback caused by the fungus *Hymenoscyphus fraxineus* could be a reason for the low *F. excelsior* numbers [61], ash is well adapted to thrive on floodplains but not able to resist this fungal infestation at present, and therefore reforestation of ash may not be suitable. *C. betulus* was numerous in the seepage water zone and tributaries, while the species' presence on the active floodplain was only apparent on the high elevated sites with lower annual flooding duration. This is most likely a consequence of the lower flood tolerance of *C. betulus*, which is not listed as typical species in the NATURA 2000 classification for riparian mixed forests. The exchange of *C. betulus* for *U. laevis* as the second most dominating species in the HF forests on the seepage water zone and tributaries may therefore be a consequence of the different hydrological conditions.

5. Conclusions

C stocks developed positively with age and the C storage function of old forests did not significantly vary with forest structure on the low active floodplain or according to different hydrological conditions. Old forests on the low active floodplain, the high active floodplain, the seepage water zone, and tributaries fulfill the same ecosystem function of C storage and the locations are therefore at first glance equally suitable for reforestation campaigns. However, C storage is only one ecosystem service among many that should be quantified and evaluated to provide decisive and suitable land management advice.

Additionally, the influence of climate change should also be considered. *Q. robur* is a good candidate for reforestations at all hydrological situations and should be accompanied by other suitable species such as *U. laevis* in all hydrological conditions and *C. betulus* in less frequently flooded conditions. *F. excelsior* is at present not a good candidate for reforestation because of the high risk of dieback.

Author Contributions: Conceptualization, H.A.S., M.S., K.J. and K.L.; Data curation, H.A.S.; Formal analysis, H.A.S.; Funding acquisition, K.J., M.S. and K.L.; Investigation, H.A.S.; Methodology, H.A.S., T.H., M.S. and K.L.; Project administration, K.L.; Resources, K.J.; Software, H.A.S.; Supervision, K.J. and K.L.; Validation, H.A.S., K.J. and K.L.; Visualization, H.A.S.; Writing—original draft, H.A.S.; Writing—review & editing, H.A.S., T.H., M.S., K.J. and K.L. All authors have read and agreed to the published version of the manuscript.

Funding: This research was conducted within the interdisciplinary project MediAN (Mechanismen der Ökosystemdienstleistungen von Hartholz-Auenwäldern: Wissenschaftliche Analyse sowie Optimierung durch Naturschutzmanagement). This research was funded by the Bundesministerium für Bildung und Forschung (BMBF 01LC1601A).

Institutional Review Board Statement: Not applicable.

Informed Consent Statement: Not applicable.

Data Availability Statement: The data presented in this study are available on request from the corresponding author. The data are not publicly available due to privacy issues.

Acknowledgments: We would like to thank the landowners who participated in this study and the Biosphere reserve Niedersächsische Elbtalaue and Brandenburg Flusslandschaft Elbe with special thanks to Hans-Jürgen Kelm and Torsten Hennig. Thanks to soil scientists Adrian Heger and Lizeth Vásconez for the soil type analysis. Many thanks are also given to field and lab assistants Lilli Hamm and Fernanda Chavez, as well as to all other MediAN colleagues.

Conflicts of Interest: The authors declare no conflict of interest. The funders had no role in the design of the study; in the collection, analyses, or interpretation of data; in the writing of the manuscript, or in the decision to publish the results.

Appendix A

Table A1. Tree stem volume equations for all species inventoried in the Middle Elbe study area are shown, taken from Zianis et. al., 2005 (Appendix C) [35]. The country where the equation originates from and the units for volume (V), diameter at breast height (D), and the height (H) of individual trees are shown.

Tree Species		Unit			Equation	Parameters					
		V	D	H		a	b	c	d	e	f
Acer spp.	NT	dm^3	cm	m	$D^a \cdot H^b \cdot exp(c)$	1.89756	0.97716	−2.94253			
Alnus glutinosa	NT	dm^3	cm	m	$D^a \cdot H^b \cdot exp(c)$	1.85749	0.88675	−2.5222			
Betula pendula	NT	dm^3	cm	m	$D^a \cdot H^b \cdot exp(c)$	1.8906	0.26595	−1.07055			
Carpinus spp.	NT	dm^3	mm	m	$a \cdot D^{(b+c)} \cdot H^d$	0.00021491	2.258957614	0.001411006	0.60291075		
Corylus avellana	NO	dm^3	cm	m	$a + b \cdot D^2 + c \cdot D^2 \cdot H + d \cdot D \cdot H^2 + e \cdot H^2$	−1.86827	0.21461	0.01283	0.0138	−0.06311	
Fagus sylvatica	NT	dm^3	cm	m	$D^a \cdot H^b \cdot exp(c)$	1.55448	1.5588	−3.57875			
Fraxinus excelsior	NT	dm^3	cm	m	$D^a \cdot H^b \cdot exp(c)$	1.95277	0.77206	−2.48079			
Picea abies	GER	m^3	m	m	$a \cdot H \cdot D^2$	0.502					
Pinus sylvestris	GER	m^3	cm	m	$a \cdot D^b \cdot H^c$	0.000056537	1.960466	0.894433			

Table A1. *Cont.*

Tree Species	Unit				Equation	Parameters					
		V	D	H		a	b	c	d	e	f
Populus spp.	NT	dm^3	mm	m	$a \cdot D^{(b+c)} \cdot H^d$	0.0009507	1.895629295	0.001650837	0.8392146		
Prunus avium	BE	m^3	cm	m	$a + b{\cdot}D + c{\cdot}D^2 + d{\cdot}D^3 + e{\cdot}H + f{\cdot}D^2{\cdot}H$	−0.002311	−0.00117728	0.000149061	-7.8058×10^{-6}	0.00033282	0.000031526
Quercus robur	NT	dm^3	cm	m	$D^a{\cdot}H^b{\cdot}\exp(c)$	2.00333	0.85925	−2.86353			
Ulmus spp.	NT	dm^3	cm	m	$D^a{\cdot}H^b{\cdot}\exp(c)$	1.942950	1.292290	−4.200640			

Table A2. Specific wood densities for tree species inventoried in the HF forests of the middle Elbe and sourced from the global wood density database [36].

Tree Species	Specific Wood Density (g cm^{-3})
Acer spp.	0.525
Alnus glutinosa	0.439
Betula pendula	0.525
Carpinus betulus	0.706
Corylus avellana	0.517
Fagus sylvatica	0.585
Fraxinus excelsior	0.560
Picea abies	0.370
Pinus sylvestris	0.422
Populus alba	0.353
Prunus avium	0.474
Quercus robur	0.560
Ulmus spp.	0.551

Table A3. Mean (±SE, n = 5), minimum and maximum carbon stocks of old hardwood floodplain forests on the active floodplain with different ages and forest structures. The total carbon stock combines five carbon pools: The above-and belowground carbon stocks (AGC and BGC) of trees, shrubs, and standing dead trees (SDT) ≥ 5 cm diameter at breast height, as well as coarse woody debris (DWD) and litter. The mean, standard error of the mean (SE), minimum (min), and maximum (max) values are shown for the five replicate plots per forest type. All carbon stocks are presented in Megagrams carbon per hectare (Mg ha^{-1}).

Forest Age and Structure		Total C Stock	Tree	Shrub	SDT	DWD	Litter
Young plantation	Mean	50.2	42.9	0.9	1.3	1.8	3.3
	SE	10.8	10.4	0.8	0.5	0.7	0.6
	Min	29.2	19.6	0.0	0.0	0.4	1.9
	Max	88.2	77.7	4.0	2.8	3.6	5.0
Old dense	Mean	140.6	108.6	3.2	14.3	11.8	2.8
	SE	11.6	8.8	1.3	7.1	3.3	0.4
	Min	116.2	82.6	0.8	0.0	5.6	1.8
	Max	181.8	133.6	7.9	32.9	22.1	4.0
Old sparse	Mean	180.4	139.9	4.0	26.9	7.2	2.4
	SE	26.6	14.8	1.9	13.3	3.3	0.5
	Min	140.1	109.0	0.0	3.4	1.7	1.5
	Max	282.4	193.7	9.7	77.5	19.3	4.0

Table A4. Mean (±SE, n = 5), minimum and maximum carbon stocks of old hardwood floodplain forests at different hydrological conditions. The total carbon stock combines five carbon pools: The above-and belowground carbon stocks (AGC and BGC) of trees, shrubs, and standing dead trees ≥ 5 cm diameter at breast height, as well as dead woody debris and litter. The mean, standard error of the mean (SE), minimum (min), and maximum (max) values are shown for the five replicate plots per hydrological condition. All carbon stocks are presented in megagrams carbon per hectare (Mg ha^{-1}).

Hydrological Condition		Total C Stock	Tree	Shrub	SDT	DWD	Litter
Low active floodplain	Mean	140.6	108.6	3.2	14.3	11.8	2.8
	SE	11.6	8.8	1.3	7.1	3.3	0.4
	Min	116.2	82.6	0.8	0.0	5.6	1.8
	Max	181.8	133.6	7.9	32.9	22.1	4.0
High active floodplain	Mean	163.5	127.7	5.7	17.6	9.7	2.8
	SE	8.3	8.7	2.0	13.0	5.3	0.3
	Min	140.0	104.4	2.6	0.0	0.8	2.4
	Max	185.1	151.8	12.6	68.2	29.7	3.8
Seepage water zone	Mean	145.3	130.3	2.6	3.0	5.0	4.4
	SE	11.8	14.7	1.1	1.6	1.9	0.3
	Min	124.4	104.5	0.0	0.0	0.4	3.9
	Max	190.0	184.1	5.5	8.7	10.9	5.5
Tributary	Mean	146.5	127.1	0.7	6.3	8.2	4.2
	SE	20.4	18.4	0.7	1.6	3.9	0.8
	Min	88.5	78.1	0.0	0.4	2.3	2.2
	Max	189.0	168.4	3.4	10.4	23.4	6.0

References

1. Brown, A.G.; Harper, D.; Peterken, G.F. European Floodplain Forests: Structure, Functioning and Management. *Glob. Ecol. Biogeogr. Lett.* **1997**, *6*, 169–178. [CrossRef]
2. Sutfin, N.A.; Wohl, E.E.; Dwire, K.A. Banking carbon: A review of organic carbon storage and physical factors influencing retention in floodplains and riparian ecosystems. *Earth Surf. Process. Landf.* **2016**, *41*, 38–60. [CrossRef]
3. Dybala, K.E.; Matzek, V.; Gardali, T.; Seavy, N.E. Carbon sequestration in riparian forests: A global synthesis and meta-analysis. *Glob. Chang. Biol* **2019**, *25*, 57–67. [CrossRef]
4. BMU. *Auenzustandsbericht: Flussauen in Deutschland*; Bundesministerium für Umwelt, Naturschutz und Reaktorsicherheit (BMU), Bundesamt für Naturschutz (BfN): Berlin, Germany, 2009.
5. Brunotte, E.; Dister, E.; Günther-Diringer, D.; Koenzen, U.; Mehl, D. Flussauen in Deutschland-Erfassung und Bewertung des Auenzustandes. *Nat. Und Biol. Vielfalt* **2009**, *87*, 141.
6. Scholz, M.; Mehl, D.; Schulz-Zunkl, C.; Kasperidus, H.D.; Born, W.; Henle, K. Ökosystemfunktionen von Flussauen—Analyse und Bewertung von Hochwasserretention, Nährstoffrückhalt, Kohlenstoffvorrat, Treibhausgasemissionen und Habitatfunktion. *Nat. Und Biol. Vielfalt* **2012**, *124*, 258.
7. Naiman, R.J.; Decamps, H.; McClain, M.E. *Riparia: Ecology, Conservation, and Management of Streamside Communities*; Elsevier Academic Press: London, UK, 2005.
8. Griscom, B.W.; Adams, J.; Ellis, P.W.; Houghton, R.A.; Lomax, G.; Miteva, D.A.; Schlesinger, W.H.; Shoch, D.; Siikamaki, J.V.; Smith, P.; et al. Natural climate solutions. *Proc. Natl. Acad. Sci. USA* **2017**, *114*, 11645–11650. [CrossRef] [PubMed]
9. Dave, R.; Saint-Laurent, C.; Murray, L.; Daldegan, G.A.; Brouwer, R.; Scaramuzza, C.A.d.M.; Raes, L.; Simonit, S.; Catapan, M.; Contreras, G.G.; et al. *Second Bonn Challenge Progress Report. Application of the Barometer in 2018*; IUCN: Gland, Switzerland, 2019.
10. Hornung, L.K.; Podschun, S.A.; Pusch, M. Linking ecosystem services and measures in river and floodplain management. *Ecosyst. People* **2019**, *15*, 214–231. [CrossRef]
11. Thomas, H.; Nisbet, T.R. An assessment of the impact of floodplain woodland on flood flows. *Water Environ. J.* **2007**, *21*, 114–126. [CrossRef]
12. Dixon, S.J.; Sear, D.A.; Odoni, N.A.; Sykes, T.; Lane, S.N. The effects of river restoration on catchment scale flood risk and flood hydrology. *Earth Surf. Process. Landf.* **2016**, *41*, 997–1008. [CrossRef]
13. Leyer, I.; Mosner, E.; Lehmann, B. Managing floodplain-forest restoration in European river landscapes combining ecological and flood-protection issues. *Ecol. Appl.* **2012**, *22*, 240–249. [CrossRef] [PubMed]
14. Mckenney, R.; Jacobson, R.B.; Wertheimer, R.C. Woody Vegetation and Channel Morphogenesis in Low-Gradient, Gravel-Bed Streams in the Ozark Plateaus, Missouri and Arkansas. *Geomorphology* **1995**, *13*, 175–198. [CrossRef]

15. Antonarakis, A.S.; Milan, D.J. Uncertainty in Parameterizing Floodplain Forest Friction for Natural Flood Management, Using Remote Sensing. *Remote Sens.* **2020**, *12*, 1799. [CrossRef]
16. Bennett, E.M.; Peterson, G.D.; Gordon, L.J. Understanding relationships among multiple ecosystem services. *Ecol. Lett.* **2009**, *12*, 1394–1404. [CrossRef] [PubMed]
17. Cierjacks, A.; Kleinschmit, B.; Babinsky, M.; Kleinschroth, F.; Markert, A.; Menzel, M.; Ziechmann, U.; Schiller, T.; Graf, M.; Lang, F. Carbon stocks of soil and vegetation on Danubian floodplains. *J. Plant. Nutr. Soil Sci.* **2010**, *173*, 644–653. [CrossRef]
18. Rieger, I.; Lang, F.; Kleinschmit, B.; Kowarik, I.; Cierjacks, A. Fine root and aboveground carbon stocks in riparian forests: The roles of diking and environmental gradients. *Plant Soil* **2013**, *370*, 497–509. [CrossRef]
19. Pretzsch, H. *Forest Dynamics, Growth, and Yield*; Springer: Berlin/Heidelberg, Germany, 2009; pp. 1–39. [CrossRef]
20. Thom, D.; Keeton, W.S. Stand structure drives disparities in carbon storage in northern hardwood-conifer forests. *For. Ecol. Manag.* **2019**, *442*, 10–20. [CrossRef]
21. Assman, E. *The Principles of Forest Yield Study. Section D- Structure, Increment and Yield of Stands in Relation to Sivicultural Treatment*; Pergamon: Oxford, UK, 1970; pp. 207–433. [CrossRef]
22. Schulze, E.D.; Beck, E.; Buchmann, N.; Clemens, S.; Müller-Hohenstein, K.; Scherer-Lorenzen, M. *Plant Ecology*, 2nd ed.; Springer: Berlin/Heidelberg, Germany, 2019. [CrossRef]
23. Bergmeier, E.; Roellig, M. Diversity, threats and conservation of European wood-pastures. In *European Wood-Pastures in Transition*; Routledge: London, UK, 2014. [CrossRef]
24. Schindler, S.; O'Neill, F.H.; Biró, M.; Damm, C.; Gasso, V.; Kanka, R.; van der Sluis, T.; Krug, A.; Lauwaars, S.G.; Sebesvari, Z.; et al. Multifunctional floodplain management and biodiversity effects: A knowledge synthesis for six European countries. *Biodivers. Conserv.* **2016**, *25*, 1349–1382. [CrossRef]
25. Reichardt, K.; Timm, L.C. *Soil, Plant and Atmosphere: Concepts, Processes and Applications*; Springer Nature: Cham, Switzerland, 2020.
26. Schwartz, R.; Gröngröft, A.; Miehlich, G. Pore Water Composition as Device for the Detection of Origin and Flow Direction of Soil Water in Alluvial Soils of the Middle Elbe River. *Acta Hydrochim. Hydrobiol.* **2003**, *31*, 423–435. [CrossRef]
27. European Commission. *Interpretation Manual of European Union Habitats—EUR28*; European Commission: Brussels, Belgium, 2013.
28. Parelle, J.; Brendel, O.; Bodénès, C.; Berveiller, D.; Dizengremel, P.; Jolivet, Y.; Dreyer, E. Differences in morphological and physiological responses to water-logging between two sympatric oak species (*Quercus petraea* [Matt.] Liebl., *Quercus robur* L.). *Ann. For. Sci.* **2006**, *63*, 849–859. [CrossRef]
29. Li, M.; Lopez, R.; Venturas, M.; Pita, P.; Gordaliza, G.G.; Gil, L.; Rodriguez-Calcerrada, J. Greater resistance to flooding of seedlings of Ulmus laevis than Ulmus minor is related to the maintenance of a more positive carbon balance. *Trees-Struct. Funct.* **2015**, *29*, 835–848. [CrossRef]
30. Urli, M.; Lamy, J.B.; Sin, F.; Burlett, R.; Delzon, S.; Porte, A.J. The high vulnerability of Quercus robur to drought at its southern margin paves the way for Quercus ilex. *Plant. Ecol.* **2015**, *216*, 177–187. [CrossRef]
31. Venturas, M.; Lopez, R.; Gasco, A.; Gil, L. Hydraulic properties of European elms: Xylem safety-efficiency tradeoff and species distribution in the Iberian Peninsula. *Trees-Struct. Funct.* **2013**, *27*, 1691–1701. [CrossRef]
32. DWD Climate Data Center (CDC). *Multi-Annual Station Means for the Climate Normal Reference Period 1981–2010, f.c.s.l.a.f.r.s.l., Version V0.x*; CDC: Atlanta, GA, USA, 2020.
33. Leyer, I. Effects of dykes on plant species composition in a large lowland river floodplain. *River Res. Appl.* **2004**, *20*, 813–827. [CrossRef]
34. Weber, A.; Hatz, M.; Rosenzweig, S. Hyd1d & hydflood—R packages for the computation of 1D water levels, flood extents and flood duration along German federal waterways Elbe and Rhine. German Federal Institute of Hydrology (BfG): Koblenz, Germany, Unpublished BfG report. 2020.
35. Zianis, D.; Muukkonen, P.; Mäkipää, R.; Maurizio, M. Biomass and Stem Volume Equations for Tree Species in Europe. *Silva Fenn.* **2005**, *4*, 1–63.
36. Zanne, A.E.; Lopez-Gonzalez, G.; Coomes, D.A.; Ilic, J.; Jansen, S.; Lewis, S.L.; Miller, R.B.; Swenson, N.G.; Wiemann, M.C.; Chave, J. Data from: Towards a worldwide wood economics spectrum. *Dryad Data Repos.* **2009**. [CrossRef]
37. UNFCCC. *Measurements for Estimation of Carbon Stocks in Afforestation and Reforestation Project Activities under the Clean Development Mechanism: A Field Manual*; UNFCCC: Bonn, Germany, 2015.
38. Mokany, K.; Raison, R.J.; Prokushkin, A.S. Critical analysis of root: Shoot ratios in terrestrial biomes. *Glob. Chang. Biol.* **2006**, *12*, 84–96. [CrossRef]
39. Böhl, J.; Brändli, U.B. Deadwood volume assessment in the third Swiss National Forest Inventory: Methods and first results. *Eur. J. For. Res.* **2007**, *126*, 449–457. [CrossRef]
40. Keeton, W.S.; Kraft, C.E.; Warren, D.R. Mature and old-growth riparian forests: Structure, dynamics, and effects on Adirondack stream habitats. *Ecol. Appl.* **2007**, *17*, 852–868. [CrossRef] [PubMed]
41. Applequist, M.B. A simple pith locator for use with off-centre increment cores. *J. For.* **1958**, *56*, 141.
42. Horner, G.J.; Baker, P.J.; Mac Nally, R.; Cunningham, S.C.; Thomson, J.R.; Hamilton, F. Forest structure, habitat and carbon benefits from thinning floodplain forests: Managing early stand density makes a difference. *For. Ecol. Manag.* **2010**, *259*, 286–293. [CrossRef]

43. Junk, W.; Bayley, P.B.; Sparks, R.E. The flood pulse concept in river-floodplain systems. *Dodge Ed. Proc. Int. Large River Symp. Can. Spec. Publ. Fish. Aquat. Sci.* **1989**, *106*, 110–127.
44. Hesse, C.; Krysanova, V. Modeling Climate and Management Change Impacts on Water Quality and In-Stream Processes in the Elbe River Basin. *Water* **2016**, *8*, 40. [CrossRef]
45. Huang, S.C.; Krysanova, V.; Hattermann, F. Projections of climate change impacts on floods and droughts in Germany using an ensemble of climate change scenarios. *Reg. Environ. Chang.* **2015**, *15*, 461–473. [CrossRef]
46. Barkmann, T.; Siebert, R.; Lange, A. Land-use experts' perception of regional climate change: An empirical analysis from the North German Plain. *Clim. Chang.* **2017**, *144*, 287–301. [CrossRef]
47. Heklau, H.; Jetschke, G.; Bruelheide, H.; Seidler, G.; Haider, S. Species-specific responses of wood growth to flooding and climate in floodplain forests in Central Germany. *Iforest-Biogeosci. For.* **2019**, *12*, 226–236. [CrossRef]
48. Dieter, M.; Elsasser, P. Carbon Stocks and Carbon Stock Changes in the Tree Biomass of Germany's Forests. Kohlenstoffvorrate und -veranderungen in der Biomasse der Waldbaume in Deutschland. *Forstwiss. Cent.* **2002**, *121*, 195–210. [CrossRef]
49. Förster, A.; Culmsee, H.; Leuschner, C. Thinned northern German Scots pine forests have a low carbon storage and uptake potential in comparison to naturally developing beech forests. *For. Ecol. Manag.* **2021**, *479*. [CrossRef]
50. Wellbrock, N.; Grüneberg, E.; Riedel, T.; Polley, H. Carbon stocks in tree biomass and soils of German forests. *Cent. Eur. For. J.* **2017**, *63*, 105–112. [CrossRef]
51. Gschwantner, T.; Lanz, A.; Vidal, C.; Bosela, M.; Cosmo, L.; Fridman, J.; Gasparini, P.; Kuliesis, A.; Tomter, S.; Schadauer, K. Comparison of methods used in European National Forest Inventories for the estimation of volume increment: Towards harmonisation. *Ann. For. Sci.* **2016**, *73*, 807–821. [CrossRef]
52. Giese, L.A.; Aust, W.M.; Trettin, C.C.; Kolka, R.K. Spatial and temporal patterns of carbon storage and species richness in three South Carolina coastal plain riparian forests. *Ecol. Eng.* **2000**, *15*, S157–S170. [CrossRef]
53. Schöngart, J.; Arieira, J.; Felfili Fortes, C.; Cezarine de Arruda, E.; Nunes da Cunha, C. Age-related and stand-wise estimates of carbon stocks and sequestration in the aboveground coarse wood biomass of wetland forests in the northern Pantanal, Brazil. *Biogeosciences* **2011**, *8*, 3407–3421. [CrossRef]
54. Keeton, W.S.; Whitman, A.A.; McGee, G.C.; Goodale, C.L. Late-Successional Biomass Development in Northern Hardwood-Conifer Forests of the Northeastern United States. *For. Sci.* **2011**, *57*, 489–505. [CrossRef]
55. van Breugel, M.; Ransijn, J.; Craven, D.; Bongers, F.; Hall, J.S. Estimating carbon stock in secondary forests: Decisions and uncertainties associated with allometric biomass models. *For. Ecol. Manag.* **2011**, *262*, 1648–1657. [CrossRef]
56. Glaeser, J.; Schmidt, P.A. Zur historischen Entwicklung des Baumartenbestandes von Hartholz-Auenwäldern—Dargestellt am Beispiel des Leipziger Auenwaldes. *Allgem. For. Jagdztg.* **2006**, *178*, 90–97.
57. Larsen, J.B. Ecological Stability of Forests and Sustainable Silviculture. *For. Ecol. Manag.* **1995**, *73*, 85–96. [CrossRef]
58. Pretzsch, H.; Bielak, K.; Block, J.; Bruchwald, A.; Dieler, J.; Ehrhart, H.P.; Kohnle, U.; Nagel, J.; Spellmann, H.; Zasada, M.; et al. Productivity of mixed versus pure stands of oak (*Quercus petraea* (Matt.) Liebl. and *Quercus robur* L.) and European beech (Fagus sylvatica L.) along an ecological gradient. *Eur. J. For. Res.* **2013**, *132*, 263–280. [CrossRef]
59. Ma, Z.L.; Chen, H.Y.H.; Bork, E.W.; Carlyle, C.N.; Chang, S.X. Carbon accumulation in agroforestry systems is affected by tree species diversity, age and regional climate: A global meta-analysis. *Glob. Ecol. Biogeogr.* **2020**, *29*, 1817–1828. [CrossRef]
60. Guyot, V.; Jactel, H.; Imbaud, B.; Burnel, L.; Castagneyrol, B.; Heinz, W.; Deconchat, M.; Vialatte, A. Tree diversity drives associational resistance to herbivory at both forest edge and interior. *Ecol. Evol.* **2019**, *9*, 9040–9051. [CrossRef] [PubMed]
61. Erfmeier, A.; Haldan, K.L.; Beckmann, L.M.; Behrens, M.; Rotert, J.; Schrautzer, J. Ash Dieback and Its Impact in Near-Natural Forest Remnants—A Plant Community-Based Inventory. *Front. Plant. Sci.* **2019**, *10*, 658. [CrossRef] [PubMed]

water

MDPI

Article

Wet Meadow Plant Communities of the Alliance *Trifolion pallidi* on the Southeastern Margin of the Pannonian Plain

Andraž Čarni [1,2], Mirjana Ćuk [3], Igor Zelnik [4], Jozo Franjić [5], Ružica Igić [3], Miloš Ilić [3], Daniel Krstonošić [5], Dragana Vukov [3] and Željko Škvorc [5,*]

1 Research Centre of the Slovenian Academy of Sciences and Arts, Institute of Biology, 1000 Ljubljana, Slovenia; andraz.carni@zrc-sazu.si
2 School for Viticulture and Enology, University of Nova Gorica, 5000 Nova Gorica, Slovenia
3 Department of Biology and Ecology, Faculty of Science, University of Novi Sad, 21000 Novi Sad, Serbia; mirjana.cuk@dbe.uns.ac.rs (M.Ć.); ruzica.igic@dbe.uns.ac.rs (R.I.); milos.ilic@dbe.uns.ac.rs (M.I.); dragana.vukov@dbe.uns.ac.rs (D.V.)
4 Department of Biology, Biotechnical Faculty, University of Ljubljana, 1000 Ljubljana, Slovenia; igor.zelnik@bf.uni-lj.si
5 Faculty of Forestry, University of Zagreb, 10000 Zagreb, Croatia; jfranjic@sumfak.hr (J.F.); dkrstonosic@sumfak.hr (D.K.)
* Correspondence: zskvorc@sumfak.hr

Citation: Čarni, A.; Ćuk, M.; Zelnik, I.; Franjić, J.; Igić, R.; Ilić, M.; Krstonošić, D.; Vukov, D.; Škvorc, Ž. Wet Meadow Plant Communities of the Alliance *Trifolion pallidi* on the Southeastern Margin of the Pannonian Plain. *Water* **2021**, *13*, 381. https://doi.org/10.3390/w13030381

Academic Editor: Thomas Hein
Received: 30 December 2020
Accepted: 26 January 2021
Published: 1 February 2021

Abstract: The article deals with wet meadow plant communities of the alliance *Trifolion pallidi* that appear on the periodically inundated or waterlogged sites on the riverside terraces or gentle slopes along watercourses. These plant communities are often endangered by inappropriate hydrological interventions or management practices. All available vegetation plots representing this vegetation type were collected, organized in a database, and numerically elaborated. This vegetation type appears in the southeastern part of the Pannonian Plain, which is still under the influence of the Mediterranean climate; its southern border is formed by southern outcrops of the Pannonian Plain and its northern border coincides with the influence of the Mediterranean climate (line Slavonsko Gorje-Fruška Gora-Vršačke Planine). Numerical analysis established four plant associations—*Trifolio pallidi–Alopecuretum pratensis*, *Ventenato dubii–Trifolietum pallidi*, *Ranunculo strigulosi–Alopecuretum pratensis*, and *Ornithogalo pyramidale–Trifolietum pallidi*. Each association was elaborated in detail: diagnostic plant species, nomenclature, geographical distribution, climatic and ecological conditions, and possible division into subassociations. Results are presented in a distribution map, figures resulting from numerical analysis, and a synoptic table. The hydrological gradient was found as the most important factor shaping the studied plant communities. The article also brings new field data on this vegetation type, which has not been sampled for decades and is in process of evaluation to be included as a special habitat type in the Habitat Directive.

Keywords: plant communities; habitat; meadow; Pannonian Plain; *Trifolion pallidi*; vegetation

1. Introduction

Wet meadows of *Trifolion pallidi* appear on clayey, mesotrophic to eutrophic soils on riverside terraces and gentle slopes along the rivers in the southeastern margin of the Pannonian Plain. Inundations of the sites during winter and spring are common, and the humidity of sites is often influenced by the high level of groundwater, which is frequently close to the surface. During the summer, the ground may dry up [1].

Hydrological factors such as seasonal floods and high water table in the soil significantly define the composition of plant communities exposed to such influences. The hydrological gradient has been generally accepted as the primary factor that determines the growth of different species and the structure of herbaceous wetland plant communities [2–5]. Water level fluctuation that is also characteristic of studied sites shapes the plant communities in different ways. The sensitivity to inundation and drying have both important roles in the distribution and abundance of species [6,7]. The sensitivity of plant species to water level fluctuations is reflected in their reduced growth and abundance [8,9].

Vegetation and ecology of the meadows on the southern margin of the Pannonian Plain were studied in relation to climatic gradient by Ilijanić [10]. He divided the wet meadow vegetation of northern Croatia into three groups. The most easterly group extends from the river Orljava eastwards and is characterized by a dry climate with low precipitation (600–800 mm). Dry periods are in April and from July to September, with an average monthly temperature of 22–23 °C and an absolute maximum in July over 40 °C. He also stated that other ecological conditions in northern Croatia are very similar; only climatic conditions are different, and this is reflected in the vegetation. He also reported that wet meadows from the area are more similar to those appearing along the rivers Sava and Morava in Serbia [11,12] than to meadows of western Croatia and established the western and northern boundaries of the elaborated vegetation.

Ilijanić [13] later classified wet meadows of the region within the alliance *Trifolion pallidi*, integrating it in the order *Trifolio–Hordeetalia*, which encompasses wet meadows under the influence of the Mediterranean climate in the eastern Mediterranean, namely, meadows of the alliances *Trifolion resupinati*, which occur in northern Macedonia and *Molinio–Hordeion* from the coastal region of Croatia. The order today also includes an alliance encompassing salty meadows in the northern Balkans (*Trifolio–Ranunculion pedati*) and wet meadows on karst poljes in the Apennines (*Ranunculion velutini*) [14]. There still exist some open questions. For instance, Purger [15] suggests the classification of the alliance *Trifolio–Ranunculion pedati* as a suballiance within the alliance *Festucion pseudovinae* of the class of salted steppe and steppic grasslands *Festuco–Puccinellietea*. There was also a discussion about the classification of the alliance *Alopecurion utriculati* [16], wet meadows distributed over the whole Balkan Peninsula. This problem was solved by Mucina and Theurillat in [14] by its lectotypification—they have selected the same association as a type of *Alopecurion* as of the alliance *Molinio–Hordeion* and in this way, *Alopecurion* became a homotypic synonym of *Molinio–Hordeion*.

A great part of wet meadows of *Trifolion pallidi* is (co)dominated by *Alopecurus pratensis*, a species of broad ecological amplitude and wide distribution range. It is a dominant species in many plant communities across Europe (e.g., [17–19]), which are assigned to various higher syntaxa due to different floristic compositions, reflecting different (macro) climatic and ecological conditions and their phytogeographic position.

In the wider Pannonian region, *Alopecurus pratensis*-dominated communities can be found in various habitats, in salted inundated meadows [20], saline steppic pastures [15], temporarily flooded and grazed pastures [21], and in wet meadows [22].

The *Trifolion pallidi* communities appear only on the southeastern margin of the Pannonian Plain. According to the numerical analysis of meadow vegetation along the southern margin of the Pannonian Plain, it was established that *Trifolion pallidi* communities form a distinct group of communities (alliance) appearing only in its southeastern margin [23], and they cannot be found in the central part of the Pannonian Plain [1,24]. This was confirmed also by recent numerical analysis of habitats on the European scale [25].

(Macro) climatic, phytogeographical, and ecological factors are often interdependent in terms of their influence on site conditions; e.g., in drier macroclimatic conditions, plant communities with a mesic character can establish in wet sites to compensate for the lack of precipitation. *Alopecurus pratens* is dominated meadows (*Alopecurion*/*Deschampsion*), widespread in west Croatia, under the increased influence of the drier continental climate toward the east (east Croatia, Serbia), thus retreat to locally wetter sites, building *Trifolion pallidi* communities [23].

Many vegetation types of meadows are included in the European Habitat Directive [26] and the Interpretation Manual of European Union Habitats [27]. The vegetation under consideration (wet meadows of *Trifolion pallidi*) has not yet been considered within the

European Habitat Directive [26] but has been proposed for consideration in the accession process of Serbia [23].

The aim of the paper is to collect all available data about *Trifolion pallidi* communities and present the appearance of wet meadows of the alliance *Trifolion pallidi* on the south-eastern margin of the Pannonian Plain. We tried to reveal the most important factors that enable the diversity of those meadows. We prepared an insight into its classification up to subassociation level taking into consideration also results of new synthetic works that have appeared recently. At the same time, we provided new field data on this habitat, which is under consideration within the European Habitat Directive [26].

2. Materials and Methods

2.1. Study Area

The study area extends along the southeastern margin of the Pannonian Basin, encompassing eastern parts of Croatia and the northern part of Serbia (Figure 1). This area is under the influence of a moderate continental climate: a moderately warm and humid climate with warm summers [28,29]. The climatic data for three meteorological stations in the region are—Osijek mean annual temperature 11.0 °C and annual precipitation 655 mm; Novi Sad 10.9 °C and 647 mm and Valjevo 11.4 °C and 787 mm [28] (http://www.hidmet.gov.rs, accessed 31 October 2020).

Figure 1. Position of the studied area (rectangle) within SE Europe.

Deciduous forests dominated by Hungarian oak (*Quercus frainetto*) appear in the area [30]. In the west, the area of research corresponds to the western localities of Hungarian oak in Slavonia (Kutjevo) [31]. In the north, the area includes a major part of Vojvodina. This area is still influenced by the Mediterranean climate, which arrives from the south (Figure 1) along rivers (e.g., Morava) [32]. The southern boundary corresponds to the geomorphological border of the Pannonian Plain [33], which corresponds to the phytogeographic border of the Pannonian Plain toward the Illyrian region proposed [34] and the border of the central European region toward the Mediterranean region [35].

2.2. Data

We collected all available relevés elaborated according to the standard central European method [36,37] dealing with wet meadows of *Trifolion pallidi* from the region under consideration. This study is based on a data set consisting of 181 relevés—146 relevés were used from the available literature sources [11,12,38–44], and 35 were newly made in the field (Figure A2). The relevés were stored in Turboveg [45] and elaborated in the Juice program [46].

We took into consideration all plant associations considered to belong to the alliance *Trifolion pallidi* according to [23]. These are *Ranunculo–Alopecuretum*, *Trifolio–Alopecuretum*, and *Ornithogalo–Trifolietum*. We added some more associations that were included within these three associations, *Oenantho banaticae–Alopecuretum*, *Ononido spinosiformis–Alopecuretum*, and *Rhinantho rumelici–Filipenduletum vulgaris*, which is usually considered within the frame of *Arrhenatherion* [40,47] and *Ventenato–Trifolietum pallidi*, which was originally considered to be part of *Molinion* [38] and later of *Deschampsion/Cnidion* [40]. On the other hand, we excluded the association *Agrostio caninae–Hordeetum secalini*, which was originally assigned to the *Trifolion pallidi* alliance [13,47] but, according to [23], belongs to the alliance *Alopecurion*.

Unweighted Ellenberg indicator values (EIV) [48] for nutrients and moisture were calculated for each relevé to facilitate ecological interpretation of clusters. Climatic conditions were estimated by climatic data extracted from the Worldclim database [49]. These variables and the number of species in relevés are presented as box–whisker diagrams. We also extracted longitude, latitude, and altitude from relevé material. We correlated all these variables with the floristic gradient presented by the first two axes of detrended correspondence analysis (DCA).

2.3. Nomenclature

Taxonomic nomenclature follows Euro+Med Plantbase [50]. Nomenclature of syntaxa is according to [23] for meadow vegetation and [14] for other syntaxa; nomenclature is adjusted according to the International Code of Phytosociological Nomenclature [51]. Throughout the contribution, we use *Trifolio–Alopecuretum* for *Trifolio pallidi–Alopecuretum pratensis*, *Ventenato–Trifolietum* for *Ventenato dubii–Trifolietum pallidi*, *Ranunculo–Alopecuretum* for *Ranunculo strigulosi–Alopecuretum pratensis*, and *Ornithogalo–Trifolietum* for *Ornithogalo pyramidale–Trifolietum pallidi*.

2.4. Numerical Analysis

To reduce noise in the data, outlier analysis was conducted using PC-ORD 5.0 (MjM Software Design, Gleneden Beach, OR, USA), and four relevés whose species composition deviated more than ±2 SD from the mean calculated Euclidean distance of all plots were excluded. Finally, 175 relevés were included in the analysis, which contained a total of 321 taxa of vascular plants.

Classification of relevés was performed using cluster analysis in PC-ORD 5.0. The OptimClass method [52] for identifying the optimal partition suggested square root transformation of cover values of species, Beta flexible ($\beta = -0.25$) for group linkage with the relative Sørensen index as the distance measure.

Diagnostic species were determined by calculating fidelity using the phi (Φ) coefficient. Only species with $\Phi > 0.4$ and a probability under the random expectation of the observed pattern of species occurrence lower than 0.001 (Fisher's exact test) were considered diagnostic for associations and subassociations [53]. Diagnostic species of subassociations were calculated only within relevés classified to the relevant association. To calculate fidelity, the number of relevés for each (sub) association was virtually standardized to equal size [54].

Within this elaboration, we treated as diagnostic for the alliance *Trifolion pallidi* and order *Trifolio–Hordeetalia* all plant species that are considered diagnostic for these syntaxa in Horvatić [55], Ilijanić [13], Šilc et al. [56], and Škvorc et al. [23], as follows: *Agrostis stolonifera* subsp. *maritima*, *Alopecurus pratensis*, *Alopecurus utriculatus*, *Anacamptis palustris* subsp. *palustris*, *Carex distans*, *Carex praecox*, *Centaurea angustifolia* subsp. *pannonica*, *Cirsium canum*, *Clematis integrifolia*, *Galium debile*, *Hordeum secalinum*, *Inula britanica*, *Lathyrus nissolia*, *Lotus tenuis*, *Lythrum virgatum*, *Medicago arabica*, *Oenanthe silaifolia*, *Poa pratensis*, *Poa trivialis* (including subsp. *sylvicola*), *Ranunculus polyanthemos*, *Ranunculus sardous*, *Ranunculus strigulous*, *Trifolium fragiferum*, and *Trifolium pallidum*. We included some of them in the synthetic table (Figure A3) as diagnostic species of associations and excluded some widely distributed species, such as *Poa pratensis* and *Poa trivialis* (in which the subspecies *sylvicola* is often not distinguished).

Detrended correspondence analysis (DCA) was performed by the Vegan program [57] in the R program environment (https://cran.wu.ac.at/). The first two axes are presented on the diagram with passively projected climatic and ecological variables that show the highest correlation with the axes calculated by Spearman correlations. We prepared a synoptic table (Figure A3) and a table of new unpublished relevés used in the analysis (Figure A2); we also showed the distribution of individual syntaxa on the map and a map presenting literature and own field data (Figure A1).

3. Results

Classification analysis Figure 2 showed that relevés can be divided into six clusters (groups). They are consistent with the existing classification scheme, with the exception of *Ventenato–Trifolietum*, which has not appeared in previous analyses. The diagram shows four associations, two of which are subdivided into subassociations (Figure 2). These groups were the subject of further analysis.

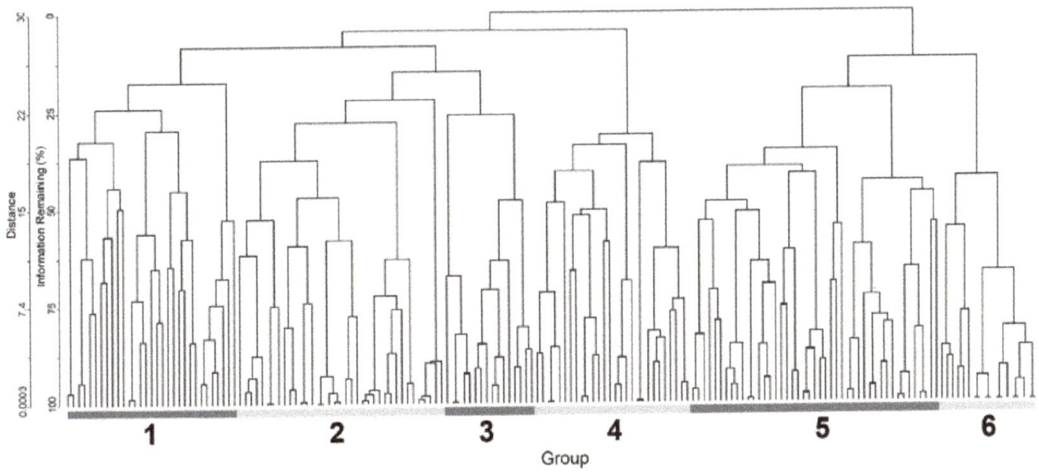

Figure 2. Dendrogram of analyzed relevés obtained with square root transformation of cover values in percentage, Beta flexible ($\beta = -0.25$) and group linkage with the relative Sørensen index. Legend: 1—*Ranunculo–Alopecuretum typicum*, 2—*Trifolio–Alopecuretum rhinanthetosum*, 3—*Trifolio–Alopecuretum typicum*, 4—*Ventenato–Trifolietum*, 5—*Ranunculo–Alopecuretum filipenduletosum*, 6—*Ornithogalo–Trifolietum*.

EIV values for moisture Figure 3 show that the wettest sites are those of *Ranunulo–Alopecuretum typicum*, *Ventenato–Trifolietum*, and *Trifolio–Alopecuretum typicum*. At the same time, it was found that the nutrient status is highest within *Ranunuculo–Alopecuretum typicum* and *Trifolio–Alopecuretum typicum*, while there are fewer nutrients within *Ventenato–Trifolietum*. It must be taken into account that the moisture of sites is a result of inundations and higher groundwater table in combination with soil characteristics (clay). Inundations also bring nutrients to the sites, but this is not always the case with high groundwater, so the two properties are not necessarily parallel. It can be seen that *Ornithogalo–Trifolietum* appears on the driest and nutrient poorest stands, while the remaining groups, such as *Ranunuculo–Alopecuretum filipenduletosum* and *Trifolio–Alopecuretum rhinanthetosum*, possess an intermediate position in respect to these ecological factors. It can be seen that altitude has a very narrow range of 100 m and does not play an important role in the differentiation of groups. This division is also shown by the ecological division of groups reflected on the first axis of DCA analysis (Table 1 and Figure 4).

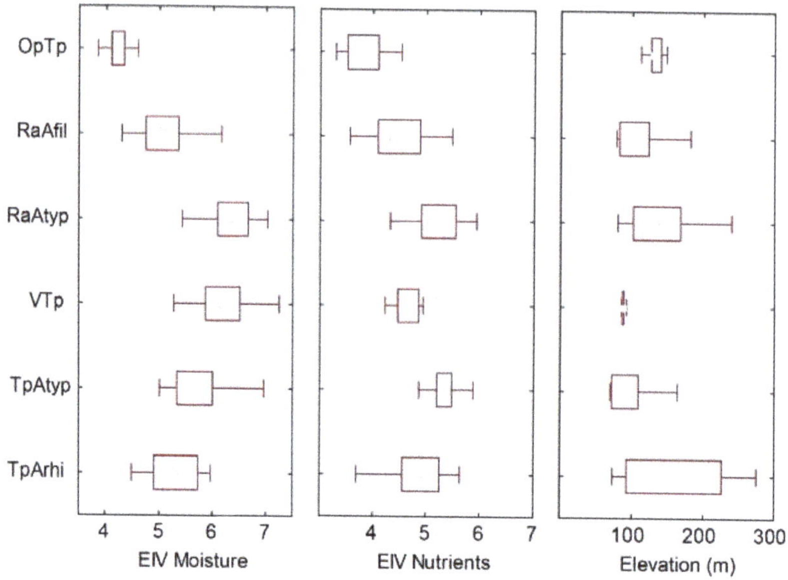

Figure 3. The box–whisker diagram of Ellenberg indicator values (EIV) bioindicator values for moisture and nutrients and altitude (elevation): EIV moisture, EIV nutrients, and elevation. Boxes show the 25–75% quartile range and median value; whiskers indicate the range of values, except outliers. Legend: OpTp—*Ornithogalo–Trifolietum*, RaAfil—*Ranunculo–Alopecuretum filipenduletosum*, RaAtyp—*Ranunculo–Alopecuretum typicum*, VTp—*Ventenato–Trifolietum*, TpAtyp—*Trifolio–Alopecuretum typicum*, TpArhi—*Trifolio–Alopecuretum rhinanthetosum*.

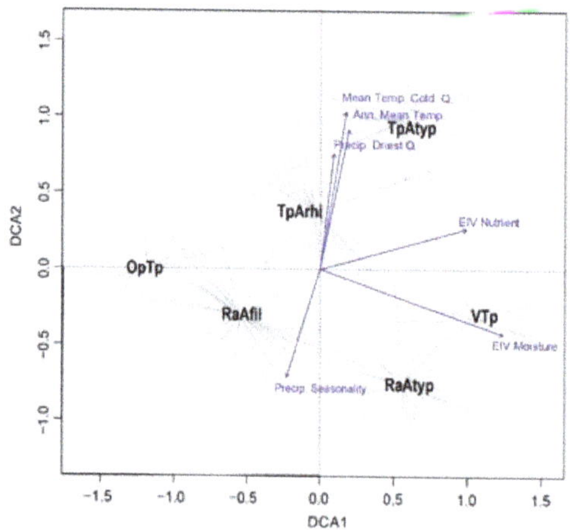

Figure 4. Diagram of Detrended Correspondence Analysis (DCA) of relevés with the centroid of groups and spider plots with passively projected EIV nutrient and EIV moisture and climatic variables—precipitation seasonality, mean temperature of the coldest quarter, mean annual temperature, precipitation in the driest quarter. Eigenvalues for the first two axes are 0.493 and 0.295, respectively. The legend is the same as in Figure 3.

Table 1. Spearman correlations of the first two detrended correspondence analysis (DCA) axes, with environmental variables. Only significant correlations are presented ($p < 0.001$).

Environmental Variable	DCA1	DCA2
Elevation		0.25
Latitude		0.66
Longitude	0.31	
Annual Mean Temperature		0.62
Temperature Seasonality		0.38
Max Temperature of Warmest Month		0.33
Min Temperature of Coldest Month		0.39
Mean Temperature of Wettest Quarter	0.30	0.50
Mean Temperature of Warmest Quarter	0.31	0.55
Mean Temperature of Coldest Quarter		0.70
Annual Precipitation		0.30
Precipitation of Wettest Month	0.41	0.44
Precipitation of Driest Month		0.34
Precipitation Seasonality		0.54
Precipitation of Wettest Quarter	0.40	
Precipitation of Driest Quarter		0.45
Precipitation of Warmest Quarter	0.34	0.20
Precipitation of Coldest Quarter		0.55
Light EIV	0.36	
Moisture EIV	0.89	0.28
Nutrients EIV	0.66	

Climatic conditions reflect the geographic position of the studied plant communities (Figures 5 and 6). There is more precipitation (mean annual, of the driest quarters) in *Trifolio–Alopecuretum* (both subassociations) appearing in the northern part of central Serbia. At the same time, the temperature regime is more propitious in this region (annual mean temperature, mean temperature of the coldest quarter). The climatic conditions become more severe (precipitation seasonality) in the regions toward the central Pannonian Plain, where *Ranunuculo–Alopecuretum* (both subassociations), *Ornithogalo–Trifolietum*, and *Ventenato–Trifolietum* can be found. Macroclimatic conditions correspond to the second DCA axis (Table 1 and Figure 4) and separate communities according to their geographical position (Figure 6). The number of plant species varies between the studied communities and reaches higher values in the subassociations (Figure 5), which occur on slightly drier sites than typical forms (Figure 3).

The DCA diagram shows the separation of groups along two gradients—axis 1 represents the ecological gradient, and axis 2 represents the macroclimatic (in our case also geographic) gradient (Figure 4 and Table 1).

Axis 1 shows the separation along EIV moisture and nutrient. *Ventenato–Trifolietum*, *Trifolio–Alopecuretum typicum* and *Ranunculo–Alopecuretum typicum* can be found in the wettest and most nutrient-rich sites; on drier and nutrient poorer sites, *Trifolio–Alopecuretum rhinanthetosum*, *Ranunuculo–Alopecuretum filipenduletosum*, and *Ornithogalo–Trifolietum* appear. Axis 2 shows the distinction between relevés appearing in the region under stronger influence of the Mediterranean climate in the northern part of central Serbia (*Trifolio–Alopecuretum* (both subassociations) and those from Vojvodina (northern Serbia) and eastern Croatia, where this influence diminishes (*Ranunculo–Alopecuretum* (both subassociations), *Ornithogalo–Trifolietum*, and *Ventenato–Trifolietum*).

Wet meadows of *Trifolion pallidi* appear along the southeastern margin of the Pannonian Plain because there is a certain influence of the Mediterranean climate from the south (e.g., along river Morava), which gradually diminishes toward the north (Table 1, Figure 6). Along this gradient, we can distinguish two associations of wet meadows that are widely distributed in the area, namely, *Trifolio–Alopecuretum* thriving in areas with a milder climate in the northern part of central Serbia and *Ranunculo–Alopecuretum* appearing northwards in the area in which the climate becomes harsher, i.e., in the area approximately till the line from Slavonsko Gorje over Fruška Gora to Vršačke Planine. Two associations have been described on small areas and represent local site conditions—*Ventenato–Trifolietum* in extremely wet conditions along the Danube and Drava River and *Ornithogalo–Trifolietum* in the driest conditions on Vršačke Planine. This distinction between the associations is predominately on a (macro)climatic scale.

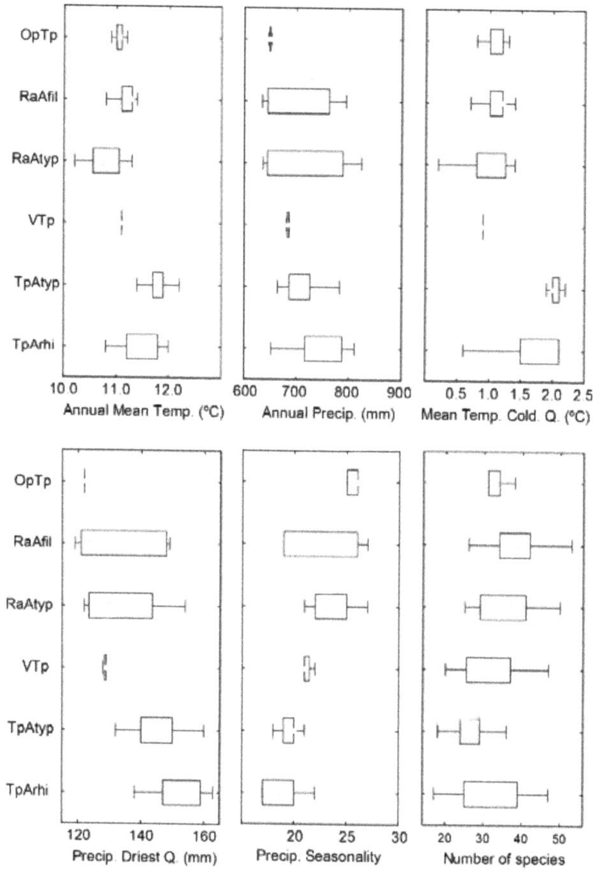

Figure 5. Macroclimatic features represented by annual mean temperature, annual precipitation, mean temperature of the coldest quarter, precipitation of the driest quarter, and precipitation seasonality and number of species. The legend is the same as in Figure 3.

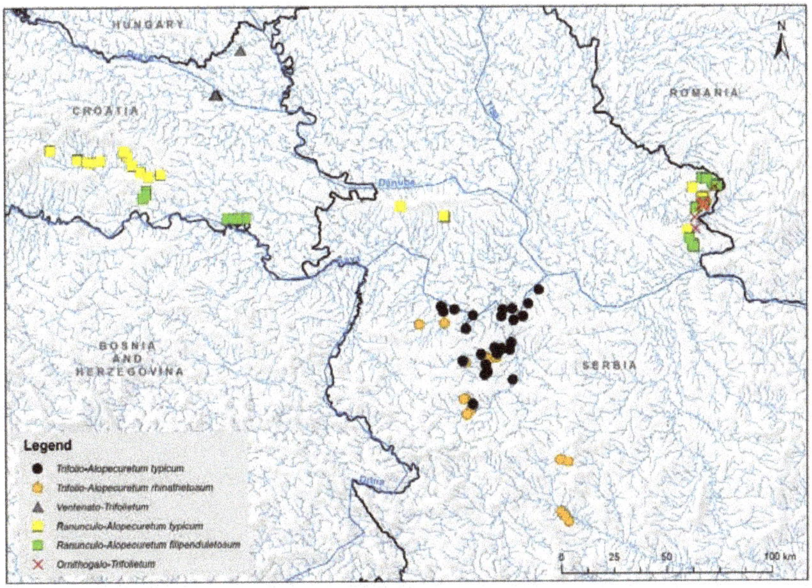

Figure 6. The geographical position of elaborated plant communities.

4. Discussion

We confirmed the findings of Ilijanić [10,13], who reported that wet meadows of *Trifolion pallidi* appear in the eastern part of Croatia and northern Serbia. Ilijanić [13] also anticipated that such vegetation could also be found in southern Romania and northern Bulgaria, but it has not yet been reported from those regions [58,59]. However, some plant communities of a transitional character have been found there (e.g., [60]).

The present analysis is based on 20% of new field data. The new data did not produce their own clusters but were merged among existing syntaxa (Figures A1 and A2). This indicates that no major change has appeared in species composition during different periods of sampling. It supports the estimation of [1], who found out that only a small percentage of the area of these communities is degraded (up to 20%), whereas the degree of degradation is slight (to moderate). We cannot judge about the extension of these communities from our data.

The *Trifolion pallidi* communities are distributed in the southeastern margin of the Pannonian Plain. The northern border coincides with the influence of the Mediterranean climate that comes from the south [32] and it extends from Slavonsko Gorje over Fruška Gora to Vršačke Planine. The transitional character of *Ventenato–Trifolietum* will be discussed below. The southern border coincides with the border of the Pannonian Plain [33].

Thermophilous deciduous oak forests of the alliances *Quercion frainetto* and *Quercion pubescenti–petraeae* and mesic oak-hornbeam forests of *Erythronio–Carpinion* appear in the area of distribution of *Trifolion pallidi*, but steppe woodlands of *Aceri–Quercion*, which appear on loess plateaus toward the north, cannot be found [30,61,62].

The hydrological gradient obtained on the base of EIV moisture values significantly correlated with axis 1 of DCA (Table 1), so we can confirm that the moisture gradient turned out as the most important factor in shaping the composition of the studied plant communities. Soil moisture and nutrient content have been shown to be the most important factors that determine the species composition of wet grasslands [63].

Any measures that would significantly change the extent or frequency of inundations of these sites or the level of groundwater would negatively influence the studied plant communities [64] and could eventually lead to changes of floristic composition and to substantial loss of regional biodiversity. The changes in hydrology would in the worst case cause the vanishing of this habitat type.

A distinction within associations on a local scale also exists, reflecting ecological site conditions. Two subassociations can be distinguished within each of the two broadly distributed associations; each reflects wetter and more nutrient-rich site conditions, on the one hand, and less wet and nutrient poorer conditions on the other. *Ranunuculo–Alopecuretum typicum* and *Trifolio–Alopecuretum typicum* can thus be found on wetter and more nutrient-rich sites and *Ranunuculo–Alopecuretum filipenduletosum* and *Trifolio–Alopecuretum rhinanthetosum* on less wet and nutrient poorer sites. The other two associations appear on a limited area and in fairly unique ecological site conditions, so they were not subdivided into subassociations. This distinction on a subassociation level is recognized on a local scale.

4.1. Description of Individual Plant Communities

4.1.1. Association *Trifolio pallidi–Alopecuretum pratensis* Cincović 1959

Diagnostic plant species: *Cichorium intybus, Mentha pulegium, Ranunculus sardous, Rorippa sylvestris, Trifolium patens, Veronica serpyllifolia.*

Nomenclatural remark. The association was lectotypified in Aćić et al. [65] (p. 46). As lectotype, they selected Cincović 1959 [12]: (Table 3, rel. 22). According to Cincović [12], this relevé belongs to the subassociation *rhinanthetosum rumelici*. Article 19a [51] stated that a lectotype must be chosen from relevés assigned to the subassociation *typicum*, if such a subassociation exists. In Cincović [12], such a subassociation does exist and this makes the lectotypification invalid. We select here as lectotype of the association *Trifolio pallidi–Alopecuretum pratensis*: Cincović 1959 [12]: Table 3, relevé 14, lectotypus hoc loco. This relevé is also the autonym of the subassociation *typicum* (Art. 13b). At the same time, we typify the subassociation *Trifolio pallidi–Alopecuretum pratensis rhinanthetosum rumelici* Cincović 1959, as lectotype: Cincović 1959 [12]: Table 3, rel. 23, lectotypus hoc loco.

This association appears in the northern part of central Serbia. It covers about half of the grassland area in the region and forms a large part of wet meadows along rivers in northwestern Serbia. These plant communities appear over fairly broad ecological conditions, from regularly inundated riverbanks to extreme sites that are barely reached by inundation water. The level of groundwater also varies significantly in these sites [12].

These meadows can easily be recognized in the landscape at the end of spring by the grey-green color of flowering *Alopecurus pratensis*. Communities are two-layered. The upper layer is formed by *Alopecurus pratensis, Festuca pratensis, Rumex crispus, Cichorium intybus*, and *Thalictrum flavum*, while in the lower layer, *Trifolium pallidum, Trifolium patens*, and *Poa pratensis* appear. The lower layer is denser than the upper one and forms a close canopy, so other plant species have difficulty appearing there [11].

The association has been divided into several subassociations (*caricetosum, rhinanthetosum rumelici, trifolietum hybridi*, and *typicum*) by Cincović [11,12], but analysis shows that only two subtypes can be distinguished at the level of subassociation—*typicum* and *rhinanthetosum rumelici*.

The subassociation *rhinanthetosum rumelici* is differentiated by *Anthoxanthum odoratum, Carex spicata, Cynosurus cristatus, Danthonia alpina, Rumex acetosa*, and *Trifolium pratense*. It appears at the limit of the inundation area on alluvial terraces and is rarely inundated. In addition, the level of groundwater is deep, between 3 m and 5 m below the surface during the vegetation season. This subassociation encompasses drier forms of communities [12].

The subassociation *typicum* is differentiated by *Lysimachia nummularia, Mentha pulegium, Trifolium patens*, and *Verbena officinalis*. It includes the majority of the other three subassociations described by Cincović [11,12] and Škvorc et al. [23], as *caricetosum* Cincović 1956 nom. inval. (Art. 4), *trifolietosum hybridi* Cincović 1959 and *lysimachietosum nummulariae* Ćuk in Škvorc et al. 2020. It appears in the inundation area along rivers and is periodically under water every year at the end of spring and beginning of summer although the duration depends on the amount of water in the particular year. The groundwater table is high—from only 10 cm to 180 cm below the surface [12].

4.1.2. Association *Ventenato dubii–Trifolietum pallidi* Ilijanić 1968

Diagnostic plant species: *Bromus arvensis, Bromus racemosus, Campanula patula, Carex acuta, Centaurium erythraea, Euphorbia palustris, Gratiola officinalis, Inula britannica, Iris sibirica, Leucanthemum vulgare agg., Lysimachia vulgaris, Moenchia mantica, Poa angustifolia, Poa palustris, Ranunculus acris, Ranunculus repens, Rhinanthus alectorolophus, Scleranthus annuus, Serratula tinctoria, Trifolium striatum, Ventenata dubia, Veronica longifolia, Vicia tetrasperma.*

Nomenclatural remark. We selected the lectotype of the association *Ventenato dubii–Trifolietum pallidi* as Ilijanić 1968 [38]: Table 3, rel. 15, lectotypus hoc loco.

The association has been described from the area along the Drava River north of Osijek. These plant communities appear on the riverbanks of the small Krašica River, on the boundary of the inundation area, but the soil is heavy and clayey and well maintains humidity. It mediates among *Molinion, Deschampsion* (syn. *Cnidion*), and *Trifolion pallidi* [23,38,40]. The floristic composition is a mixture of the diagnostic species of all three alliances: *Molinion* (*Iris sibirica, Gentiana pneumonanthe*), *Deschampsion* (*Serratula tinctoria, Gratiola officinalis*), and *Trifolion pallidi* (*Trifolium pallidum, Oenanthe silaifolia*). Three therophytic species (*Trifolium pallidum, Ventenata dubia*, and *Trifolium striatum*) can be observed, which were considered to be characteristic species of this association [38]. Communities are (co)dominated by *Alopecurus pratensis*. This association has been divided into two subassociations—*poetosum palustris* and *lotetosum corniculati* Ilijanić [38].

In a broader analysis [23], the relevés of this association were partly classified among outliers and partly classified within *Trifolio–Alopecuretum*. The results of this large-scale analysis and the appearance of many *Trifolion pallidi* species persuaded us to include these transitional relevés in the analysis, where this association occurred. Additional analysis should be done in the future to clarify its synsystematic position. The present analysis does not support its further division into subassociations.

4.1.3. Association *Ranunuculo strigulosi–Alopecuretum pratensis* Vučković ex Aćić et al. 2013 mut. Ćuk in Čarni et al. 2021

Diagnostic plant species: *Betonica officinalis, Briza media, Carex distans, Centaurea jacea* agg., *Cirsium canum, Cruciata laevipes, Galium debile, Ranunculus strigulosus, Symphytum officinale*.

Nomenclatural remark. This association was invalidly described by Vučković [43] as *Ranunculus stevenii–Alopecuretum* but was later validated by Aćić et al. [65] as *Ranunculo acris–Alopecuretum pratensis* with *Ranunculus acris* L. 1753 as the name-giving taxon. We think that the species *Ranunculus strigulosus* (formerly misinterpreted as *Ranunculus steveni* Andrz. 1814) from the *Ranunculus acris* complex better describes the association because *Ranunculus strigulosus* Schur 1866 is a typical species of wet meadows in southeastern Europe. This concept is accepted in various regional floras [66] (p. 142) [67] (p. 552). We therefore propose mutation of the association name *Ranunculo acris–Alopecuretum pratensis* Vučković ex Aćić et al. 2013 [65] (p. 46) to *Ranunculo strigulosi–Alopecuretum pratensis* Vučković ex Aćić et al. 2013 mut. Ćuk in Čarni et al. *nomen mutatum novum* according to Art. 45 [51].

The species *Ranunculus strigulosus* has been misinterpreted in several regional floras as *Ranunculus steveni* or *R. stevenii* (e.g., [68,69]) [70]. The species in these communities (also in the table published by Vučković [43]) is thus *Ranunculus strigulosus*.

The association appears over a large area between Slavonsko Gorje and Vršačke Planine. It appears on deep soils that are wet, fertilized, and periodically inundated. According to an analysis made by Vučković [43], a comparison of this association with the *Trifolio–Alopecuretum*, thriving in the northern part of central Serbia, shows more hemicryptophyes and fewer therophytes, as well as more Eurasian and fewer sub-Mediterranean species; the differences are still more pronounced in comparison with associations appearing further toward the south (classified within *Trifolion resupinati*).

We have also integrated *Oenanthno banaticae–Alopecuretum pratensis* Parabućski et Stojanović ex Aćić at al. 2013, *Ononido spinosifolii–Alopecuretum pratensis* Butorac ex Aćić et al. 2013 and *Rhinantho rumelici–Filipenduletum vulgaris* Ilijanić 1969 (phantom name in Rauš et al. [40] (p. 321)) within this association.

Two subassociations, *typicum* and *filipenduletosum* Škvorc et al. 2020, can be distinguished, based on humidity and nutrient availability. The results of the analysis are not in accordance with [23], who integrated the whole association *Oenantho banaticae–Alopecuretum pratensis* within *Ranunculo–Alopecuretum pratensis*, as a drier subassociation *filipenduletosum*. The present analysis shows that only the drier subassociation (*Oenanthno–Alopecuretum filipenduletosum* Parabućski et Stojanović 1988 nom. inval. (Art. 5)) can be classified within the drier subassociation *Ranunculo–Trifolietum filipenduletosum*, whereas the subassociation from more humid sites (*Oenantho–Alopecuretum bolboschenetosum* Parabućski et Stojanović 1988 nom. inval. (Art. 5)) is classified within the typical subassociation due to the presence of many hygrophilous species, such as *Gratiola officinalis*, *Bolboschoenus maritimus*, and *Trifolium hybridum*, and the lack of thermophilous species, such as *Filipendula vulgaris*, *Leucanthemum vulgare*, *Rhinanthus rumelicus*, *Festuca valesiaca*, etc. appearing in the drier subassociation [42].

The subassociation *typicum* is differentiated by *Agrostis stolonifera*, *Carex hirta*, *Carex vulpina*, *Equisetum palustre*, *Galium mollugo*, *Holcus lanatus*, and *Lythrum salicaria*. It appears on wetter and more nutrient-rich sites. We classified *Ranunculo–Alopecuretum oenanthetosum banaticae* Vučković 1991 nom. inval. (Art. 5, Art. 13 b) (the nomenclatural type of the association is within this subassociation), *Oenathae–Alopecuretum bolboschenetosum* and *Ononido–Alopecuretum* within this subassociation.

The subassociation *filipenduletosum vulgaris* is differentiated by *Briza media*, *Filipendula vulgaris*, *Fragaria viridis*, *Galium verum*, *Leontodon hispidus*, *Rhinanthus rumelicus*, and *Tragopogon pratensis*. It appears on drier sites that are nutrient poorer. Within this subassociation are also classified *Ranunculo–Alopecuretum rumicetosum acetosae* Vučković 1991 nom. inval. (Art. 5), *Ranunulo–Alopecuretum stahyetosum officinalis* Vučković 1991 nom. inval. (Art. 5), *Oenantho–Alopecuretum filipenduletosum*, and *Rhinantho rumelici–Filipenduletum*.

4.1.4. Association *Ornithogalo pyramidale–Trifolietum pallidi* Vučković ex Aćić et al. 2013

Diagnostic plant species: *Achillea millefolium agg.*, *Dianthus armeria*, *Eryngium campestre*, *Euphorbia cyparissias*, *Festuca valesiaca*, *Galium verum*, *Medicago falcata*, *Ornithogalum pyramidale*, *Pilosella bauhinii*, *Pimpinella saxifraga*, *Plantago media*, *Potentilla argentea*, *Ranunculus polyanthemos*, *Rhinanthus rumelicus*, *Stellaria graminea*, *Thymus pulegioides*, *Trifolium medium*.

The association possesses a very limited distribution area, being found only on the southern slopes of Vršačke Planine. It is named after two sub-Mediterranean plant species, *Trifolium pallidum* and *Ornithogalum pyramidale*. The sites are drier and nutrient poorer than those of the previous associations and it is transitional to dry grasslands of *Festuco–Brometea* [43]. Two subassociations have been described within this association—*Ornithogalo–Trifolietum trifolietosum pallidi* Vučković 1991 nom. inval. (Art. 5) and *Ornithogalo–Trifolietum rhinanthetosum rumelici* Vučković 1991 nom. inval. (Art. 5), but our analysis did not confirm the division into lower syntaxa.

4.2. *Syntaxomomic Scheme of the Studied Plant Communities*

- *Molinio–Arrhenatheretea Tx. 1937*
- *Trifolio–Hordeetalia Horvatić 1963*
- *Trifolion pallidi Ilijanić 1969*
- *Trifolio pallidi–Alopecuretum pratensis* Cincović 1959
    - *typicum*
    - *rhinathetosum rumelici* Cincović 1959
- *Ventenato–Trifolietum pallidi* Ilijanić 1968
- *Ranunculo strigulosi–Alopecuretum pratensis* M. Vučković ex Aćić et al. 2013 mut. Ćuk in Čarni et al. 2021
    - *typicum*

- *filipenduletosum* Škvorc et al. 2020
- *Ornithogalo pyramidale–Trifolietum pallidi* M. Vučković ex Aćić et al. 2013

5. Conclusions

The article presents species-rich plant communities of the wet meadows that harbor a major part of regional biodiversity. The continuation of mowing and safeguarding against inappropriate hydrological interventions are key factors for maintaining these communities [71,72]. The main threats to these habitats are changes of traditional management practice, which could be abandonment leading to reforestation, or intensification of agriculture through fertilization and drainage [1]. Changes in hydrological conditions such as changes in the extent or frequency of inundations or changes in the level of groundwater on these sites would negatively influence these plant communities. The changes in hydrology would in the worst-case cause the vanishing of this habitat type [64].

We have provided new field data about this habitat type that has not been sampled for decades. They prove that it is still of good quality, worth protecting, and it deserves integration in the Habitat Directive.

Author Contributions: Conceptualization, A.Č., M.Ć., and Ž.Š.; Data curation, Ž.Š.; Formal analysis, J.F., D.K., and Ž.Š.; Funding acquisition, A.Č., I.Z., J.F., R.I., D.V., and Ž.Š.; Investigation, A.Č., M.Ć., J.F., D.K., and Ž.Š.; Methodology, A.Č.; Project administration, Ž.Š.; Resources, A.Č., M.Ć., and Ž.Š.; Software, Ž.Š.; Supervision, A.Č.; Validation, A.Č., I.Z., R.I., M.I., and Ž.Š.; Visualization, D.K. and Ž.Š.; Writing—original draft preparation, A.Č., M.I., and D.V.; Writing—review and editing, A.Č., I.Z., R.I., M.I., D.V., and Ž.Š. All authors have read and agreed to the published version of the manuscript.

Funding: This paper resulted from bilateral projects between Serbia and Croatia "Diversity of Vegetation of Wet Meadows in a Climatic Gradient along the Southern Edge of the Pannonian Plain" financed by the Ministry of Education, Science, and Technological Development of the Republic of Serbia and the Ministry of Science and Education of the Republic of Croatia. A. Čarni and I. Zelnik were supported by projects of the Slovenian Research Agency (ARRS, P1-0236 and ARRS, P1-0212). M. Ćuk, R. Igić, M. Ilić, and D. Vukov were supported by the Ministry of Education, Science, and Technological Development of the Republic of Serbia (grant no. III 43002).

Institutional Review Board Statement: Not applicable.

Informed Consent Statement: Not applicable.

Data Availability Statement: Study is based on published material [11,12,38–44] and new field data Figure A2.

Conflicts of Interest: The authors declare no conflict of interest. The funders had no role in the design of the study; in the collection, analyses, or interpretation of data, in the writing of the manuscript, or in the decision to publish the results.

Appendix A

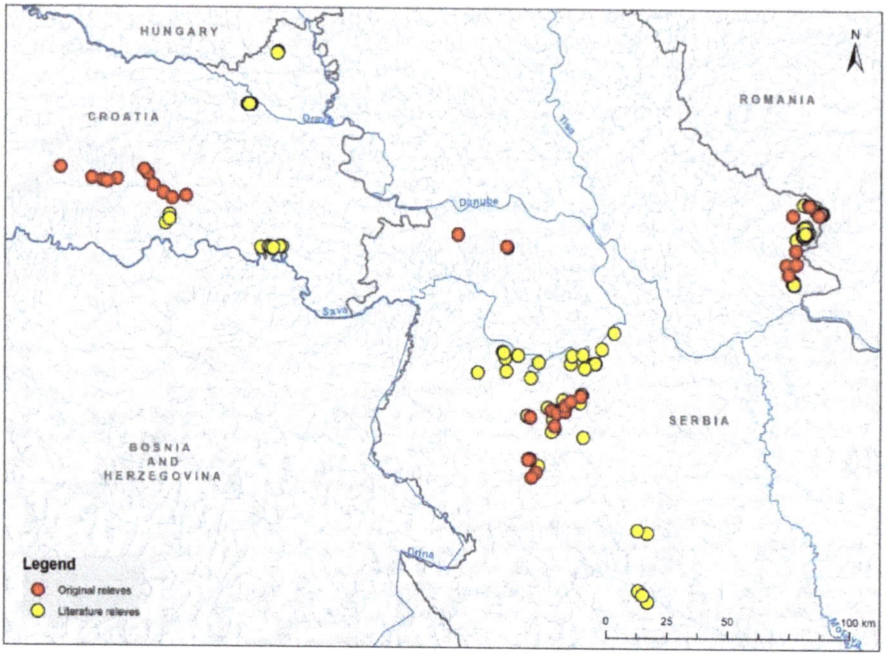

Figure A1. The geographical position of original relevés from fieldwork sampling and relevés collected from the literature.

Relevé number	1111	1111112222222222333	333
	1234567890123	4567890123456789012	345
Cichorium intybus	.+++.++++++.1	.........+...+.....+	...
Rorippa sylvestris	.+++.+.+.++++	...+.........+....+.	...
Ranunculus sardous	.+.....+..+.+	....+..............	...
Veronica serpyllifolia	.+.......+...+	.++...+............	...
Cirsium canum	....	2+1+21++1112+21.1..	...
Ranunculus strigulosus	1.++1........	2122234+2233..+.2..	...
Centaurea jacea agg.	.+.+.++++++++	++..+1+..+++...+.+2	+..
Carex distans	....	..1...1.++1+..2....	...
Galium debile	+............	....+...+.++...+...	...
Symphytum officinale	....+........	...+.......+1++.+..	...
Betonica officinalis	+.........+...	+1+..+.2...........	...
Cruciata laevipes	..+...++.....	+.........+........+	...
Achillea millefolium agg.	..1+.+..++...	.......+..+...1+..1+	21+
Ranunculus polyanthemos	.+.112+.1++1+	..........+..+..+.++	++1
Stellaria graminea	+..++........	+++...+.+.......+..++	+++
Rhinanthus rumelicus	....	+...........1.+...2+	+1+
Galium verum	........++...	+...++.+..........+.	111
Alopecurus pratensis	+.44444334323	2233123234331++2+22	223
Trifolium pallidum	.5332++524+52	.......+.........+43	.42
Oenanthe silaifolia	+...+22+++..+	.+.+++...++1+..2...	...
Lathyrus nissolia	+.+.+++++..++	................+.+.	++.
Medicago arabica	.+.+.1++.1++1	...+.+..2...+1+....	...
Poa trivialis	++.++++++...+	11+11+++222221113+2	.++
Schedonorus pratensis	13+.1.+.+.+.3	12.23++1+1.+2423223	21+
Trifolium pratense	+1222+124+1+.	1+.1+.++++.221+...1	++.
Carex hirta	+++++1+++...+	+1122+1+1+1+1+1.1.+	...
Poa pratensis	2113222+11331	+1+..+1+.++....+.+.	+.+
Lathyrus pratensis	1.+.1+.++..+.	1++++++++1++..+.+++	+..
Potentilla reptans	++++2+21+2+2+	++...+1..+++..+211.	...
Plantago lanceolata	++.11.+.+.2+1	2+++1++1+1+1..+..++	...
Rumex crispus	+++++++1+++++	.+.1++++..+.1..+.+.	...
Galium mollugo	+.+...+++.++.	+++.+++++++++++...+	...
Lychnis flos-cuculi	+.++.++++...+	121+2+++1+12.......+	...
Holcus lanatus	1.+3++..+...+	332213123+21..1....	...
Rumex acetosa	+.++1+1.+..++	+21...+.++1+..1...+	+..
Trifolium campestre	.2+..311++3.3	2+..2.....++...+.++	++.
Convolvulus arvensis	.++++.+++.+1.	.++1+..+.........++	+++
Erigeron annuus	..++..+++.++.	++.....++.+.+.+++++	+..
Lotus corniculatus	.++..++.+.++.	.+..+.++.+.+...2.+1	+.+
Vicia sativa	+++++++++++++	...........++......	+.+
Carex spicata	++.+.+.+++..+	+..++++++.+...+....	...
Anthoxanthum odoratum	.++++...+++.+	21+..1+31.........1.	2..
Carex vulpina	1......+....+	++.++++.1++++.1.+..	...
Bromus hordeaceus	.+.......1++.	+..++++.+..+1..1.+1	.1.
Ranunculus repens	.......++.....	+++21++.++++21+....	...
Lysimachia nummularia	+1...++......	.++++.+.+...2.1++..	...
Cynosurus cristatus	.++1++..1+...	+++.+.+...+........	+..
Lythrum salicaria	...++.+++....	+..+.++...+..++++..	...
Crepis biennis	.+..+.+++1++.	...........+++...++	...
Trifolium repens	++.+.++.+.+..	.+.......+...+.....+	+..
Ononis spinosa	+....+.......	+...+++++111.......	+..
Geranium dissectum	.+++++11.121+	....	.+.
Prunella vulgaris	++....+.+..++	++..1...........+1.	...
Daucus carota	.+.......++.++	.+.......+....+++..	+..
Bellis perennis	.+.........+..	+++.+++.++.+.......	...
Elytrigia repens	...+.+...++1+	...+.........1.1...	++.
Veronica arvensis	.......+++.++.	+......++...+......	.++
Ajuga reptans	....	+++.+++++.++....1..	...
Myosotis arvensis	.++.+.+.+++++	........+..........	...
Filipendula vulgaris	+.1+3...1....	..+.....+..........	.r+
Leucanthemum vulgare agg.	..++....+..+.	++.......++....+....	...
Dactylis glomerata	..+.+.....+..	..........++.....+.2	.++
Veronica chamaedrys	...+.........	+.1.......+++......++	+..
Vicia tetrasperma	....++2+++...	................+.++	...
Mentha aquatica	.......+.+..+	..+.....1..+1.+..+..	...
Taraxacum officinale	.........+++.+	..............++.1++.	...
Equisetum palustre	....	..+..++..++2.+.+..	...
Glechoma hederacea	....	......+...+..++..+++	+.+
Danthonia alpina	++...1+++..++	....	...
Lysimachia vulgaris	+.......+.....	++++.....+.+.......	...
Vicia hirsuta	+........++.2.	.........+.........+	++.
Eleocharis palustris	+............	....+.+..+.+++.+....	...
Cerastium fontanum	....+.+.+....	..+....+..+.+....+..	...
Mentha longifolia	......++....+	....+..+++.+.......	...
Agrostis stolonifera	....	..++.+..+.+..++1...	...
Calystegia sepium	+............	..........+.+++1+....	...
Moenchia mantica	.........+...+	+....+.++..+.......	...
Taraxacum palustre	....	.....+++.++++.......	...
Gratiola officinalis	+.......++....	......1....+....1...	...
Lolium perenne	..++..11.....	....+.........+......	...
Juncus compressus	.....+.......	.....++.......2.++...	...
Cnidium silaifolium	....	+++..+..++.........	...
Trifolium hybridum	....	....+.....1++.3......	+..
Cirsium arvense	+.......+...+.	................+..+..	...
Plantago major	........++....	...............++.+...	...
Cardamine amara	....	.....+++....++........	...
Medicago lupulina	....	.....++.+1+.........	...
Pastinaca sativa	....	............++.++...+.	...
Carex filiformis	2...+........	............+....+....	...
Juncus effusus	.+....+...+...	...............+......	...
Leontodon hispidus	.+...........	..+...+...+.........	...
Bromus racemosus	.......+.....+	....	.++
Carex brizoides	....	+.2...1.+..........	...
Luzula campestris	....	+.+.....+..........	+..
Inula salicina	....	+.............+.+.+..	...
Angelica sylvestris	....	..+.........+.+.2....	...
Festuca rupicola	....	..+................+.	1.2
Thalictrum lucidum	....	...............+......+.+	..+
Carex muricata	....	...............++++	...
Allium vineale	....	...............+++.	..+

Other species. *Agrimonia eupatoria* 33: +, 35: 1; *Agrostis capillaris* 33: +, *Ajuga genevensis* 3: +, 4: +, *Alopecurus rendlei* 18: 2; *Althaea officinalis* 27: +; *Anacamptis palustris* 28: +, 29: +; *Arrhenatherum elatius* 23: 1, *Bolboschoenus maritimus* 27: 2, *Bothriochloa ischaemum* 34: +, 35: +; *Bromopsis inermis* 5: +; *Bromus commutatus* 8: 1, 12: +, 13: +; *Calamagrostis epigejos* 29: +; *Campanula patula* 9: +, 15: +, 21: +; *Campanula rapunculus* 3: +; *Cardamine pratensis agg.* 1: +; *Carex flacca* 8: +; *Carex lepidocarpa* 22: 1; *Carex nigra* 30: 2; *Carex pallescens* 16: +, 24: +; *Centaurea macroptilon* 15: +, 19: +, 20: +; *Centaurium erythraea* 33: +; *Cerastium brachypetalum* 22: +, 23: +, 24: +, *Cerastium glomeratum* 11: +, 12: 1, 13: +; *Cirsium vulgare* 11: +, 34: +, 35: +; *Clinopodium vulgare* 21: +, 33: 1, *Colchicum autumnale* 14: +, 21: +; *Crataegus monogyna* 33: +; *Cruciata glabra* 14: +; *Dactylorhiza incarnata* 24: +, 25: +, 28: +; *Deschampsia cespitosa* 9: +, 20: +; *Dipsacus fullonum* 7: +; *Dorycnium pentaphyllum* 33: +; *Elymus caninus* 13: +; *Eupatorium cannabinum* 30: +; *Euphorbia esula* 3: +, 4: +, 5: +; *Filago germanica* 33: 1, *Fragaria viridis* 3: +, 9: 1, 33: +; *Galega officinalis* 8: +, 29: +; *Galium aparine* 35: +; *Galium palustre* 7: 1, 9: +, *Heracleum sphondylium* 16: +, 21: +, *Hypericum perforatum* 10: +; *Hypochaeris radicata* 2: +, 10: +; *Inula helenium* 23: +; *Iris pseudacorus* 8: +, 25: +; *Juncus articulatus* 16: +, *Juncus conglomeratus* 13: +, 16: +, 22: +; *Juncus inflexus* 24: +, 26: +, 28: 3; *Knautia drymeia* 16: +; *Lactuca serriola* 26: +, 27: +, 34: +, *Lathyrus aphaca* 27: +; *Lathyrus tuberosus* 7: +, 8: +, 32: +, *Leucojum aestivum* 18: +, 19: +, *Lycopus europaeus* 7: +, 8: +, 28: +; *Medicago sativa* 11: +; *Myosotis scorpioides* 22: +; *Oenanthe fistulosa* 24: +; *Picris hieracioides* 3: +, 23: +, 33: +; *Poa compressa* 1: +, *Prunus spinosa* 33: +, 35: +; *Rosa arvensis* 33: +; *Rosa gallica* 4: +, 7: +, 9: +; *Rubus caesius* 8: +, 30: +, 33: +; *Rumex obtusifolius* 30: +; *Sanguisorba officinalis* 5: +, *Scabiosa ochroleuca* 33: +; *Scirpus sylvaticus* 16: +, *Scorzoneroides autumnalis* 19: +, 22: +; *Scutellaria galericulata* 1: +, 4: +, 7: +; *Scutellaria hastifolia* 29: +, 30: +; *Securigera varia* 4: +; *Serratula tinctoria* 14: +, *Sherardia arvensis* 3: +, 5: +; *Silene latifolia* 3: +, 30: +; *Solidago canadensis* 26: +; *Solidago gigantea* 14: +, 24: +; *Sonchus arvensis* 29: +; *Stachys palustris* 30: +; *Succisa pratensis* 16: +; *Thalictrum simplex* 24: +; *Torilis arvensis* 34: +, 35: +; *Tragopogon pratensis* 31: +, 32: +; *Trifolium incarnatum* 3: +, *Trifolium montanum* 33: +; *Trisetum flavescens* 14: +, 21: +, 35: +; *Tussilago farfara* 30: +; *Valerianella locusta* 2: +, 21: +; *Veronica persica* 12: +; *Vicia grandiflora* 7: +; *Vicia lutea* 33: +; *Vicia pannonica* 27: +, *Vicia sepium* 15: +, 22: +; *Viola tricolor* 11: +; *Xeranthemum annuum* 34: +, 35: +.

Localities, as number of relevé, date (yearmonthdate), aspect, slope (if exist), latitude, longitude (both in degrees), description of locality: 1. 20160518, 167m, 44.282830, 19.944640, RS, W Serbia, surroundings of Valjevo, village Popučke, Kolubara-Bare; 2. 20160519, 280m, 44.328380, 19.906900, RS, W Serbia, surroundings of Valjevo, village Zabrdica, Reke; 3. 20160519, 267m, E, 5°, 44.330880, 19.913960, E, RS, W Serbia, surroundings of Valjevo, village Zabrdica, Reke; 4. 20160519, 237m, E, 2°, 44.329320, 19.918180, RS, W Serbia, surroundings of Valjevo, village Zabrdica, Reke; 5. 20160519, 182m, 44.264900, 19.921420, RS, W Serbia, surroundings of Valjevo, 6. 20160520, 94m, 44.479530, 19.925980, RS, W Serbia, surroundings of Ub, village Novaci; 7. 20160520, 97m, 44.501200, 20.036830, RS, W Serbia, surroundings of Ub, village Vrelo; 8. 20160520, 94m, 44.490020, 20.064810, RS, W Serbia, surroundings of Ub, village Takovo; 9. 20160520, 89m, 44.495510, 20.107920, RS, W Serbia, surroundings of Ub, village Crvena Jabuka, 10. 20160520, 93m, 44.514690, 20.111090, RS, W Serbia, surroundings of Ub, village Milorci; 11. 20160520, 91m, 44.528940, 20.141330, RS, W Serbia, surroundings of Ub, village Brović, 12. 20160520, 87m, 44.545550, 20.193950, RS, W Serbia, surroundings of Ub, village Liso Polje; 13. 20160520, 104m, 44.444150, 20.049900, RS, W Serbia, surroundings of Ub, village Gunjevac; 14. 20150514, 160m, 45.366080, 17.667100, HR, Slavonia, village Mihaljevci, 15. 20150514, 145m, 45.357190, 17.726570, HR, Slavonia, village Eminovci, along a brook; 16. 20150514, 149m, E, 3°, 45.361020, 17.803100, HR, Slavonia, village Rajsavac; 17. 20150514, 177m, S, 1°, 180m, 45.392750, 17.945870, HR, Slavonia, village Gradiste; 18. 20150514, 130m, 45.337400, 17.993570, HR, Slavonia, village Čaglin; 19. 20150514, 128m, 45.337410, 17.993570, HR, Slavonia, between villages Čaglin and Migalovci; 20. 20150515, 206m, 45.405860, 17.502710, HR, Slavonia, betwen villages Orjavac and Kujnik; 21. 20150515, 143m, 45.351813, 17.750320, HR, Slavonia near to the village Svetinja; 22. 20150515, 167m, 45.377420, 17.964310, HR, Slavonia izmedju Vukojevica i Milanluga, used as pasture. 23. 20150515, 190m, 45.309320, 18.043870, HR, Slavonia, village Sovski dol; 24. 20150515, 166m, 45.290070, 18.091500, HR, Slavonia, village Paka; 25. 20150515, 130m, 45.296830, 18.164560, HR, Slavonia, village Music, in front of entrance in the village, 26. 20170521, 140m, 45.083180, 19.839680, RS, Fruška gora, village Rivica. 27. 20170521, 150m, 45.086410, 19.835610, RS, Fruška gora, village Rivica. 28. 20170521, 224m, SE, 2°, 45.135760, 19.580400, RS, Fruška gora, village Ležimir, 29. 20170530, 89m, 45.147660, 21.336480, HR Vršac, village Kozluk, 30. 20170531, 84m, 44.975010, 21.289140, Vršačke planine, village Straža. 31. 20170530, 97m, 45.179520, 21.424270, RS, Vršačke planine, between vilages Veliko Središte and Gudurica; 32. 20170531, 70m, 44.940850, 21.300180, RS, village Jasenovo; 33. 20170530, 166m, E, 2°, 45.144190, 21.470990, RS, Vršačke planine, village Markovac, near to the graveyard; 34. 20170531, 84m, NW, 2°, 44.977360, 21.341880, RS, village Banatska Subotica. 35. 20170531, 122m, 45.024690, 21.341880, RS, village Vojvodinci.

Figure A2. Analytic table of the new field data used in the analysis. The relevés are classified in accordance with Figure A3, diagnostic species of the associations are framed. The following associations were sampled: 1–13 *Trifolio pallidi–Alopecuretum pratensis*, 14–32 *Ranunculo strigulosi–Alopecuretum pratensis*, and 33–35 *Ornithogalo pyramidale–Trifolietum pallidi*.

Group No.	1	2	3	4	5	6
No. of relevés	31	37	16	28	45	18
Cichorium intybus	81	57	.	7	27	22
Trifolium patens	32	76*	19	.	36	.
Ranunculus sardous	19	54	13	4	.	.
Rorippa sylvestris	42	30	.	7	2	6
Veronica serpyllifolia	26	30	.	11	2	.
Mentha pulegium	3	43*	.	.	2	.
Trifolium pratense	71	30	.	54	67	56
Rumex acetosa	52	8	63	32	69	61
Anthoxanthum odoratum	65	11	31	32	69	56
Cynosurus cristatus	61	14	.	21	58	6
Carex spicata	45	8	.	32	.	11
Danthonia alpina	32	.	.	.	.	.
Lysimachia nummularia	13	65	.	54	27	.
Verbena officinalis	.	30	.	.	4	.
Iris sibirica	.	.	94	14	2	.
Rhinanthus alectorolophus	.	.	69	.	.	.
Poa angustifolia	.	.	63	.	.	.
Euphorbia palustris	.	3	63	4	.	.
Ventenata dubia	10	.	63	.	.	.
Veronica longifolia	.	.	56	11	2	.
Serratula tinctoria	.	8	69	29	13	.
Poa palustris	.	3	50	7	.	.
Ranunculus acris	.	.	56	7	18	.
Carex acuta	.	.	38	.	.	.
Moenchia mantica	23	3	56	18	4	.
Ranunculus repens	10	14	69	68	9	.
Vicia tetrasperma	19	5	81	7	27	39
Scleranthus annuus	.	.	31	.	.	.
Campanula patula	13	.	56	7	36	.
Centaurium erythraea	.	.	38	.	4	6
Bromus racemosus	10	32	63	.	18	11
Gratiola officinalis	16	35	75	46	42	.
Leucanthemum vulgare agg.	29	24	81	18	67	11
Lysimachia vulgaris	6	.	38	21	.	.
Trifolium striatum	.	.	31	.	.	6
Bromus arvensis	3	.	44	.	2	22
Cirsium canum	.	.	.	89	36	.
Ranunculus strigulosus	23	3	.	82	58	.
Carex distans	.	.	.	36	22	.
Centaurea jacea agg.	65	38	31	64	93	6
Galium debile	3	.	.	46	16	.
Briza media	.	.	6	7	64*	11
Symphytum officinale	3	24	.	54	27	.
Cruciata laevipes	10	.	.	14	33	.
Betonica officinalis	26	16	.	21	58	6
Carex hirta	45	16	6	75	31	.
Lythrum salicaria	16	11	31	50	11	
Carex vulpina	13	14	.	71	7	.
Agrostis stolonifera	13	.	25	64	7	.
Galium mollugo	23	.	6	54	11	.
Holcus lanatus	23	.	6	54	7	.
Equisetum palustre	.	.	31	32	.	.
Filipendula vulgaris	39	8	.	11	89	67
Tragopogon pratensis	.	.	19	4	51	56
Leontodon hispidus	10	5	38	11	49	.
Fragaria viridis	6	.	.	.	31	6
Festuca valesiaca	23	.	.	.	16	100
Thymus pulegioides	.	.	.	.	11	89
Euphorbia cyparissias	.	.	.	.	4	83
Eryngium campestre	.	.	.	.	.	67
Ornithogalum pyramidale	3	.	.	4	13	78
Pimpinella saxifraga	.	.	.	.	22	72
Potentilla argentea	10	8	.	.	.	67
Rhinanthus rumelicus	29	3	.	11	87*	100
Stellaria graminea	13	.	.	39	53	83
Trifolium medium	3	.	.	.	2	44
Plantago media	.	5	.	.	42	67
Pilosella bauhinii	.	.	.	.	20	50
Achillea millefolium agg.	58	49	13	25	71	100
Medicago falcata	.	.	.	.	9	33
Ranunculus polyanthemos	84	65	13	14	62	94
Galium verum	35	5	44	18	82*	89
Dianthus armeria	.	.	.	.	18	33
Alopecurus pratensis	97	95	94	96	82	94
Trifolium pallidum	74	78	100	21	71	94
Oenanthe silaifolia	32	54	50	32	13	.
Lotus corniculatus	65	62	50	25	53	67
Clematis integrifolia	10	70	.	32	29	.
Lathyrus nissolia	32	.	.	14	22	39
Inula britannica	6	5	63*	14	16	17
Medicago arabica	39	5	.	21	.	17
Lythrum virgatum	.	19	.	21	18	.
Carex praecox	.	32	19	.	2	6
Plantago lanceolata	81	92	94	46	44	6
Schedonorus pratensis	58	59	38	82	96	83
Potentilla reptans	77	84	50	68	51	28
Rumex crispus	77	81	63	50	36	17
Lychnis flos-cuculi	71	54	63	75	49	.
Lathyrus pratensis	45	22	31	71	56	56
Taraxacum officinale	58	76	69	18	22	28
Poa pratensis	94	86	.	54	71	50
Poa trivialis	55	78	.	86	29	11
Trifolium campestre	39	.	56	36	49	67
Daucus carota	39	16	25	21	49	22
Trifolium repens	58	46	6	11	13	33
Ononis spinosa	10	14	.	32	49	39
Prunella vulgaris	39	27	44	18	18	17
Trifolium hybridum	13	19	.	43	18	22
Veronica chamaedrys	3	.	31	21	40	17
Medicago lupulina	.	11	.	18	33	6
Bellis perennis	16	16	13	32	4	.
Ajuga reptans	.	3	13	39	16	.
Lolium perenne	23	5	.	7	.	11
Convolvulus arvensis	58	65	6	18	4	44
Erigeron annuus	23	.	44	39	47	6
Bromus hordeaceus	19	22	38	39	20	22
Elytrigia repens	32	41	.	18	16	22
Vicia hirsuta	23	16	25	11	22	50
Vicia sativa	42	14	19	7	4	44
Glechoma hederacea	3	.	.	29	29	39
Scutellaria hastifolia	13	38	.	21	9	.
Geranium dissectum	48	16	.	.	.	11
Galium palustre	10	19	31	.	4	.
Inula salicina	.	16	38	14	16	.
Carex filiformis	6	16	31	7	9	.
Leucojum aestivum	3	19	31	7	.	.
etc.						

Figure A3. Percentage synoptic table of the *Trifolion pallidi* meadows. Diagnostic species of syntaxa are framed or indicated by asterisk. Diagnostic species of the *Trifolion pallidi* and *Trifolio–Hordeetalia* are from literature. Legend to columns: 1—*Trifolio pallidi–Alopecuretum rhinanthetosum rumelici*, 2—*Trifolio pallidi–Alopecuretum typicum*, 3—*Ventenato–Trifolietum pallidi*, 4—*Ranunculo acris–Alopecuretum typicum*, 5—*Ranunculo acris–Alopecuretum filipenduletosum*, 6—*Ornithogalo pyramidale–Trifolietum pallidi*.

References

1. Janssen, J.A.M.; Rodwell, J.S.; García Criado, M.; Gubbay, S.; Haynes, T.; Nieto, A.; Sanders, N.; Landucci, F.; Loidi, J.; Symank, A.; et al. *European Red List of Habitats. Part 2. Terrestrial and Freshwater Habitats*; European Union: Luxembourg, 2016; p. 44. [CrossRef]
2. Dwire, K.A.; Kauffman, J.B.; Brookshire, E.N.J.; Baham, J.E. Plant biomass and species composition along an environmental gradient in montane riparian meadows. *Oecologia* **2004**, *139*, 309–317. [CrossRef] [PubMed]
3. Luo, W.B.; Song, F.B.; Xie, Y.H. Trade-off between tolerance to drought and tolerance to flooding in three wetland plants. *Wetlands* **2008**, *28*, 866–873. [CrossRef]

4. Chen, X.; Li, X.; Xie, Y.; Li, F.; Hou, Z.; Zeng, J. Combined influence of hydrological gradient and edaphic factors on the distribution of macrophyte communities in Dongting Lake wetlands (China). *Wetl. Ecol. Manag.* **2015**, *23*, 481–490. [CrossRef]
5. Gaberščik, A.; Krek, J.L.; Zelnik, I. Habitat diversity along a hydrological gradient in a complex wetland results in high plant species diversity. *Ecol. Eng.* **2018**, *118*, 84–92. [CrossRef]
6. Moran, J.; Skeffington, M.S.; Gormally, M. The influence of hydrological regime and grazing management on the plant communities of a karst wetland (Skealoghan turlough) in Ireland. *Appl. Veg. Sci.* **2008**, *11*, 13–24. [CrossRef]
7. Dolinar, N.; Rudolf, M.; Šraj, N.; Gaberščik, A. Environmental changes affect ecosystem services of the intermittent Lake Cerknica. *Ecol. Complex.* **2010**, *7*, 403–409. [CrossRef]
8. Grime, J.P. Competitive exclusion in herbaceous vegetation. *Nature* **1973**, *242*, 344–347. [CrossRef]
9. Grime, J.P. *Plant Strategies Vegetation Processes and Ecosystem Properties*, 2nd ed.; Wiley: Chichester, UK, 2001; p. 417.
10. Ilijanić, L. Typologisch-geographische Gliederung der Niederungswiesen Nordkroatiens im klimatischen Zusammenhang. *Acta Bot. Croat.* **1963**, *22*, 119–132.
11. Cincović, T. Tipovi livada u Posavini. *Zb. Rad. Poljopr. Fak. Beograd.* **1956**, *4*, 31–56.
12. Cincović, T. Livadska vegetacija u rečnim dolinama zapadne Srbije. *Zb. Rad. Poljopr. Fak. Beograd.* **1959**, *7*, 1–62.
13. Ilijanić, L. Das *Trifolion pallidi*, ein neuer Verband der Ordnung *Trifolio-Hordeetalia* H-ić. *Acta Bot. Croat.* **1969**, *28*, 151–160.
14. Mucina, L.; Bültmann, H.; Dierßen, K.; Theurillat, J.P.; Raus, T.; Čarni, A.; Šumberová, K.; Willner, W.; Dengler, J.; Gavilán, R.G.; et al. Vegetation of Europe: Hierarchical floristic classification system of vascular plant, bryophyte, lichen, and algal communities. *Appl. Veg. Sci.* **2016**, *19* (Suppl. 1), 3–264. [CrossRef]
15. Purger, D. Subterranean clover (*Trifolium subterraneum* L.) in communties in the Great Hungraian Plain: Questions of coenology, nomenclature and syntaxonomy. *Acta Bot. Hung.* **2006**, *46*, 399–413. [CrossRef]
16. Zeidler, H. Das *Alopecurion utriculati*, ein neuer Verband Balkanischer Wiesengesellschaften. *Vegetatio* **1954**, *5–6*, 292–301. [CrossRef]
17. Jovanović, R. Tipovi dolinskih livada Jasenice. *Arch. Biol. Sci.* **1957**, *9*, 1–14.
18. Rodríguez-Rojo, M.P.; Jiménez-Alfaro, B.; Jandt, U.; Bruelheide, H.; Rodwell, J.S.; Schaminée, J.H.J.; Perrin, P.M.; Kącki, Z.; Willner, W.; Fernández-González, F.; et al. Diversity of lowland hay meadows and pastures in Western and Central Europe. *Appl. Veg. Sci.* **2017**, *20*, 702–719. [CrossRef]
19. Goncharenko, I.; Kozyr, M.; Senchylo, O. Classification of the floodplain meadows of the Seym and the Dnieper river valleys in the north-eastern part of Ukraine. *Biologia* **2020**, *75*, 53–70. [CrossRef]
20. Deák, B.; Valkó, O.; Török, P.; Tóthmérész, B. Solonetz meadow vegetation (*Beckmannion eruciformis*) in East-Hungary—An alliance driven by moisture and salinity. *Tuexenia* **2014**, *34*, 187–203. [CrossRef]
21. Zaliberová, M.; Škodová, I. Potentillion anserinae. In *Rastlinné Spoločenstvá Slovenska. 5. Travinno-Bylinná Vegetácia*; Hegedüšová Vantarová, K., Škodová, I., Eds.; Veda: Bratislava, Slovakia, 2014; pp. 331–350.
22. Botta-Dukát, Z.; Chytrý, M.; Hájková, P.; Havlová, M. Vegetation of lowland wet meadows along a climatic continentality gradient in Central Europe. *Preslia* **2005**, *77*, 89–111.
23. Škvorc, Ž.; Ćuk, M.; Zelnik, I.; Franjić, J.; Igić, R.; Ilić, M.; Krstonošić, D.; Vukov, D.; Čarni, A. Diversity of wet and mesic grasslands along a climatic gradient on the southern margin of the Pannonian Basin. *Appl. Veg. Sci.* **2020**, *23*, 676–697. [CrossRef]
24. Borhidi, A.; Kevey, B.; Lendvai, G. *Plant Communities of Hungary*; Akadémiai Kiadó: Budapest, Hungary, 2012; p. 544.
25. Chytrý, M.; Tichý, L.; Hennekens, S.M.; Knollová, I.; Janssen, J.A.; Rodwell, J.S.; Peterka, T.; Marcenò, C.; Landucci, F.; Danielka, J.; et al. EUNIS Habitat Classification: Expert system, characteristic species combinations and distribution maps of European habitats. *Appl. Veg. Sci.* **2020**, *23*, 648–675. [CrossRef]
26. European Council. Council Directive 92/43/EEC of 21 May 1992 on the Conservation of natural habitats and of wild fauna and flora. *Off. J. Eur. Union* **1992**, *206*, 7–50.
27. European Commission. *Interpretation Manual of European Union Habitats*; European Commission, DG Environment: Brussels, Belgium, 2013; p. 144.
28. Zaninović, K.; Gajić-Čapka, M.; Perčec Tadić, M.; Vučetić, N.; Milković, J.; Bajić, A.; Cindrić, L.; Cvitan, L.; Katušin, Z.; Kaučić, D.; et al. *Climate Atlas of Croatia 1961–1990, 1971–2000*; Državni Hidrometeorološki Zavod: Zagreb, Croatia, 2008; p. 200.
29. Milovanović, B. Climate regionalization of Serbia according to Köppen climate classification. *J. Geogr. Inst. Jovan Cvijic SASA* **2017**, *67*, 103–114. [CrossRef]
30. Stupar, V. Mesophilous *Quercus frainetto* dominated forests from Western Balkans. *South-East Eur. For.* **2020**, *11*, 97–110. [CrossRef]
31. Vukelić, J. *Šumska Vegetacija Hrvatske*; Sveučilište u Zagrebu: Zagreb, Croatia, 2012; p. 403.
32. Pantović, J.; Grdović, S.; Veljić, M.; Sabovljević, M.S. Biogeographic traits of the bryophyte flora of Serbia. *Bot. Serbica* **2019**, *43*, 143–150. [CrossRef]
33. Ćalić, J.; Gaudenyi, T.; Milošević, M.V.; Štrbac, D.; Milivojević, M. Geomorphometrical method for delineation of plains-case study of the south-eastern (Serbian) segment of the Pannonian Plain. *Carpathian J. Earth Environ. Sci.* **2012**, *7*, 239–248.
34. Adamović, L. *Die Vegetationsverhältnisse der Balkanländer (Mösische Länder): Umfassend Serbien, Altserbien, Bulgarien, Ostrumelien, Nordthrakien und Nordmazedonien*; Vegetation der Erde 11; W. Engelmann: Leipzig, Germany, 1909; p. 567.
35. Oberdorfer, E. Gliederung und Umgrenzung der Mittelmeervegetation auf der Balkanhalbinsel. *Ber. Geobot. Forsch. Rübel* **1948**, *1974*, 84–111.
36. Braun-Blanquet, J. *Pflanzensoziologie, Grundzüge der Vegetationskunde*, 3rd ed.; Springer Verlag: Wien, Austria, 1964; p. 631.

37. Westhoff, V.; van der Maarel, E. The Braun-Blanquet approach. In *Handbook of Vegetation Science 5, Classification and Ordination of Communities*; Whittaker, R.H., Ed.; Dr. W. Junk: The Hague, The Netherlands, 1973; pp. 287–399.
38. Ilijanić, L. Die Ordnung *Molinetalia* in der Vegetation Nordostkroatiens. *Acta Bot. Croat.* **1968**, *26–27*, 161–179.
39. Cindrić, M. Biljni Pokrov Podnožja Dilj-Gore Sjeveroistočno od Slavonskog Broda. Master's Thesis, University of Zagreb, Zagreb, Croatia, 1974.
40. Rauš, Đ.; Šegulja, N.; Topić, J. Vegetacija sjeveroistočne Hrvatske. *Glas. šumske pokuse* **1985**, *23*, 223–355.
41. Šegulja, N.; Topić, J. Prilog poznavanju pašnjaka istočne Slavonije. *Poljopr. Znan. Smotra.* **1978**, *46*, 91–94.
42. Parabućski, S.; Stojanović, S. *Oenanthe (banatica)-Alopecuretum pratensis* ass. nova. *Zb. Matice Srp. Prir. Nauke.* **1988**, *74*, 71–78.
43. Vučković, M. *Livadska i Livadsko-Stepska Vegetacija Vršačkih Planina*; Matica Srpska, Odeljenje za Prirodne Nauke: Novi Sad, Serbia, 1991; p. 95.
44. Butorac, B. *Vegetacija Fruškogorskog Lesnog Platoa, Monografija Fruške Gore*; Matica Srpska, Odeljenje za Prirodne Nauke: Novi Sad, Serbia, 1992; p. 163.
45. Hennekens, S.M.; Schaminée, J.H.J. TURBOVEG, a comprehensive data base management system for vegetation data. *J. Veg. Sci.* **2001**, *12*, 589–591. [CrossRef]
46. Tichý, L. JUICE, software for vegetation classification. *J. Veg. Sci.* **2002**, *13*, 451–453. [CrossRef]
47. Trinajstić, I. *Biljne Zajednice Republike Hrvatske*; Akademija Šumarskih Znanosti: Zagreb, Croatia, 2008; p. 179.
48. Ellenberg, H.; Weber, H.E.; Düll, R.; Wirth, V.; Werner, W.; Paulißen, D. *Zeigerwerte von Pflanzen in Mitteleuropa*; Erich Goltze: Göttingen, Germany, 1992; p. 248.
49. Fick, S.E.; Hijmans, R.J. Worldclim 2: New 1-km spatial resolution climate surfaces for global land areas. *Int. J. Climatol.* **2017**, *37*, 4302–4315. [CrossRef]
50. Euro + Med Plantbase. Available online: https://www.emplantbase.org (accessed on 20 November 2020).
51. Theurillat, J.P.; Willner, W.; Fernández-González, F.; Bültmann, H.; Čarni, A.; Gigante, D.; Mucina, L.; Weber, H.E. International Code of Phytosociological Nomenclature, 4th edition. *Appl. Veg. Sci.* **2020**, *24*. [CrossRef]
52. Tichý, L.; Chytrý, M.; Hájek, M.; Talbot, S.S.; Botta-Dukát, Z. OptimClass: Using species-to-cluster fidelity to determine the optimal partition in classification of ecological communities. *J. Veg. Sci.* **2010**, *21*, 287–299. [CrossRef]
53. Chytrý, M.; Tichý, L.; Holt, J.; Botta-Dukát, Z. Determination of diagnostic species with statistical fidelity measures. *J. Veg. Sci.* **2002**, *13*, 79–90. [CrossRef]
54. Tichý, L.; Chytrý, M. Statistical determination of diagnostic species for site groups of unequal size. *J. Veg. Sci.* **2006**, *17*, 809–818. [CrossRef]
55. Horvatić, S. Vegetacijska karta otoka Paga s općim pregledom vegetacijskih jedinica Hrvatskog primorja. *Prir. Istraž. Jugosl. Akad.* **1963**, *33*, 1–187.
56. Šilc, U.; Aćić, S.; Škvorc, Ž.; Krstonošić, D.; Franjić, J.; Dajić Stevanović, Z. Grassland vegetation of the *Molinio-Arrhenatheretea* class in the NW Balkan Peninsula. *Appl. Veg. Sci.* **2014**, *17*, 591–603. [CrossRef]
57. Oksanen, J.; Blanchet, F.G.; Friendly, M.; Kindt, R.; Legendre, P.; McGlinn, D.; Minchin, P.R.; O'Hara, R.B.; Simpson, G.L.; Solymos, P.; et al. Package "vegan". Community Ecology Package. Available online: https://cran.rproject.org/web/packages/vegan (accessed on 20 January 2020).
58. Sanda, V.; Öllerer, K.; Burescu, P. *Fitocenozele din Romania-Sintaxonomie, Structura, Dinamica si Evolutie*; Ars Docendi: Bucureşti, Romania, 2008; p. 570.
59. Tzonev, R.T.; Dimitrov, M.A.; Roussakova, V.H. Syntaxa according to the Braun-Blanquet approach in Bulgaria. *Phytol. Balc.* **2009**, *15*, 209–233.
60. Karácsonyi, C. Two new meso-hygrophytic plant communities in the northern part of Dealurile de Vest (Western Hills) area (Romania). *Contrib. Bot.* **2011**, *46*, 7–15.
61. Čarni, A.; Košir, P.; Karadžić, B.; Matevski, V.; Redžić, S.; Škvorc, Ž. Thermophilous deciduous forests in Southeastern Europe. *Plant Biosyst.* **2009**, *143*, 1–13. [CrossRef]
62. Purger, D.; Lengyel, A.; Kevey, B.; Lendvai, G.; Horváth, A.; Tomić, Z.; Csiky, J. Numerical classification of oak forests on loess in Hungary, Croatia and Serbia. *Preslia* **2014**, *87*, 47–66.
63. Zelnik, I.; Čarni, A. Plant species diversity and composition of wet grasslands in relation to environmental factors. *Biodivers. Conserv.* **2013**, *22*, 2179–2192. [CrossRef]
64. Stromberg, J.C. Restoration of riparian vegetation in the south-western United States: Importance of flow regimes and fluvial dynamism. *J. Arid. Environ.* **2002**, *49*, 17–34. [CrossRef]
65. Aćić, S.; Šilc, U.; Lakušić, D.; Vukojičić, S.; Dajić Stevanović, Z. Typification and correction of syntaxa from the class *Molinio-Arrhenatheretea* Tx. 1937 in Serbia. *Hacquetia* **2013**, *12*, 39–54. [CrossRef]
66. Martinčič, A. *Mala Flora Slovenije. Ključ za Določanje Praprotnic in Semenk*, 4th ed.; Tehniška Založba Slovenije: Ljubljana, Slovenia, 2007; p. 967.
67. Nikolić, T. *Flora Croatica–Vaskularna flora Republike Hrvatske*; Alfa d.d.: Zagreb, Croatia, 2019; Volume 4, p. 664.
68. Hayek, A. *Prodromus Florae Peninsulae Balcanicae. 1. Band: Pteridophyta, Gymnospermae, Dicotyledoneae (Apetalae et Choripetalae). Repertorium Speciorum Novarum Regni Vegetabilis*; Verlag des Repertoriums: Dahlem bei Berlin, Germany, 1927; p. 1193, Beihefte 30.
69. Josifović, M. *Flora SR Srbije*; Srpska Akademija Nauka i Umetnosti: Beograd, Serbia, 1970; Volume 1, p. 326.

70. Hörandl, E.; von Raab-Straube, E. Ranunculeae. In Euro + Med Plantbase. Available online: http://www.europlusmed.org/cdm_dataportal/taxon/562ee98a-4851-4af1-acef-126d2a5e450f (accessed on 20 November 2020).
71. Szépligeti, M.; Kőrösi, Á.; Szentirmai, I.; Házi, J.; Bartha, D.; Bartha, S. Evaluating alternative mowing regimes for conservation management of Central European mesic hay meadows: A field experiment. *Plant Biosyst.* **2018**, *152*, 90–97. [CrossRef]
72. Török, P.; Janišová, M.; Kuzemko, A.; Rūsiņa, S.; Dajić Stevanović, Z. Grasslands, their threats and management in Eastern Europe. In *Grasslands of the World: Diversity, Management and Conservation*; Squires, V.R., Dengler, J., Hua, L., Feng, H., Eds.; CRC Press: Boca Raton, FL, USA, 2018; pp. 64–88.

water

MDPI

Article

Distribution of Charophyte Oospores in the Curonian Lagoon and their Relationship to Environmental Forcing

Vaiva Stragauskaitė [1,*], Martynas Bučas [1] and Georg Martin [1,2]

[1] Marine Research Institute, Klaipeda University, H. Manto 84, 92294 Klaipeda, Lithuania; martynas.bucas@jmtc.ku.lt (M.B.); georg.martin@ut.ee (G.M.)
[2] Estonian Marine Institute, University of Tartu, Mäealuse 14, 12618 Tallinn, Estonia
* Correspondence: vaivastragauskaite@gmail.com

Abstract: Lack of knowledge about distribution of charophyte fructifications and importance of environmental conditions in the Baltic Sea coastal waters fostered us to assess the spatial-temporal patterns of oospore bank in relationship with environmental factors in the Curonian Lagoon (Lithuanian part). We mapped the distribution of oospores in 2017–2019. The importance of environmental factors was determined by the cluster analysis and boosted regression trees. Four oospores species were recorded up to 4 m depth. The highest mean densities (58,000 ind·m^{-2}) of viable fructifications were found along the eastern shore, where the densest charophyte stands were recorded. Viable fructifications showed a clear pattern of filling the oospore bank after the vegetation season and a depletion during the summer as they germinated. The distance from charophyte stands, salinity, bottom slope aspect, and wave exposure were the most important environmental variables. Full fructifications mostly occurred within <0.5 km distance from the charophyte stands restricted to flat and sheltered areas exposed to the northern and eastern slopes. Empty fructifications were mostly found within <2 km distance from the charophyte stands but their high density was limited to <1 km distance from the charophyte stands and on the northeastern bottom slopes and >1.5 salinity.

Keywords: oospores; gyrogonites; Baltic Sea; wave exposure; charophyte stands; seafloor topography; salinity gradient

Citation: Stragauskaitė, V.; Bučas, M.; Martin, G. Distribution of Charophyte Oospores in the Curonian Lagoon and their Relationship to Environmental Forcing. *Water* **2021**, *13*, 117. https://doi.org/10.3390/w13020117

Received: 23 November 2020
Accepted: 31 December 2020
Published: 6 January 2021

Publisher's Note: MDPI stays neutral with regard to jurisdictional claims in published maps and institutional affiliations.

1. Introduction

A propagule bank of charophytes consists of sexual propagules oospores and vegetative propagules [1]. Fertilized oogonia form oospores, which for many species continue their development into calcified gyrogonites. After a decay of charophyte thalli, sexual propagules (fructifications) can remain in sediments for a long period and stay viable up to 300 years [2]. Therefore, for many charophyte species, dispersion, colonization, and maintenance of populations depend entirely on sexual reproduction [1].

Despite the widely recognized importance of oospore bank for charophyte ecology and life cycle, there are few studies that have been focused on distribution of oospores. In the Baltic Sea, the studies on oospore bank have been conducted only along the German coastline [3–5], where in contrast to the recent vegetation (angiosperm dominated), more oospores than seeds of angiosperms have been found in the sediment samples. Extensive studies on the distribution of charophytes have been performed in Swedish, Polish, and Estonian coastal waters [6–9], but obtained results have not been related to an oospore bank.

Mechanisms of spatial variation in potential diaspore banks were also only considered in a few studies [1]. Charophyte stands are considered as a primary source of oospores [10]. Although the density of oospores correlated with the coverage of charophytes in the shallow lakes [11], the oospore density in the shallow brackish lagoon of the Baltic Sea did not correspond to the cover of charophyte vegetation [10]. Another important factor for a spread of oospores is hydrodynamics, which transports oospores to other places and causes their resuspension or borrowing. Water birds are recognized as responsible for a long-distance dispersal of oospores, which if passed through a digestive system are viable and able to germinate [1].

Regarding the water framework directive (2000/60/EC), an understanding of a distribution of oospores in estuarine lagoons is important for an ecological status assessment and a development of macrophythobenthos indicators. For the transitional waters (estuaries and lagoons), about one-third of macrophythobenthos indicators have not been tested for pressure–impact relationships, being the least validated [12]. Therefore, a knowledge development on the ecology of charophytes and other macrophytes in transitional waters of the Baltic Sea is at an early stage compared to coastal and inland waters. There are no data about a status of a charophyte oospore bank and factors determining an oospore distribution in the biggest estuarine lagoons of the Baltic Sea. In the Curonian Lagoon, only the long-term changes of charophyte habitats have been investigated [13], and the empirical relationships between the distribution of charophytes and environmental factors have been assessed previously [14]. The relatively low importance in explaining the variation in the distribution (<28%) of charophytes by environmental factors highlights the need to look into other important factors such as spatial and temporal patterns of charophyte fructifications.

Thus, the aim of the present study was to determine distribution and seasonal patterns of a charophyte oospore bank and its relationship with charophyte stands and abiotic factors in the Curonian Lagoon. We estimated ripening stages of oospores in sediments, hypothesizing that an absence of charophytes in some areas of the lagoon can be explained by a lack of viable fructifications and by a strong hydrodynamic forcing, with both being the most important factors for a distribution of oospores in the estuarine part of the lagoon. We also tried to assess the oospore densities required to maintain charophyte stands for a next generation.

2. Study Area

The Curonian Lagoon is the largest estuarine lagoon in the Baltic Sea, with an area of 1.584 km^2. The northern (estuarine) part of the lagoon belongs to the state of Lithuania (area 413 km^2) and extends from the Nemunas River inflow to the Baltic Sea via the Klaipeda Strait. The mean depth of the lagoon is 3.8 m, and only the Klaipeda Straight has a maximum 14.5 m depth [15]. The lagoon is almost freshwater, with a salinity range of 0.1–7, depending on an inflow of brackish waters from the Baltic Sea through the narrow and 10 km long strait during cyclonic periods [16]. The inflow of brackish waters mainly occurs along the deeper western shore of the lagoon, whereas freshwater masses move along the eastern shore [15,17].

These water masses strongly effect the water clarity in the lagoon; moreover, the water transparency is largely determined by an algal density and a periodic sediment resuspension during periods of high winds and wave activity [18]. The Curonian Lagoon is considered as eutrophic or hyper-eutrophic with recurring spring diatom blooms followed by summer cyanobacteria blooms, a phenomenon that has been reported for several decades [19,20]. Bottom sediments in the lagoon up to the depth of 2.7 m consist mainly of fine sand, whereas coarse silt and fine silty mud prevail at the depth over 3 m [21].

The diversity of charophytes is relatively low (seven species) in the study area, with *Chara contraria* being the most frequent, followed by *Chara aspera, Nitellopsis obtusa, Chara baltica, Tolypella nidifica, Chara globularis*, and *Chara canescens* [13]. The charophytes form stands shallower than the 2 m depth together with the dominant angiosperms such as *Potamogeton perfoliatus, Potamogeton rutilus*, and *Stuckenia pectinata* [14].

3. Materials and Methods

3.1. Field Sampling

An oospore bank was assessed in the Lithuanian part of the Curonian Lagoon in the period of 2017–2019 (Figure 1). The locations of the sampling sites were mainly based on the previous studies of the distribution of charophytes [14], i.e., to sample different depths across the whole study area. In November 2017, a preliminary survey was carried out along the eastern and western shores of the study area in order to estimate a general extent of charophyte fructifications in the upper littoral part (<1 m depth), where most of the charophyte stands are restricted [13]. The depth extent and distribution of charophyte fructifications were surveyed along the depth gradient (0.5, 1, 1.5, 2, 3, and 4 m) in the 7 major sections (profiles) between the western and the eastern shores of the study area (Figure 1). In several profiles (1, 3, and 6), surveys were performed in June 2018 (after the start of the vegetation season), in October 2018 (after the end of the vegetation season), and in May 2019 (before the vegetation season), attempting to represent the main seasonal patterns of a charophyte life cycle.

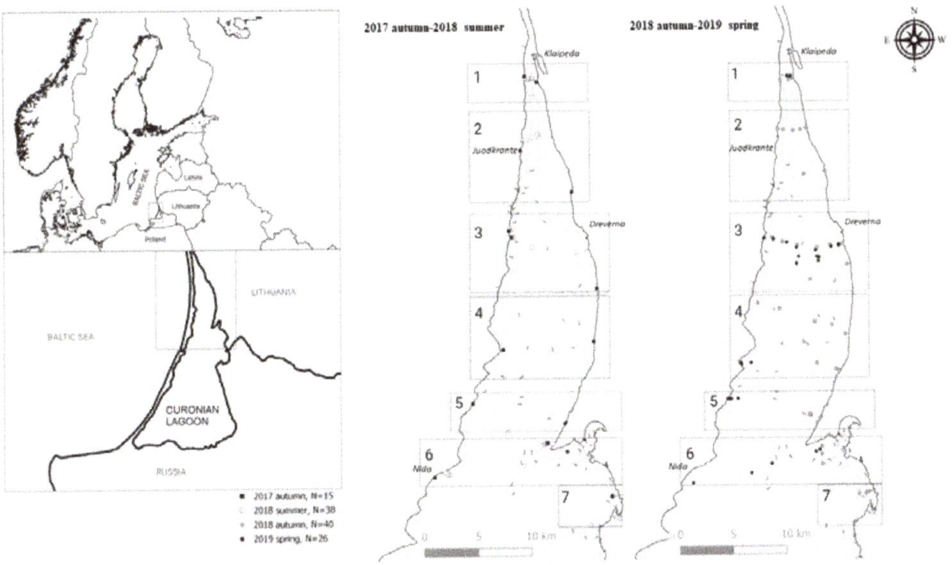

Figure 1. The study area with the sampling sites (indicated by symbols) and profiles (indicated by squares with numbers) in the Curonian Lagoon for the period of 2017–2019. N—number of the study sites.

In each study site, 3 sediment cores (replicates) were sampled within <5 m distance from each other using a Plexiglass tube (the diameter of 4.5 cm; with a plastic valve on top), which was operated with a push rod (extending to 6 m long) from a boat. The upper 5 cm sediment layer was collected from each core and the samples were stored in separate plastic containers in the dark at 5 °C. In the laboratory, the sampled sediments were sieved (at the mesh sizes of 1, 0.5, and 0.2 mm), and charophyte oospores were collected from the latter 2 mesh sizes. Oospores in each sample were counted and identified using the determination keys [22–24] and compared with live specimens under a microscope. A viability of oospores was tested by clicking them with a needle. Fructifications were classified to the 4 types: empty oospores and empty gyrogonites (empty fructifications), and full oospores and full gyrogonites as viable fructifications (see detailed description of oospores and gyrogonites provided by [25]). In each sample, a density of the 4 types of fructifications was calculated (expressed as ind. m^{-2}).

3.2. Data Analysis

For the assessment of the importance of environmental factors explaining the distribution of charophyte fructifications, we tested 1 biotic and 6 abiotic variables: a distance from charophyte stands (biotic), a wave exposure (relative exposure index—REI), a bottom slope, its aspect, a topographic position index (TPI), a depth, and a water salinity. Using QGIS 3.4.1 [26], the distance from charophyte stands was assessed by Euclidean distance from the sites where charophytes were recorded in 2014–2015 [14]. REI was estimated for 16 sectors (fetch rays at increment of 11.25°) according to Malhotra and Fonseca [27]: $REI = \left(\sum_{i=1}^{8} \left(\sum_{j=1}^{16} F_j cos(\theta_j) / \sum_{j=1}^{16} cos(\theta_j) \right)_i V_i D_i \right) / 8$, where F_j = length for the *jth* direction fetch ray after clipping to shoreline and interrogating bathymetry (the maximum length of each ray was fixed at 10 km), θ_j = angle between the *ith* fetch ray and the *jth* ray, V_i = wind speed for the *ith* direction, and D_i = wind duration for the *ith* direction. The REI value of 0 indicated absolutely sheltered from wave effect, REI < 50 corresponded to very sheltered areas in the study area (in the shallow northeastern shore), while REI > 300 showed the most exposed study sites in the southern part of the study area. The bathymetric data were obtained from the hydrological model [17]. The model was coupled with the wind data (velocity and direction) for the vegetation period (May–September) during 2000–2017, which was measured in the coastal hydrometeorological stations located in Nida and Klaipeda. REI was calculated in R 4.0.3 [28] applying the "raster" [29], the "rgdal" [30], the "geosphere" [31], and the "rgeos" [32] packages. In this study, the seafloor geomorphological parameters (the bottom slope, its aspect, and TPI) were considered as proxies for effects of water currents. The bottom slope and its aspect were derived from the bathymetric data (the spatial resolution of 48 × 48 m) using terrain function with the "raster" package in R. Using the same data, TPI was calculated by the tpi function and the focal window size of 59 (in order to represent large-scale topographic patterns) with the "spatialEco" package [33] in R. The TPI value of 0 indicates flat relief (plain), positive TPI values show elevated relief forms (convexities), while negative values represent concave surfaces. The depth was considered as a proxy of a light and wave gradient. The mean near bottom salinity data for the vegetation period during 2013–2017 was obtained from the same hydrological model developed for the Curonian Lagoon [17]. For each sediment sampling site in 2019, we extracted all environmental data from their raster layers in QGIS or R.

For the analysis of seasonal dynamics of charophyte fructifications, we selected the third profile, where the similar locations of study sites were sampled in the main seasons (summer, autumn, and spring).

3.3. Statistical Analysis

The Bray–Curtis dissimilarity [34] was applied to quantify differences in the densities of the fructifications between the profiles (the averaged density of each type of fructifications at three depths: 1, 2, and 3 m), which was inspected by the hierarchical cluster analysis using the "vegan" package [35] in R. The one factor analysis-of-similarities was used to test differences between groups of clusters.

The importance of 7 environmental factors was assessed by indirect and direct approaches. The indirect approach was based on a comparison of means of each environmental factor among several groups of the study sites, i.e., the sites with and without charophyte fructifications and delineated clusters (groups of the profiles) in the cluster analysis. The Welch ANOVA with the Games–Howell post hoc test in the "userfriendlyscience" package [36] in R were used for the comparisons of means due to the heterogeneity of variance and the unequal sample size between the groups. The same tests were applied for the comparisons of mean densities of charophyte fructifications among the different seasons and depths (1 and 2 m).

The statistical modelling and machine learning methods were chosen as the direct approach for the assessment of the importance of 7 environmental factors (explanatory variables) on the distribution (occurrence) and the density of charophyte fructifications (4 response variables). Firstly, the generalized additive models were fitted for each response variable; however, the residuals did not meet the assumption of homoscedasticity of variance and multicollinearity of several explanatory variables, e.g., between TPI and

REI ($r_{Spearman} = -0.62$), between the salinity and the distance to the charophyte stands ($r_{Spearman} = -0.61$). Therefore, the boosted regression trees (BRT) method was selected. BRT was originally developed to predict data in computer science [37], and has recently been used for predicting species density, distribution, and diversity [38]. This method does not assume that an explanatory variable has the same relationship with a response across the entire range of the environmental factor, as BRT uses decision trees to classify explanatory variables and predict the response by minimizing a loss function [39]. BRT produces a partial dependence plot to illustrate the relationship between an explanatory variable and a response, after controlling for all other explanatory variables in a model. This technique improves a prediction accuracy because it adjusts weights on the basis of explanatory variables through a stagewise learning of the data. Moreover, BRT addresses a multicollinearity issue because it considers interactions among explanatory variables—a response to an explanatory variable depends on the values of other explanatory variables at the higher levels of trees.

BRT was performed using the package "H_2O" [40] in R. Two models with the Bernoulli distribution were performed for the occurrence of full and empty fructifications, and two models with the Poisson distribution for the density of full and empty fructifications. For each response, multiple models were run varying either the model learning rate (between 0.1 and 0.001) and the number of trees (between 50 and 5000). Then, the optimum model parameters were selected on the basis of the model performance (the area under the receiver operating characteristic curve for the models with the Bernoulli distribution, and R^2 for the models with the Poisson distribution): the learning rate of 0.005, the number of trees at 100, and the bag fraction of 0.75. The tree complexity was set to 2 in order to represent two-way interactions between environmental factors. For each model, the explained deviance with all environmental factors was assessed by $Explained\ Deviance = \left(\frac{\text{null deviance} - \text{residual deviance}}{\text{null deviance}}\right)$. The relative importance of explanatory variables in predicting the response variable (i.e., the empirical improvement in reducing a squared error relative to other explanatory variables) was estimated. Partial dependence plots were derived for each model, which showed a marginal effect of explanatory variables on the predicted outcome of a model [36].

4. Results

4.1. Structure and Distribution of Oospore Bank

Fructifications (oospores and gyrogonites) of four charophyte species were identified in the study area. In some cases, *C. contraria* and *C. aspera* calcified gyrogonites were hardly distinguished from each other, and also there were many oospores damaged by abrasion which were determined as the *Chara contraria/aspera* group. The species composition was similar among the second and seventh profiles, where oospores of *C. contraria/aspera* dominated (Figure 2). The oospores of *Chara baltica* and *Tolypella nidifica* were only found in the first profile and accounted for 0.5% of the total density.

Occurrence of viable (full) and empty fructifications together were recorded in 59% of sampled sites during the study period (Figure 2), while full fructifications were found in only 33%. Viable fructifications were recorded in all sites along the eastern shore, whereas they were recorded only at the second and third profiles along the western shore. Empty fructifications were distributed further from the eastern shore than viable ones, except for the second and the third profiles, where fructifications occurred across both shores.

The highest density of empty oospores was observed in the fourth profile at 1 m depth (up to 88,000 ind. m^{-2} near the eastern shore) and in the third profile near the western shore at 2.5 m depth at the beginning of June 2018. Empty gyrogonites had the highest density (20,000 ind. m^{-2}) in the third profile at 1.5 m depth. The maximum number of full oospores (11,875 ind. m^{-2}) was recorded in the first profile at 1 m depth. Viable gyrogonites were most abundant in the profiles 1–3 at depths of 1–2 m (up to 78,000 ind. m^{-2}) in October 2018.

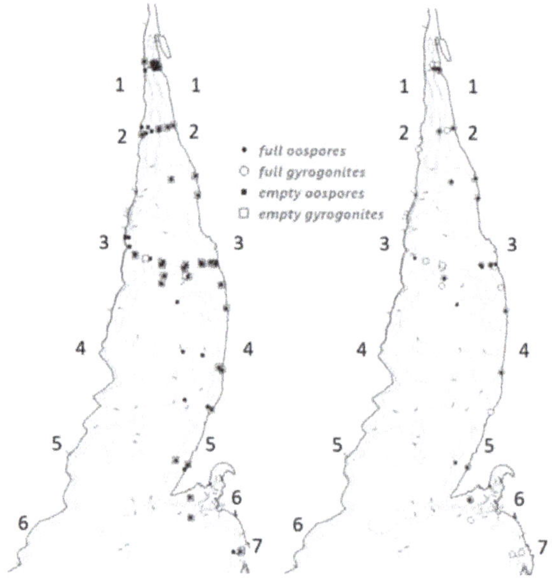

Figure 2. The distribution of charophyte fructifications in the study area: empty oospores and gyrogonites (**left**), viable (full) oospores, and gyrogonites (**right**). Numbers along the shores indicate the profiles.

The empty oospores made up most of the oospore bank, and their percentage increased with a depth (Table 1). The percentage of other fructifications decreased with depth, and there were no records of full fructifications deeper than 2.5 m.

Table 1. The oospore bank structure in different depths.

Depth, m	Empty Oospores, %	Empty Gyrogonites, %	Full Oospores, %	Full Gyrogonites, %
0.5–1	72.5	10.7	6.9	9.8
1.5–2	75.0	16.1	0.5	8.3
2–2.5	87.4	6.1	1.6	4.9
>2.5	96.4	3.6	0.0	0.0

The two groups (clusters) of the profiles were identified ($R = 0.98$, $p < 0.05$) after the cluster analysis (Figure 3). The first group consisted of the profiles 1–4, where charophyte fructifications were found on both the eastern and western shores (Figure 4A). The second group consisted of the profiles 5–7, where fructifications were found only on the eastern shore (Figure 4B). The mean density of fructifications in the first cluster was >80% higher than in the second one.

4.2. Seasonal Patterns

There were significant seasonal differences in the density of *C. contraria/aspera* fructifications (Figure 5), and they were dependent on the shore and the depth. On the eastern shore at the 1 m depth, the mean density of empty gyrogonites was statistically significantly (Games–Howell post hoc test, $p < 0.05$) higher in the summer of 2018 than in the spring of 2019. There was no significant ($p > 0.05$) difference in the mean density of empty or full oospores. The mean density of full gyrogonites in the autumn of 2018 was significantly ($p < 0.05$) higher than in the summer of 2018 and the spring of 2019.

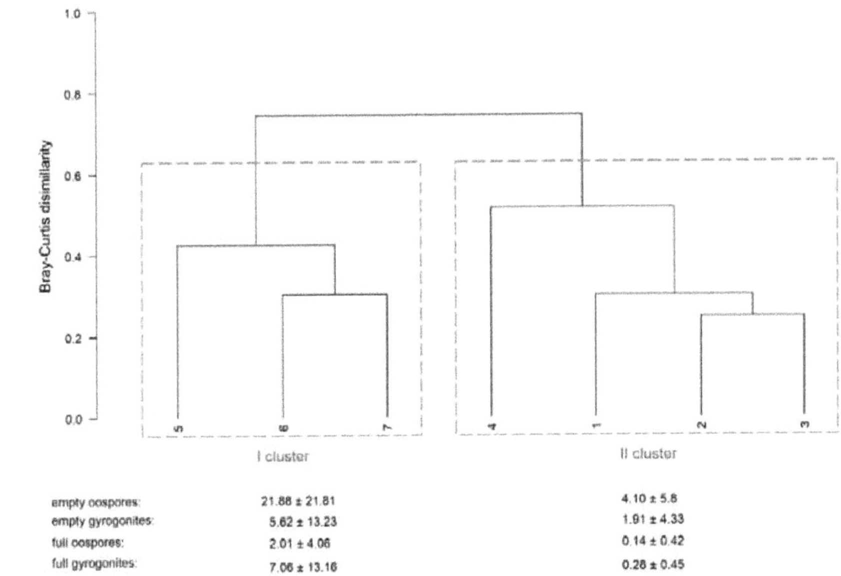

Figure 3. The dendrogram derived from the density of charophyte fructifications found in the seven profiles in the study area. The two main clusters (marked with Roman numerals) of the profiles (marked with Arabic numerals), indicated by dashed lines, and the mean (±standard deviation) density (10^3 ind. m^{-2}) of empty and full fructifications are provided below.

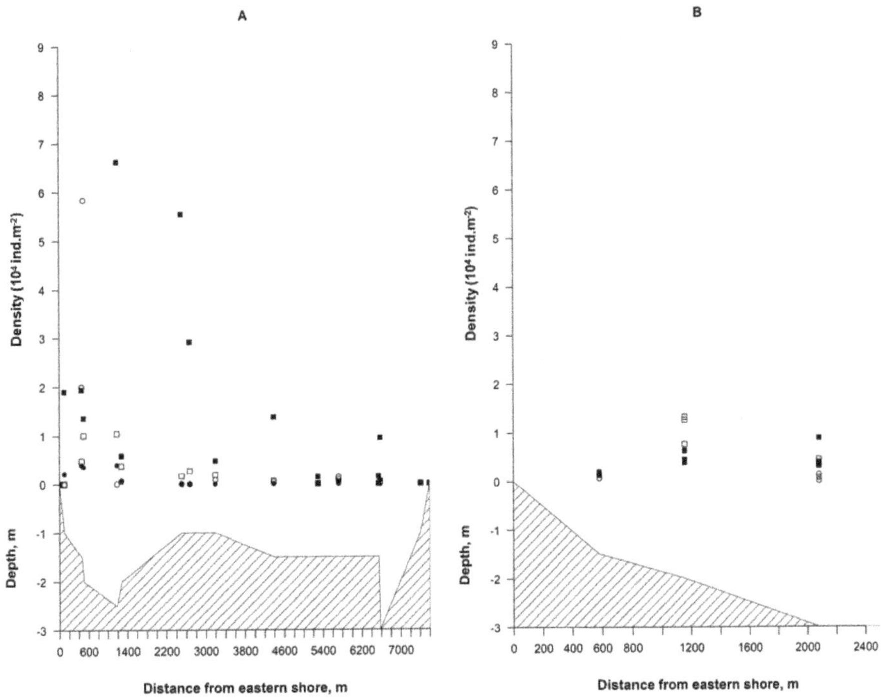

Figure 4. The distribution of the density of charophyte fructifications in the most representative profiles (third and sixth) of the first cluster (**A**) and the second cluster (**B**) during the period of 2017–2019. Full circles—full oospores, empty circles—full gyrogonites, full squares—empty oospores, and empty squares—empty gyrogonites.

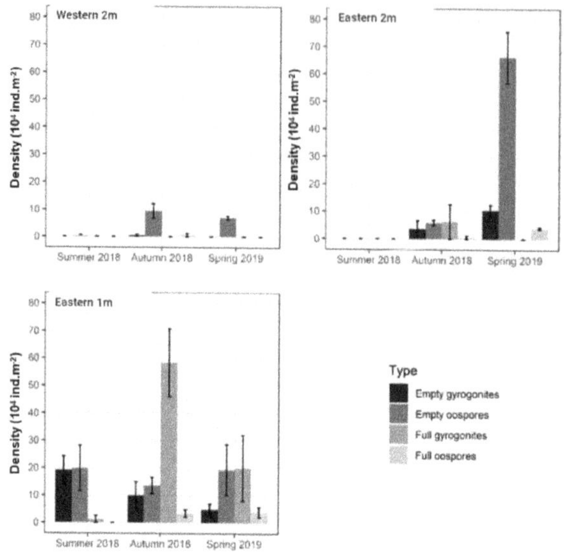

Figure 5. The seasonal dynamics of the mean (±standard deviation) density of *Chara contraria/contraria group* fructifications in the third profile of the study area. For the comparison, the profile was divided into sections according to the side of the lagoon shore (eastern and western) and the depth zone (1 and 2 m).

We did not find charophyte fructifications on the western shore at the 1 m depth during all seasons. Empty fructifications prevailed at the 2 m depth, but no significant differences ($p > 0.05$) were observed in the mean density between the seasons for any type of fructification. At the same depth on the eastern shore, the highest mean density of empty oospores and gyrogonites were estimated in the spring 2019, while no fructifications were recorded in the summer 2018. The mean density of full gyrogonites in the autumn of 2018 was significantly ($p < 0.05$) higher than in the spring of 2019. The mean density of full oospores in the spring of 2019 was significantly ($p < 0.05$) higher than in the autumn of 2018.

4.3. Environmental Factors Explaining Patterns in Oospore Bank

The magnitude of the environmental variables assessed varied throughout the study area (Table 2). Charophyte fructifications were found in the depth from 0.5 to 4 m, where the relative wave exposure index (REI) varied from 44 to 335. The highest distance between the fructification presence and the charophyte stands was determined at the level of >6.5 km. The maximum near bottom salinity of 1.8 indicated occasional intrusions of brackish Baltic Sea waters to the investigation area. Fructifications occurred on 0.003–0.907° slopes with 0.4–351.8° aspect and bottom places where topographic position index (TPI) ranged from −0.339 to 0.626.

REI, the distance from the charophyte stands, the salinity, and partly the aspect of bottom slopes were the most important environmental factors for characterizing differences between the cluster groups and the areas with presence and absence of charophyte fructifications. The statistically significant differences in the mean REI were determined between the sites with the presence and absence of fructifications (the mean ± standard deviation, respectively, 150 ± 67 and 205 ± 82; $p < 0.01$), between the first and the second cluster groups (respectively, 144 ± 80 and 202 ± 76; $p < 0.01$), between the first cluster group and the sites with absence of fructifications ($p < 0.05$), and between the presence of fructifications and the second cluster group ($p < 0.001$). There were significant differences in the mean distance from the charophyte stands among the same groups—between the sites with the presence and absence of fructifications (respectively, 1.2 ± 1.5 km and 5.2 ± 3.9 km;

$p < 0.001$), between the first and the second cluster groups (respectively, 1.2 ± 1.4 km and 6.1 ± 3.6 km; $p < 0.001$), between the sites with the presence of fructifications and the second cluster ($p < 0.001$), between the first cluster and the sites with the absence of fructifications ($p < 0.001$), and between the sites with the absence of fructifications and the second cluster ($p < 0.01$). The statistical differences in the mean salinity were determined among these groups—between the sites with the presence and absence of fructifications (respectively, 0.7 ± 0.6 and 0.3 ± 0.4; $p < 0.001$), between the first and the second cluster groups (respectively, 0.8 ± 0.6 and 0.2 ± 0.05; $p < 0.001$), between the first cluster group and the sites with the absence of fructifications ($p < 0.001$), between the sites with the presence of fructifications and the second cluster group ($p < 0.001$), and between the second cluster group and the sites with the absence of fructifications ($p < 0.05$). The statistical differences in the mean aspect of bottom slopes were determined only between the sites with the presence and absence of fructifications (respectively, 134.8 ± 102.3° and 203.4 ± 109.9°; $p < 0.01$).

Table 2. The statistics of environmental variables in the sites with presence and absence of charophyte fructifications, and in the two clusters delineated in the cluster analysis (Figure 3).

Variable, unit	Minimum	Maximum	Mean	Median	Standard Deviation
Sites with Presence of Fructifications in the Study Area					
Depth, m	0.5	4.0	1.7	1.5	0.8
REI	44	271	150	140	67
Distance from the charophyte stands, km	0.01	6.5	1.2	0.6	1.5
Salinity	0.1	1.7	0.7	0.5	0.6
Bottom slope,°	0.003	0.907	0.127	0.051	0.214
Aspect,°	0.4	351.8	134.8	100.2	102.3
TPI	−0.339	0.626	0.205	0.118	0.334
Sites without Fructifications in the Study Area					
Depth, m	0.5	5.0	1.8	1.5	1.0
REI	71	335	205	218	82
Distance from the charophyte stands, km	0.4	13.3	5.2	4.8	3.9
Salinity	0.1	1.8	0.3	0.1	0.4
Bottom slope,°	0.003	1.171	0.135	0.107	0.179
Aspect,°	1.7	357.2	203.4	225.8	109.9
TPI	−0.511	1.231	0.128	0.063	0.305
First Cluster					
Depth, m	0.5	4.0	1.7	1.5	0.7
REI	44	285	144	137	80
Distance from the charophyte stands, km	0.01	6.9	1.2	0.6	1.4
Salinity	0.1	1.8	0.8	0.5	0.6
Bottom slope,°	0.003	0.171	0.137	0.036	0.252
Aspect,°	0.4	357.2	159.3	119.1	113.1
TPI	−0.339	0.626	0.178	0.082	0.349
Second Cluster					
Depth, m	0.5	3.0	1.8	2.0	0.8
REI	91	335	202	199	76
Distance from the charophyte stands, km	2.1	13.3	6.1	6.3	3.6
Salinity	0.1	0.2	0.2	0.2	0.051
Bottom slope,°	0.009	0.308	0.119	0.102	0.084
Aspect,°	1.7	356.6	160.7	118.3	108.7
TPI	−0.512	0.626	0.124	0.132	0.244

In the BRT models, the explained variance (deviation) by the used environmental factors was relatively low for the occurrence of full and empty charophyte fructifications (respectively, 14% and 20%). The distance to the charophyte stands and the aspect of bottom slopes were the most important factors (relative importance > 30%) in explaining the presence/absence of full fructifications (Figure 6). The highest probability (≥40%) to find viable fructifications was less than 0.5 km away from the charophyte stands and on the northern, eastern, and southern bottom slopes (aspect of 0–225°). Less important were the wave exposure and the bottom slope (relative importance, respectively, 21% and 11%), where the highest probability to find full fructifications was up to 150 REI and relatively flat bottom areas (slope < 0.05°). The importance of TPI, the depth, and the salinity was minor (≤5%). The highest probability to find viable fructifications was on shallow (<1.2 m) and elevated areas (TPI > 0.1) affected by brackish waters (salinity > 0.1).

The distance to the charophyte stands was the most important factor (relative importance = 67%) in explaining the occurrence of empty fructifications (Figure 6). The distance from the charophytes stands had negative relationship with the response variable. Empty fructifications were mostly found (probability > 60%) less than 2 km away from the charophyte stands. The bottom aspect and slope were less important (relative importance, respectively, 16% and 8%), indicating the highest probability (>60%) of empty fructifications on flat bottom slopes (<0.15°) exposed to the northern, eastern, or southern directions (aspect of 0–160°). The other factors were of less importance (<6%). The higher probability of empty fructifications was up to 200 REI and salinity > 0.31. The effect of TPI and depth was not important for the occurrence of empty fructifications.

The explained variance by the used environmental factors was relatively low for the density of full and empty charophyte fructifications (respectively, 23% and 21%). The distance to the charophyte stands was the most important for the full fructifications (relative importance = 89%), where their high density (>15,000 ind. m^{-2}) was in less than 0.5 km away from the charophyte stands (Figure 6). Less important (relative importance = 6%) was the wave exposure, where the high density (>15,000 ind. m^{-2}) was in less than 150 REI. The other factors were even less important (<3%), where the density of fructifications was higher on relatively flat (slope < 0.1°) or elevated (TPI ≥ 0) areas. The effect of the bottom aspect, the salinity, and the depth was not important for the density of full fructifications.

The salinity was the most important factor (relative importance = 61%) in explaining the density of empty fructifications (Figure 6), where their highest density (>30,000 ind. m^2) was from the salinity >1.5. The second important (29%) factor was the distance from the charophyte stands. Relatively high density (>20,000 ind. m^{-2}) of empty fructifications was in the distance of <1 km from the charophyte stands. The importance of other factors was minor (<5%). The density of empty fructifications was higher in sheltered areas (REI < 200) on relatively flat slopes (<0.1°) or elevated (TPI of 0–100) areas exposed to the northern, southern, and western directions (120–360°). The effect of the depth was not important for the density of empty fructifications.

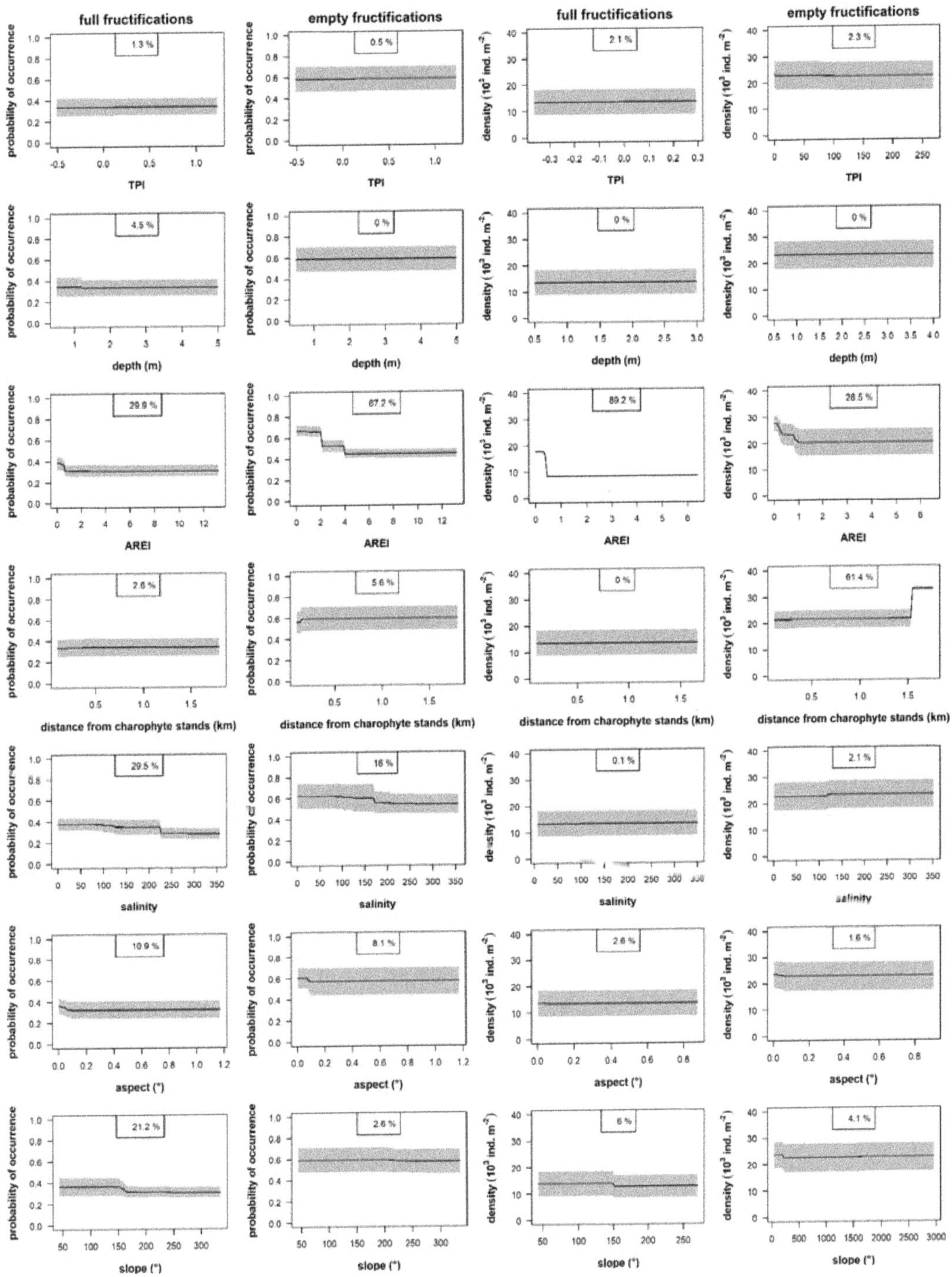

Figure 6. The partial dependence curves between the occurrence and density of full (viable) charophyte fructifications and empty charophyte fructifications and the environmental factors. The relative importance of environmental factors is provided in the top legend for each plot.

5. Discussion

5.1. Species Composition

The current paper describes spatial patterns of generative reproduction of charophytes and the potential viability of charophyte habitats in the Curonian Lagoon. We determined the oospores of four species of charophytes, whereas in total seven charophyte species have been recorded in the study area previously [13], and a total of 24 species have been recorded in the Baltic Sea [41].

Oospores of *C. contraria* and *C. aspera* were the most abundant, and these species are dominant in the study area [14]. The production of oospores is important for a long-distance dispersal [42]; therefore, a generative reproduction strategy may explain the spread of oospores of *C. contraria* in deeper areas and away from primary habitats in the upper littoral part of the Curonian Lagoon [13]. Oospores of brackish charophyte species (*C. baltica* and *T. nidifica*) were very rare in the study area and restricted to the northeastern part, which is frequently affected by inflows of the Baltic Sea waters. According to Doege et al. [43], *T. nidifica* is common in the Baltic Sea with salinity range of 8–25; however, salinities in the northern part of our study area during the summer can reach only 7.5, with the average salinity of 2 [17].

5.2. Density and Distribution

The differences in the density and distribution of charophyte fructifications were found between the eastern and western shores of the Curonian Lagoon. The highest mean densities of all oospore types were found on the eastern shore (the profiles 1–4), where the dense charophyte stands were recorded [14]. The mean density of viable fructifications at the 1 m depth was 17,500 ind. m^{-2}, which corresponds to the mean density of oospores (about 20,000 ind. m^{-2}) reported in the Westrugensche Bodden lagoon system at the same depth [4]. Although some differences in the oospore density between these waterbodies and the Curonian Lagoon could be due to species composition (*C. aspera*, *C. baltica*, and *C. canescens* were recorded in the Grieben Bay), one common pattern was observed for both lagoons—the highest densities of oospores were observed in the areas with >90% of charophyte cover at 1–1.5 m depth.

Along the western shore of the Curonian Lagoon, the distribution of fructifications was restricted to the littoral from Klaipeda to Juodkrante (Figure 1), where their mean density was the highest in vicinity of Juodkrante. This corresponded to the presence of charophyte stands [14]. We found empty fructifications down to the 2 m depth, where charophyte stands have not been recorded before. The reason of these findings could be that fructifications are carried from charophyte stands along steep bottom slopes towards the deepest part (i.e., navigation channel) by waves and currents; however, environmental conditions (mainly a light climate and sedimentation) most likely are unfavorable for germination, development, and growth there.

A colonization of unvegetated areas by charophytes may take several years and may require the sufficient number of a production and accumulation of oospores [11]. Other studies also confirm high densities of fructifications in densely vegetated areas, for example, about 20,000 ind. m^{-2} for *C. contraria* in the Grieben Bay and for *C. aspera* in the Windebyer Noor at the 0.5 m depth [44]. The mean density of full fructifications was relatively low (<10,000 ind. m^{-2}) along the western shore of the study area, as the mean coverage of charophytes was $\leq$30% [14]. This density of oospores seems to be too low to establish dense charophyte stands in this area and in the southern part of the study area along the eastern shore (the profiles 5–7).

We found the seasonality of generative reproduction in our study corresponding to a well-documented pattern [45]. In the Curonian Lagoon, viable fructifications (especially gyrogonites) showed a clear pattern of filling the oospore bank in the autumn (after vegetation season) and depletion during the summer as they germinate. Such seasonality is strongly expressed along the eastern shore at the 1 m depth. Empty gyrogonites, prevailing in the 2 m depth along both shores, had an opposite trend. The explanation for this pattern could be that empty fructifications in shallow zones are transported after a germination to deeper zones by currents and waves. Several viable fructifications were found up to the 2.5 m depth, but an insufficient light regime might have a negative impact on a charophyte development, growth, and reproduction [46,47]. We hypothesize that charophytes in the 2 m depth germinate, but they cannot produce a sufficient number of viable fructifications limiting a spread of charophyte stands to deeper areas.

Its apparent that seasonal measurements of viable and empty gyrogonites indicate a status of the charophyte habitats. On the basis of results of this study, we conclude that the mean density of full gyrogonites in autumn ≥61,000 ind. m^{-2} should be sufficient for the establishment and development of *C. contraria* in incoming spring. Certainly, the threshold requires testing under experimental conditions in order to develop the indicator of biological status for transitional waters.

5.3. Importance of Environmental Factors

In this study, the distance from the charophyte stands, the salinity, the aspect of bottom slope, and the wave exposure (REI) were revealed as statistically significant important environmental factors for explaining the distribution of charophyte fructifications. The significance of these factors was confirmed by the two approaches, enhancing the confidence that this observation was not random.

The explained deviance by the tested environmental variables was relatively low for the occurrence and the density (respectively, ≤20% and ≤23%) of fructifications in the BRT models. There could be multiple reasons for this, where most likely the main issue was the difference in the spatial and temporal scale of the measurements of the environmental factors and the charophyte fructifications. For instance, the estimates of the distance from the charophyte stands were based on the charophyte mapping data in 2014–2015 [14]. The indirectly measured factors also could affect the poor fit of the models, where REI was a surrogate of a wave impact and the bottom topographic variables were proxies for an effect of water movement. Another possible reason for the relatively low correlation between the distribution of charophyte fructifications and the environmental factors is an interaction of environmental factors in transitional type of waterbodies such as estuaries and lagoons. Due to the estuarine quality paradox [48], it is more difficult to distinguish pure effects of environmental factors than in lakes. It has been suggested that a large part of the variation in abundance models can be explained by including interactions of biotic and abiotic parameters [49]. However, data of biological parameters is usually spatially and temporally limited, and models cannot accurately predict outside the range of their sampling [50].

5.3.1. Distance from Charophyte Stands

Charophyte habitats are the main source of the oospore bank [1], and were therefore the one of the most important factors explaining the distribution of oospores in our study. The widest stands of charophytes were located in the northeastern part of the study area (the profiles 1–4) [14], where oospores were abundant. Whereas, in the remaining part of the study area (the profiles 5–7), the narrower spread of oospores and charophytes was restricted only to the eastern shore. The spatial-temporal patterns of oospores suggest that their transport to uncolonized areas was relatively low and resulted in a sharp decrease of oospore densities at the border of dense charophyte vegetated areas (from 30,000 ind. m^{-2} at the highly vegetated area in the eastern shore to 2500 ind. m^{-2} at the 2 km distance to the west). Similar distribution was observed for the colonization of *C. aspera* in the shallow lake, where no oospores were recorded in the >3 km distance [11]. As it was mentioned in the previous chapter, the relatively high number of empty oospores found in the charophyte stands indicates an intensive reproduction within this habitat.

5.3.2. Hydrodynamics and Transport of Fructification

The wave exposure was important explanatory variable for the occurrence and partly for the density of full fructifications. In general, hydrodynamics is highlighted as an important factor for the forming of macrophyte communities [8,10]. Waves also affect a distribution and spread of charophyte fructifications by a mechanical disturbance (resuspension) of fructifications [51], and by reducing water transparency (i.e., light availability) for a germination of charophytes [44].

In this study, the mean wave exposure was higher in the southeastern part (the profiles 5–7) than in the northeastern part (the profiles 1–4), which corresponds to the distribution of denser stands of charophytes in the first part than in the second [14]. It is most likely that the wave exposure limits the distribution of charophyte stands and consequently affects the density of their fructifications. There is some research performed on the seed dispersal of angiosperms, e.g., [52]; however, mechanisms of transport of oospores is poorly studied. In the Curonian Lagoon, the cyclonic storms prevail [16], and therefore wave-induced water movements should transport oospores towards the eastern shore. This could be confirmed by the observed high amount of detached charophytes with oospores on the shore after stormy periods (personal observation). It also explains why nearshore zones are relatively abundant in fructifications. However, there are no data showing at which depths charophyte thalli are usually teared and what amount of their vegetation and fructifications is transported towards a shore and to deeper areas. Thus, sheltered areas may have a higher generative recolonization potential than exposed areas, but this hypothesis was not addressed in this study.

Environment of the Curonian Lagoon is hydrodynamically active, and the water turbidity can increase during windy periods [15]. Several full fructifications were found in the wave-exposed areas (REI > 150), located in the delta part of the lagoon and along the western shore, where a concentration of suspended sediments can increase in spring and summer up to 25 mg L^{-1}. Such high turbidity may reduce a light penetration near the bottom, which may restrict an establishment of charophytes deeper as they use more resources for an elongation and less for a reproduction under low light conditions [46].

5.3.3. Salinity

Salinity is considered a major factor limiting a distribution of many macrophyte species in estuarine ecosystems [53]. However, the salinity gradient in the Curonian Lagoon did not correlate with the distribution of *C. contraria* and *C. aspera* [14]. It was demonstrated that the salinity influenced the production of *C. contraria* oospores, where its reproductive effort decreased in the salinities higher than 2, while *C. aspera* was tolerant [54]. Our results show a relative high importance (61%) of the salinity only for the density of empty fructifications, which increased in the northern part of study area. This could be explained as indirect effect of salinity, i.e., the dominant direction of riverine waters flow by the eastern shore towards the Baltic Sea [15,17], and may transport high amounts of oospores to the northern part of the lagoon. The direct effect of salinity was found only on the distribution of brackish water species (*T. nidifica* and *C. baltica*), but their fructifications were only 0.5% from the total oospore density and were restricted to the first profile.

5.3.4. Depth

The importance of water depth was relatively low (<5%) for the occurrence and distribution of charophyte fructifications, whereas other studies mention that a distribution of oospores along a depth gradient can reflect former zonation of charophytes [4]. Charophytes can be found up to the 4 m depth in the coastal waters of the Baltic Sea [6,9,10], but there are no records of maximum depth distribution of their fructifications. In the Curonian Lagoon, the maximum depth limit of full fructifications of charophytes was described at the 2.5 m depth, which is deeper than the recorded one in 2014–2015 [14]. However, we found empty fructifications down to the depth of 4 m, and their density was higher from the 1.5 to 2 m depth than in the shallower part, most likely due to a transport of oospores by waves and currents.

5.3.5. Topography of the Seafloor (Proxy of Intensity of Water Movement)

In this study, we did not directly assess the effect of water currents, which hypothetically should be important for the spread and accumulation of charophyte fructifications. However, we analyzed the importance of currents by surrogates quantitative geomorphological seafloor parameters (i.e., the slope, aspect, and TPI). The relative importance of geomorphological bottom parameters was not high (for the bottom aspect < 30%, slope < 11%, for the TPI < 3%). Nevertheless, we think that they indirectly revealed the effect of water currents since the high density of empty fructifications corresponded to the relatively flat western slopes. This can be explained by downwards transport of empty oospores to deeper areas, i.e., navigation channel, which is situated along the western shore of the study area. With respect to the TPI, the presence of full fructifications was related to lower seafloor locations (valleys) rather than to higher places (ridges). Viable oospores or gyrogonites are heavier than empty ones and probably are more often accumulated in seafloor depressions.

5.3.6. Waterfowl

We found several viable fructifications in the areas where charophytes were not recorded before, which could also be transported by waterfowl since they have a high potential in the dispersal of aquatic macrophytes [55]. In the Curonian Lagoon, the highest total bird number (30,000–48,000 ind.) was registered during the June–September 2018 [56], wherein up to 25% of them belonged to herbivorous waterfowl (coots, mute swans, and mallards). The data from transmitters attached to mute swans has provided preliminary information that the main feeding, nesting, and molting areas correspond to the areas vegetated by charophytes in the lagoon (Morkūnė et al., unpublished). These observations are useful for the assessment of a potential role of herbivorous birds on a dispersion of oospores in the study area, which requires an additional research.

6. Conclusions

The mayor number of charophyte oospores found in the estuarine part of the Curonian Lagoon belong to the dominant charophyte species (*C. contraria* and *C. aspera*). The highest mean density of all fructification types (full and empty oospores and gyrogonites) was found on the eastern shore, where the dense charophyte stands are established. Viable fructifications extended up to the 2.5 m depth, showing the improving status of charophyte stands. The temporal changes of oospore density followed the seasonal pattern—the density in autumn represented the quantity of matured oospores during summer, while the density in spring indicated how many viable fructifications remained after winter and how many had an opportunity to germinate.

The explained variation (deviance) in the occurrence and density of charophyte fructifications by the environmental factors was relatively low. The main important environmental variables were the distance to the charophyte stands, the salinity, the aspect, and the wave exposure, which have also been determined as significant factors in other studies in the Baltic Sea. In the lagoon, full fructifications were abundant in sheltered nearshore areas in a proximity of around 1 km from the charophyte stands. The relatively lower importance of the salinity and geomorphological seafloor parameters probably indirectly revealed the effect of water currents. All these findings confirm our hypothesis that absence of charophytes in some areas (especially western and southern parts of the study area) are due to unfavorable environmental conditions (mainly due to the wave exposure and water transparency) and lack of viable oospores. The mean density of viable fructifications in spring was similar to the results recorded in other studies in the Baltic Sea and can be considered as sufficient to maintain charophyte stands in the Curonian Lagoon.

Author Contributions: Conceptualization, V.S. and M.B.; methodology, M.B. and. V.S.; software, V.S. and M.B.; validation, V.S. and M.B.; formal analysis, V.S. and M.B.; investigation, V.S. and M.B.; resources, V.S. and M.B.; data curation, V.S., M.B., and G.M.; writing—original draft preparation, V.S. and M.B.; writing—review and editing, V.S., M.B., and G.M.; visualization, V.S. and M.B.; supervision, M.B. All authors have read and agreed to the published version of the manuscript.

Funding: This research was funded by the Research Council of Lithuania (no. S-MIP-19-29). Work was supported by the Doctorate Study program in Ecology and Environmental Sciences, Klaipeda University (for V.S.).

Institutional Review Board Statement: Not applicable.

Informed Consent Statement: Not applicable.

Data Availability Statement: The data presented in this study are available on request from the corresponding author. The data are not publicly available due to the thesis that is being prepared from these data.

Acknowledgments: We are grateful to J. Gintauskas and E. Tiškus for assistance in sediment collection, and J. Mėžinė for providing hydrometeorological data for analysis. Valuable comments of the two anonymous reviewers improved the manuscript.

Conflicts of Interest: The authors declare no conflict of interest.

References

1. Bonis, A.; Grillas, P. Deposition, germination and spatio-temporal patterns of charophyte propagule banks: A review. *Aquat. Bot.* **2002**, *72*, 235–248. [CrossRef]
2. Stobbe, A.; Gregor, T.; Röpke, A. Long-lived banks of oospores in lake sediments from the Trans-Urals (Russia) indicated by germination in over 300 years old radiocarbondated sediments. *Aquat. Bot.* **2014**, *119*, 84–90. [CrossRef]
3. Steinhardt, T.; Selig, U. Comparison of recent vegetation and diaspore banks along abiotic gradients in brackish coastal lagoons. *Aquat. Bot.* **2009**, *91*, 20–26. [CrossRef]
4. Blindow, I.; Dahlke, S.; Dewart, A.; Flügge, S.; Hendreschke, M.; Kerkow, A.; Meyer, J. Long-term and interannual changes of submerged macrophytes and their associated diaspore reservoir in a shallow southern Baltic Sea bay: Influence of eutrophication and climate. *Hydrobiologia* **2016**, *778*, 121–136. [CrossRef]
5. Nowak, P.; Steinhardt, T.; von Ammon, U.; Rohde, H.; Schoor, A.; Holzhausen, A.; Schaible, R.; Schubert, H. Diaspore bank analysis of Baltic coastal waters. *Bot. Lett.* **2018**, *165*, 159–173. [CrossRef]
6. Blindow, I. Distribution of charophytes along the Swedish coast in relation to salinity and eutrophication. *Int. Rev. Hydrobiol.* **2000**, *85*, 707–717. [CrossRef]
7. Brzeska, P.; Woźniczka, A.; Pełechaty, M.; Blindow, I. New records of Chara connivens P. Salzmann ex A. Braun1835—An extremely rare and protected species in Polish brackish waters. *Acta Soc. Bot. Pol.* **2015**, *84*, 143–146. [CrossRef]
8. Torn, K.; Kovtun-Kante, A.; Herkül, K.; Martin, G.; Mäemets, H. Distribution and predictive occurrence model of charophytes in Estonian waters. *Aquat. Bot.* **2015**, *120*, 142–149. [CrossRef]
9. Kovtun, A.; Torn, K.; Martin, G.; Kullas, T.; Kotta, J.; Suursaar, Ü. Influence of abiotic environmental conditions on spatial distribution of charophytes in the coastal waters of West Estonian Archipelago, Baltic Sea. *J. Coast. Res.* **2011**, 412–416. [CrossRef]
10. Steinhardt, T.; Selig, U. Spatial distribution patterns and relationship between recent vegetation and diaspore bank of a brackish coastal lagoon on the southern Baltic Sea. *Estuar. Coast. Shelf Sci.* **2007**, *74*, 205–214. [CrossRef]
11. Van den Berg, M.S.; Coops, H.; Simons, J. Propagule bank buildup of Chara aspera and its significance for colonization of a shallow lake. *Hydrobiologia* **2001**, *462*, 9–17. [CrossRef]
12. Birk, S.; Bonne, W.; Borja, A.; Brucet, S.; Courrat, A.; Poikane, S.; Solimini, A.; Van De Bund, W.; Zampoukas, N.; Hering, D. Three hundred ways to assess Europe's surface waters: An almost complete overview of biological methods to implement the Water Framework Directive. *Ecol. Indic.* **2012**, *18*, 31–41. [CrossRef]
13. Sinkevičienė, Z.; Bučas, M.; Ilginė, R.; Vaičiūtė, D.; Kataržytė, M.; Petkuvienė, J. Charophytes in the estuarine Curonian Lagoon: Have the changes in diversity, abundance and distribution occurred since the late 1940s? *Oceanol. Hydrobiol. Stud.* **2017**, *46*, 186. [CrossRef]
14. Bučas, M.; Sinkevičienė, Z.; Kataržytė, M.; Vaičiūtė, D.; Petkuvienė, J.; Stragauskaitė, V.; Ilginė, R. How much can the occurrence and coverage of charophytes in an estuarine lagoon (Curonian Lagoon) be explained by environmental factors? *Estuar. Coast. Shelf Sci.* **2019**, *216*, 128–138. [CrossRef]
15. Christian, F.; Arturas, R.; Saulius, G.; Georg, U.; Lina, B. Hydraulic regime-based zonation scheme of the Curonian Lagoon. *Hydrobiologia* **2008**, *611*, 133–146. [CrossRef]
16. Dailidienė, I.; Davulienė, L. Long-term mean salinity in the Curonian Lagoon in 1993–2005. *Acta Zool Lit.* **2007**, *17*, 172–181. [CrossRef]

17. Zemlys, P.; Ferrarin, C.; Umgiesser, G.; Gulbinskas, S.; Bellafiore, D. Investigation of saline water intrusions into the Curonian Lagoon (Lithuania) and two-layer flow in the Klaipėda Strait using finite element hydrodynamic model. *Ocean Sci.* **2013**, *9*. [CrossRef]
18. Lesutiene, J.; Bukaveckas, P.A.; Gasiunaite, Z.R.; Pilkaityte, R.; Razinkovas-Baziukas, A. Tracing the isotopic signal of a cyanobacteria bloom through the food web of a Baltic Sea coastal lagoon. *Estuar. Coast. Shelf Sci.* **2014**, *138*, 47–56. [CrossRef]
19. Bresciani, M.; Adamo, M.; De Carolis, G.; Matta, E.; Pasquariello, G.; Vaičiūtė, D.; Giardino, C. Monitoring blooms and surface accumulation of cyanobacteria in the Curonian Lagoon by combining MERIS and ASAR data. *Remote Sens. Environ.* **2014**, *146*, 124–135. [CrossRef]
20. Vaičiūtė, D.; Bresciani, M.; Bartoli, M.; Giardino, C.; Bučas, M. Spatial and temporal distribution of coloured dissolved organic matter in a hypertrophic freshwater lagoon. *J. Limnol.* **2015**, *74*, 572–583. [CrossRef]
21. Trimonis, E.; Gulbinskas, S.; Kuzavinis, M. The Curonian Lagoon bottom sediments in the Lithuanian water area. *Baltica* **2003**, *16*, 13–20.
22. Krause, W. Zur Bestimmungsmöglichkeit subfossiler Characeen-Oosporen an Beispielen aus Schweizer Seen. *Vierteljahrsschr. Naturforsch. Ges. Zürich.* **1986**, *141*, 295–313.
23. Haas, J.N. First identification key for charophyte oospores from central Europe. *Pflugers. Arch.* **1994**, *29*, 227–235. [CrossRef]
24. Vedder, F. Morphologie und Taxonomie rezenter und subfossiler Characeen-Oosporen aus der Ostsee. *Rostock. Meeresbiolog. Beitr.* **2004**, *13*, 43–54.
25. Soulié-Märsche, I.; García, A. Gyrogonites and oospores, complementary viewpoints to improve the study of the charophytes (Charales). *Aquat. Bot.* **2015**, *120*, 7–17. [CrossRef]
26. QGIS Development Team. QGIS Geographic Information System. v 2.18.12- Las Palmas. *Open Source Geospatial Found. Proj.* **2020**. Available online: https://doi.org/http://www.qgis.org/ (accessed on 18 April 2020).
27. Malhotra, A.; Fonseca, M.S. *WEMo (Wave Exposure Model): Formulation, Procedures and Validation*; NOAA Technical Memorandum NOS NCCOS 65: Beaufort, NC, USA, 2007.
28. R Core Team. *R: A Language and Environment for Statistical Computing*; R Foundation for Statistical Computing: Vienna, Austria, 2020. Available online: https://www.R-project.org/ (accessed on 20 June 2020).
29. Hijmans, R.J. *Raster: Geographic Data Analysis and Modeling*; R Package Version 3.3-13; R Foundation for Statistical Computing: Vienna, Austria, 2020. Available online: https://CRAN.R-project.org/package=raster (accessed on 20 June 2020).
30. Bivand, R.; Keitt, T.; Rowlingson, B. *rgdal: Bindings for the 'Geospatial' Data Abstraction Library*; R Package Version 1.5-17; R Foundation for Statistical Computing: Vienna, Austria, 2020. Available online: https://CRAN.R-project.org/package=rgdal (accessed on 20 June 2020).
31. Hijmans, R.J.; Williams, E.; Vennes, C. *Geosphere: Spherical Trigonometry*; R Package Version 1.3-11; R Package Version 1.5-17; R Foundation for Statistical Computing: Vienna, Austria, 2014. Available online: https://cran.r-project.org/web/packages/geosphere (accessed on 20 June 2020).
32. Bivand, R.; Rundel, C. *Rgeos: Interface to Geometry Engine—Open Source (GEOS)*; R Package Version 0.3-11; R Package Version 1.5-17; R Foundation for Statistical Computing: Vienna, Austria, 2015. Available online: https://cran.r-project.org/web/packages/rgeos/i (accessed on 20 June 2020).
33. Evans, J.S. *SpatialEco*; R Package Version 1.3-1; R Package Version 1.5-17; R Foundation for Statistical Computing: Vienna, Austria, 2020. Available online: https://github.com/jeffreyevans/spatialEco (accessed on 20 June 2020)
34. Zuur, A.F.; Ieno, E.N.; Smith, G.M. Ordination—First encounter. *Analysing Ecological Data.* **2007**, 189–192. [CrossRef]
35. Oksanen, J.; Blanchet, F.G.; Friendly, M.; Kindt, R.; Legendre, P.; McGlinn, D.; Minchin, P.R.; O'hara, R.B.; Simpson, G.L.; Solymos, P.; et al. *Vegan: Community Ecology Package*; R Package Version 2.4-3; R Foundation for Statistical Computing: Vienna, Austria, 2016. Available online: https://cran.r-project.org/web/packages/vegan (accessed on 20 June 2020).
36. Peters, G. *Userfriendlyscience: Quantitative Analysis made Accessible*; R Package Version 0.7.0; R Foundation for Statistical Computing: Vienna, Austria, 2017. Available online: https://cran.r-project.org/web/packages/userfriendlyscience (accessed on 20 June 2020).
37. Friedman, J.H. Greedy function approximation: A gradient boosting machine. *Ann. Stat.* **2001**, 1189–1232. [CrossRef]
38. Torn, K.; Peterson, A.; Herkül, K.; Suursaar, U. Effects of climate change on the occurrence of charophytes and angiosperms in a brackish environment. *Webbia.* **2019**, *74*, 167–177. [CrossRef]
39. Elith, J.; Leathwick, J.R.; Hastie, T. A working guide to boosted regression trees. *J Anim. Ecol.* **2008**, *77*, 802–813. [CrossRef]
40. LeDell, E.; Gill, N.; Aiello, S.; Fu, A.; Candel, A.; Click, C.; Kraljevic, T.; Nykodym, T.; Aboyoun, P.; Kurka, M.; et al. *h2o: R Interface for the 'H2O' Scalable Machine Learning Platform*; R Package Version 3.30.1.3; R Foundation for Statistical Computing: Vienna, Austria, 2020. Available online: https://cran.r-project.org/web/packages/h2o/h2o.pdf (accessed on 20 June 2020).
41. Schubert, H.; Blindow, I. *Charophyte of the Baltic Sea*; The Baltic Marine Biologists Publication, No. 19; A.R.G. Gantner Verlag Kommanditgesellschaft: Ruggell, Liechtenstein, 2003; pp. 27–36.
42. Santamaría, L. Why are most aquatic plants widely distributed? Dispersal, clonal growth and small-scale heterogeneity in a stressful environment. *Acta oecol.* **2002**, *23*, 137–154. [CrossRef]
43. Doege, A.K.; van de Weyer, K.; Becker, R.; Schubert, H. Bioindikation mit Characeen [Bioindication with Characeae]. In *Armleuchteralgen. Die Characeen Deutschlands [Stoneworts. Characeae of Germany]*; Chara, A.G., Ed.; Springer: Berlin/Heidelberg, Germany, 2016; Volume 8, pp. 97–138.

44. Steinhardt, T.; Selig, U. Influence of salinity and sediment resuspension on macrophyte germination in coastal lakes. *J. Limnol.* **2011**, *70*, 11–20. [CrossRef]
45. Casanova, M.T.; Brock, M.A. Life histories of charophytes from permanent and temporary wetlands in eastern Australia. *Aust. J. Bot.* **1999**, *47*, 383–397. [CrossRef]
46. De Winton, M.D.; Casanova, M.T.; Clayton, J.S. Charophyte germination and establishment under low irradiance. *Aquat. Bot.* **2004**, *79*, 175–187. [CrossRef]
47. Sanjuan, J.; Vicente, A.; Flor-Arnau, N.; Monleón, T.; Cambra, J.; Martín-Closas, C. Effects of light and temperature on Chara vulgaris (Charophyceae) gyrogonite productivity and polymorphism–palaeoenvironmental implications. *Phycologia* **2017**, *56*, 204–212. [CrossRef]
48. Elliott, M.; Quintino, V. The estuarine quality paradox, environmental homeostasis and the difficulty of detecting anthropogenic stress in naturally stressed areas. *Mar. Pollut. Bull.* **2007**, *54*, 640–645. [CrossRef]
49. Elith, J.; Leathwick, J.R. Species distribution models: Ecological explanation and prediction across space and time. *Annu. Rev. Ecol. Evol.System.* **2009**, *40*, 677–697. [CrossRef]
50. Bertrin, V.; Boutry, S.; Alard, D.; Haury, J.; Jan, G.; Moreira, S.; Ribaudo, C. Prediction of macrophyte distribution: The role of natural versus anthropogenic physical disturbances. *Appl. Veg. Sci.* **2018**, *21*, 395–410. [CrossRef]
51. Van Zuidam, B.G.; Peeters, E.T. Wave forces limit the establishment of submerged macrophytes in large shallow lakes. *Limnol. Oceanogr.* **2015**, *60*, 1536–1549. [CrossRef]
52. Ruiz-Montoya, L.; Lowe, R.J.; Kendrick, G.A. Contemporary connectivity is sustained by wind-and current-driven seed dispersal among seagrass meadows. *Mov. Ecol.* **2015**, *3*, 9. [CrossRef]
53. Schubert, H.; Feuerpfeil, P.; Marquardt, R.; Telesh, I.; Skarlato, S. Macroalgal diversity along the Baltic Sea salinity gradient challenges Remane's species-minimum concept. *Mar. Pollut. Bull.* **2011**, *62*, 1948–1956. [CrossRef]
54. Bonis, A.; Grillas, P.; van Wijck, C.; Lepart, J. The effect of salinity on the reproduction of coastal submerged macrophytes in experimental communities. *J. Veg. Sci.* **1993**, *4*, 461–468. [CrossRef]
55. Van Leeuwen, C.H.; Van der Velde, G.; van Groenendael, J.M.; Klaassen, M. Gut travellers: Internal dispersal of aquatic organisms by waterfowl. *J. Biogeogr.* **2012**, *39*, 2031–2040. [CrossRef]
56. Morkūnė, R.; Petkuvienė, J.; Bružas, M.; Morkūnas, J.; Bartoli, M. Monthly Abundance Patterns and the Potential Role of Waterbirds as Phosphorus Sources to a Hypertrophic Baltic Lagoon. *Water* **2020**, *12*, 1392. [CrossRef]

water

MDPI

Article

Water Level Fluctuations and Air Temperatures Affect Common Reed Habitus and Productivity in an Intermittent Wetland Ecosystem

Alenka Gaberščik *, Mateja Grašič, Dragan Abram and Igor Zelnik

Department of Biology, Biotechnical Faculty, University of Ljubljana, Večna pot 111, SI-1000 Ljubljana, Slovenia; mateja.grasic@bf.uni-lj.si (M.G.); dragan.abram@gmail.com (D.A.); igor.zelnik@bf.uni-lj.si (I.Z.)

* Correspondence: alenka.gaberscik@bf.uni-lj.si; Tel.: +386-1-320-33-43

Received: 30 August 2020; Accepted: 6 October 2020; Published: 10 October 2020

Abstract: Lake Cerknica is an intermittent wetland ecosystem with extreme water level fluctuations. It hosts extensive reed stands that have colonized different habitat types. Two different stands were compared: a lake stand not directly influenced by the intermittent River Stržen and a riparian stand near River Stržen. Reed productivity (growth and assimilate allocation) was monitored for these reed stand types over 13 years (2007–2019), and this measurement was compared to monthly water levels and air temperatures. Reeds from the lake reed stand were significantly shorter with a lower shoot density, overall biomass production, and ratio of flowering plants. A correlation analysis revealed stronger and more numerous significant correlations between environmental and reed productivity parameters for the lake reed stand compared to the riparian reed stand. The variabilities of the growth and assimilate allocation parameters in the lake reed stand were both mostly explained by the combined water levels for June and July, which explained 47% and 52% of the variability, respectively. The most influential temperatures were in May, which explained 29% and 19% of the variability of growth and assimilate allocation parameters, respectively. For the riparian reed stand, water levels and temperatures out of the vegetation season appeared more important. Therefore, habitats with permanent water are more suitable for reeds than those with fluctuating water. However, fluctuating water conditions are expected to become more common due to climate change.

Keywords: *Phragmites australis*; water level fluctuations; temperature; morphometric parameters; biomass allocation

---

1. Introduction

Water level fluctuations directly affect ecosystems through their effects on the aquatic organisms, and they indirectly affect them by changing the physical and chemical conditions [1] that control different levels of the community structure [2]. The common reed (*Phragmites australis* (Cav.) Trin. and Steud) is a highly productive and cosmopolitan perennial helophyte that colonizes different wetland habitats [3]. It forms dense monospecific stands and is the dominant plant species in many ecosystems [4,5]. It shows high intraspecific diversity and phenotypic plasticity that enable its acclimation to adverse environmental conditions [6]. This is why the common reed is widely distributed across a variety of habitats [7].

One of the most important factors that significantly alters the habitat conditions of the common reed is the water level [1,5,8,9]. This defines the availability of water and nutrients [10], as well as the presence of oxygen in the root zone [11], and thus influences the structure of the reed stands. This is a collective result of different processes that include photosynthesis, respiration, mortality, and assimilate translocation between shoots and below-ground plant organs [12]. Li et al. [13] showed that water depth is a key driver in the shaping of common reed biomass production, plant height, basal diameter, and density. The common reed can tolerate water depths of up to 2 m. A decrease in water level positively affects the plants, which reflects in an increase in their assimilation area [14,15]. High water levels might also prevent rhizosphere self-oxidation via radial oxygen loss and might thus result in hypoxia [16]. Like many other helophytes, the common reed is only moderately tolerant to oxygen shortage [17,18]. High water levels are especially unfavorable at the beginning of the growing season because they slow down the early development of the plants [5,19].

Lake Cerknica is an intermittent wetland ecosystem that has extensive reed stands [20]. These reed stands cover the shallower parts of the lake area, the riparian zones of the lake tributaries, and the riparian zone of the intermittent River Stržen, which meanders across the exposed lake-bed. These habitats offer different conditions for reed stands in terms of the duration and extent of the lake floodwaters [21]. In addition to the spatial differences of these floodwaters, reed stands are also greatly influenced by water level fluctuations throughout the year. This pattern depends on the temporal and spatial distributions of precipitation over the watershed, which extends over 575 km^2 [22]. These results in water level fluctuations that are more pronounced in the lake habitat in comparison to the riparian habitat, which is permanently conditioned by water from the watercourse. These water level fluctuations do not always harmonize with the developmental cycles of the plants, including the common reed [23]. Any deviations in the normal rhythm will affect the habitus and productivity of these reeds [5]. The best conditions for the common reed are during 'normal' floods, when the water reaches the highest level in spring and then gradually decreases until summer, when the lake usually dries out. However, with autumnal rainfalls, the water level can increase again and then slowly decrease during the winter period [22]. Along with the water levels, the temperatures also have an important influence because they can significantly affect the beginning of the growing season and its length [24].

There have been many studies, which included the measurements of reed productivity parameters in relation to water level fluctuations. However, the majority of them have only presented short-term results, e.g., [13–15]. In the present study, we analyzed datasets of reed productivity parameters from the monitoring of two reed stand types at Lake Cerknica over 13 years (2007–2019)—namely the lake reed stand and the riparian reed stand. The productivity parameters were divided into two groups—the growth and assimilate allocation parameters—and were related to the environmental conditions in terms of the monthly water levels and air temperatures. These two parameters are also expected to fluctuate due to global climate change, and, therefore, this study not only provides an insight into the conditions of reed stands at Lake Cerknica but also shows the trends that will be relevant for other reed stands in other water bodies considering the likely future climate changes. We hypothesized that the most critical environmental conditions for these reeds are those in spring, when the reeds emerge from the rhizomes, and those in the autumn, as these might affect the length of the growing season. We also hypothesized that these critical periods would be of different importance relative to the two reed stand types.

2. Materials and Methods

2.1. Site Description

Lake Cerknica is located at the bottom of the Cerkniško polje depression. Polje is a karst feature that is characterized by an intermittent river that originates at one margin of the polje and sinks at the other side (Figure 1). At the Cerkniško polje, the river is called River Stržen, and the polje is its floodplain. Lake Cerknica is a Ramsar and Natura 2000 site, and it is part of the Notranjska Regional Park. The majority of the water that enters the lake is of karst origin, which reaches the lake from underground. In spring and usually in late autumn, the water surface of Lake Cerknica reaches an altitude of 550 m a.s.l., with flooding over an area of 26 km^2. The flooding phase of the lake lasts over a relatively long period, usually about 9 months of each year; thus, the floodplain area is dry for 2–3 months of each year. The pattern of water level fluctuations varies significantly over different years [22]. The extent and duration of these floodwaters result in a variety of habitats that change over time [5,21,25]. Different plant communities show clear zonal distributions that depend on the extent, duration, and frequency of these floodwaters [20].

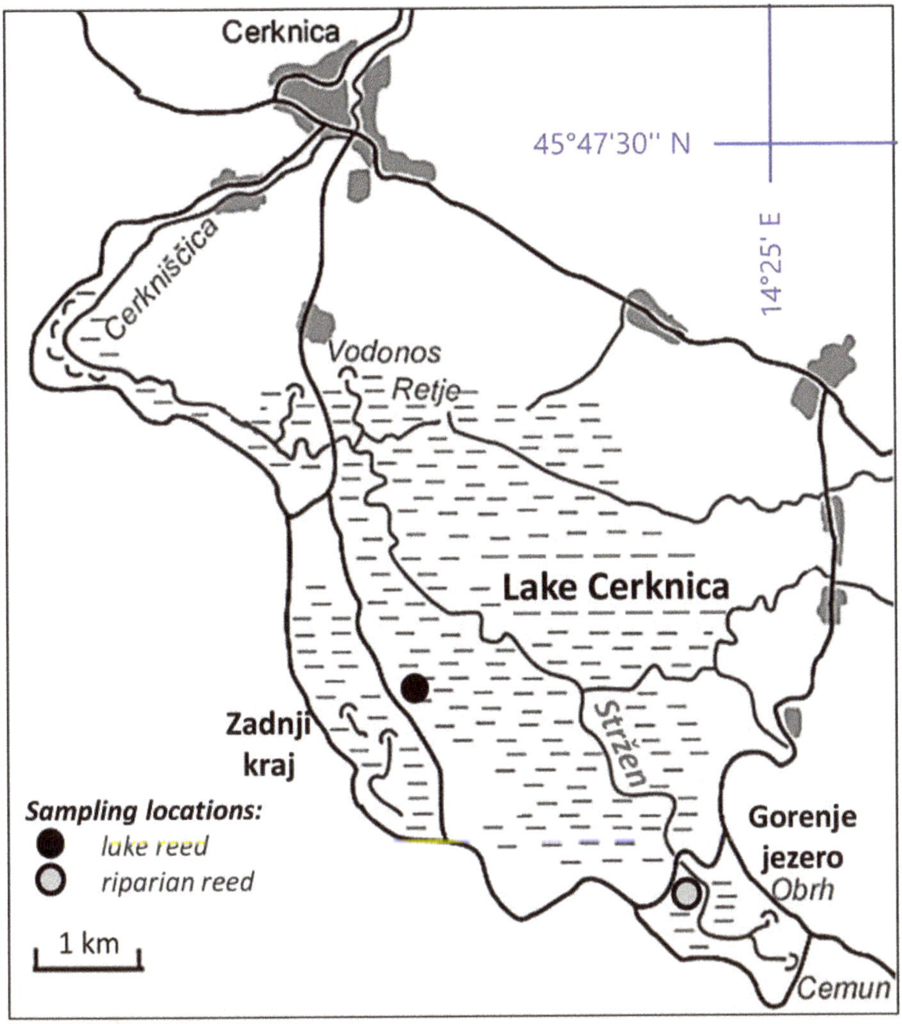

Figure 1. Map of the study area showing the two sampling locations.

Reed productivity parameters were measured here for two different reed stand types: (a) in a lake reed stand at Zadnji Kraj (45°44′27″ N; 14°22′13″ E), which experiences extreme water level fluctuations (from ~2 m high to completely dry), and (b) in a riparian reed stand in the vicinity of the intermittent River Stržen at Gorenje Jezero (45°43′40″ N; 14°24′17″ E) at the edge of the polje, where the water levels are usually < 0.5 m and where the soil water-table remains high even during dry periods (Figure 1).

2.2. Productivity Parameters

The harvesting of the above-ground parts of the reed plants from 2007 to 2019 was performed during the fruiting phase at the end of the growing season, in September, following the work of Květ et al. [26] and Cronk and Fennessy [27]. Five 0.5 × 0.5 m plots were harvested from each location at each sampling. In addition to the plant above-ground biomass, the following parameters were determined: shoot density, flowering shoots, internodes and leaves per plant, shoot height, and shoot basal diameter.

2.3. Environmental Parameters

The water level data were obtained from the Stržen–Gorenje Jezero monitoring station of the Slovenian Environment Agency (Figure 2) and the air temperatures from the nearest weather station (Postojna), as available from http://www.meteo.si/ (Figure 3).

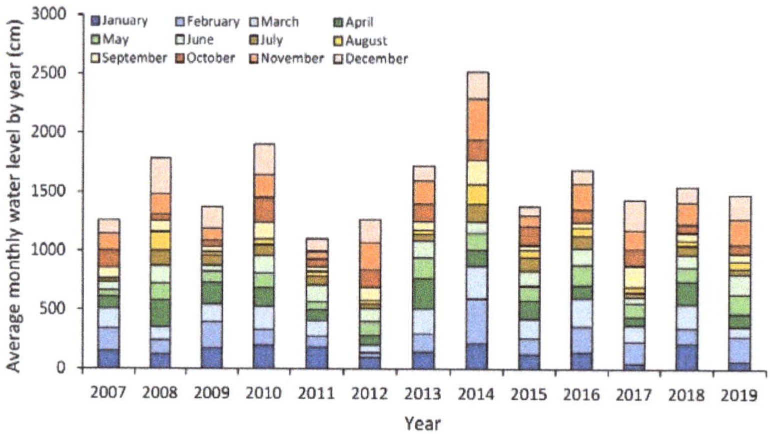

Figure 2. Average monthly water levels at the Stržen–Gorenje Jezero monitoring station (obtained from the Slovenian Environment Agency).

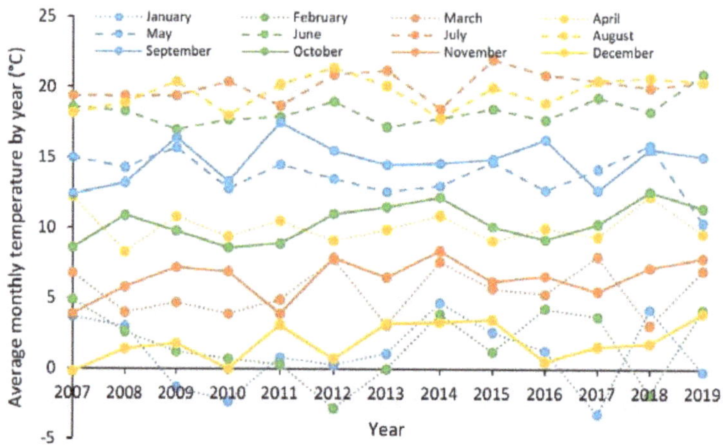

Figure 3. Average monthly air temperatures at the Postojna weather station (obtained from http://www.meteo.si/).

2.4. Statistical Analysis

Using the SPSS statistic software, version 22.0 (IBM Corp., Armonk, NY, USA), statistically significant differences between the lake and riparian reed stand productivities were evaluated using Mann–Whitney non-parametric tests, and the relationships between different productivity and environmental parameters were calculated using Spearman's correlation analysis.

To examine the impact of water level and air temperature on reed plant productivity, we considered the data from the corresponding year (January to August) and from the end of the growing season from the year before (September to December), which might have affected the accumulation of reserves in the reed rhizomes.

A detrended correspondence analysis was used for the exploratory data analysis using the Canoco for Windows 4.5 program package (Microcomputer Power: Ithaca, NY, USA). Due to the obtained gradient lengths (<3 standard deviations) [28], a redundancy analysis (RDA) was used to determine whether the variations in the productivity parameters were related to monthly water levels and/or air temperatures. Productivity was determined as two groups: the growth parameters (i.e., shoot density, internodes, leaves per plant, shoot height, and stem basal diameter) and the assimilate allocation parameters (i.e., flowering shoots, biomass of leaves, stems, and flowers). The significance of the effects of the variables was determined using Monte Carlo tests with 999 permutations. The forward selection of the explanatory variables was used to avoid collinearity. All of the variables used in the analysis were standardized.

3. Results

Common reeds from the lake and riparian reed stands significantly differed for the productivity parameters (Table 1). Reeds from the lake reed stand were significantly shorter, with significantly lower shoot density, overall biomass production, and ratio of flowering plants. The habitus of single plants was similar regarding stem basal diameter and the number of internodes and leaves but not regarding height and biomass per plant (lake reed stand: 8.8 g per plant; riparian reed stand: 12.0 g per plant). The samples from the lake reed stand generally showed a higher variability in the measured parameters in comparison to the riparian reed stand samples.

Table 1. Productivity parameters for the common reed from the lake (Zadnji Kraj) and riparian (Gorenje Jezero) reed stands from 2007 to 2019.

Parameter	Units	Reed Stand		p
		Lake	Riparian	
		Mean ± S.D.	Mean ± S.D.	
Height	cm	148.05 ± 30.59	183.27 ± 26.49	≤0.001
Basal diameter	mm	5.25 ± 1.00	5.15 ± 0.78	ns
Internodes/plant	n	16.33 ± 1.70	16.01 ± 1.68	ns
Leaves/plant	n	9.46 ± 3.20	9.89 ± 2.40	ns
Plants	n/m^2	50.40 ± 20.89	64.86 ± 26.36	≤0.001
Leaf dry mass	g/m^2	105.28 ± 46.77	159.26 ± 72.37	≤0.001
Rest dry mass	g/m^2	338.67 ± 192.44	619.62 ± 270.72	≤0.001
Flowering	n/m^2	22.98 ± 15.23	36.36 ± 14.32	≤0.001
Flowering	%	47.59 ± 26.63	58.71 ± 18.09	≤0.05

ns: not significant (Mann–Whitney non-parametric tests).

A correlation analysis revealed much stronger and more numerous significant correlations between the environmental and reed productivity parameters for the lake reed stand than for the riparian reed stand. For the lake stand, the reed productivity was mostly affected by the water levels at the end of the previous growing season, in October and November of the previous year, and the water levels at the beginning of the current season, in May and June. Almost all of these relations were negative (Table 2).

May and June were also the most important months regarding the air temperatures. However, the effects here were generally positive.

Table 2. Spearman's correlations between growth and biomass allocation parameters, as well as water levels and air temperatures, for the common reed plants from the lake reed stand (Zadnji Kraj).

Parameter	Previous Year				Current Year							
	Sept.	Oct.	Nov.	Dec.	Jan.	Feb.	Mar.	Apr.	May.	Jun.	Jul.	Aug.
Water Level												
Height	0.13	−0.30	−0.29	0.15	0.10	0.01	0.07	0.09	−0.47	−0.28	0.07	−0.17
Basal diameter	−0.16	−0.70	−0.38	0.06	−0.16	0.06	−0.29	0.05	−0.40	−0.34	−0.35	−0.48
Internodes/plant	0.15	−0.28	−0.26	0.08	−0.15	0.30	−0.08	−0.08	−0.33	−0.21	−0.22	−0.18
Leaves/plant	−0.28	−0.27	−0.37	−0.40	−0.42	−0.33	−0.47	−0.02	0.20	0.58	−0.31	0.09
Plants/m^2	0.03	−0.22	−0.02	0.27	0.25	0.11	0.06	0.17	−0.03	−0.19	0.00	−0.32
Leaf dry matter/m^2	−0.16	−0.55	−0.48	0.01	−0.00	−0.29	−0.18	0.02	−0.22	−0.06	−0.51	−0.61
Rest dry matter/m^2	0.00	−0.54	−0.38	0.23	0.23	−0.04	−0.01	0.08	−0.48	−0.47	−0.12	−0.51
Flowering (n/m^2)	0.17	−0.19	−0.21	0.21	0.22	0.09	0.07	0.08	−0.50	−0.44	0.19	−0.21
Flowering (%)	0.15	−0.01	−0.17	−0.02	0.03	−0.06	0.11	0.01	−0.41	−0.28	0.16	−0.00
Air Temperature												
Height	0.00	0.05	0.03	0.22	−0.03	0.22	−0.13	0.00	0.56	0.00	−0.02	−0.20
Basal diameter	0.25	0.13	−0.09	0.09	−0.30	0.13	0.08	−0.02	0.33	0.11	−0.02	0.11
Internodes/plant	0.07	0.52	0.18	0.46	0.06	0.51	0.18	0.21	0.36	0.33	0.04	0.01
Leaves/plant	0.32	−0.13	−0.01	−0.14	−0.14	0.07	−0.05	−0.40	−0.38	0.45	0.13	0.03
Plants/m^2	−0.03	0.05	−0.08	0.03	−0.13	−0.27	−0.09	0.16	0.03	−0.46	−0.08	0.05
Leaf dry matter/m^2	0.41	−0.21	−0.05	−0.05	−0.29	−0.29	−0.23	−0.03	0.06	−0.09	0.11	0.11
Rest dry matter/m^2	0.16	−0.02	−0.14	0.20	−0.17	−0.16	−0.05	0.09	0.49	−0.22	−0.07	−0.01
Flowering (n/m^2)	−0.16	0.03	−0.11	0.20	0.05	0.16	−0.02	0.09	0.61	−0.19	−0.21	−0.22
Flowering (%)	−0.06	−0.03	0.02	0.18	0.17	0.27	−0.02	−0.07	0.52	0.11	−0.03	−0.28

Shaded fields contain statistically significant coefficients (light grey: $0.01 \leq p \leq 0.05$; dark grey: $p \leq 0.01$.

In the riparian reed stand, the parameters most positively correlated to the water level were plant basal diameter and the height of the plants, while basal diameter was negatively correlated to air temperatures (Table 3). The correlation analysis showed the importance of the winter months, especially December and January, when water level negatively correlated with the number of internodes.

Table 3. Spearman's correlations between growth and biomass allocation parameters, as well as water levels and air temperatures, for the common reed plants from the riparian reed stand (Gorenje Jezero).

Parameter	Previous Year				Current Year							
	Sept.	Oct.	Nov.	Dec.	Jan.	Feb.	Mar.	Apr.	May.	Jun.	Jul.	Aug.
Water Level												
Height	−0.04	0.16	−0.12	−0.30	−0.21	0.20	0.04	0.02	0.17	0.14	0.36	0.46
Basal diameter	0.32	−0.11	−0.13	0.26	0.31	0.02	0.07	0.45	−0.10	0.22	0.44	0.06
Internodes/plant	−0.23	0.02	−0.00	−0.35	−0.34	0.36	0.04	0.03	0.35	0.01	−0.11	0.20
Leaves/plant	−0.25	−0.03	−0.06	−0.30	−0.27	0.18	0.08	0.10	0.24	−0.19	−0.19	−0.09
Plants/m^2	−0.07	−0.26	−0.19	0.09	0.13	−0.06	−0.12	−0.10	−0.19	0.03	−0.12	−0.14
Leaf dry matter/m^2	−0.02	−0.12	−0.32	−0.03	0.10	0.05	0.01	0.07	−0.13	−0.06	0.04	−0.09
Rest dry matter/m^2	0.13	−0.02	−0.16	0.07	0.22	0.13	0.11	0.11	−0.06	0.11	0.30	0.14
Flowering (n/m^2)	−0.01	−0.05	−0.07	−0.06	0.06	0.13	0.07	−0.02	0.09	0.07	0.12	0.16
Flowering (%)	0.05	0.30	0.11	−0.22	−0.09	0.12	0.13	−0.03	0.25	0.05	0.22	0.35
Air Temperature												
Height	−0.10	0.22	0.02	0.33	0.23	0.53	0.17	−0.20	−0.05	0.19	0.08	−0.32
Basal diameter	−0.31	0.11	0.10	−0.03	0.16	0.15	−0.37	−0.02	0.20	−0.36	−0.34	−0.56
Internodes/plant	−0.02	0.16	−0.10	−0.03	0.10	0.40	0.13	0.03	−0.23	0.21	0.10	0.08
Leaves/plant	−0.05	−0.15	−0.31	−0.21	0.02	0.05	−0.05	0.03	−0.08	−0.03	0.18	0.25
Plants/m^2	0.15	−0.05	0.05	−0.01	−0.18	−0.01	−0.01	0.06	0.02	−0.00	−0.21	−0.08
Leaf dry matter/m^2	−0.13	−0.14	−0.19	−0.05	0.05	0.13	−0.14	0.11	0.17	−0.14	−0.22	−0.14
Rest dry matter/m^2	−0.12	0.15	0.10	0.19	0.14	0.27	−0.09	0.07	0.08	−0.16	−0.23	−0.43
Flowering (n/m^2)	0.14	0.30	0.17	0.35	0.10	0.20	0.15	0.00	−0.13	0.07	−0.00	−0.24
Flowering (%)	−0.01	0.29	0.09	0.36	0.34	0.21	0.24	−0.08	−0.14	0.18	0.17	−0.14

Shaded fields contain statistically significant coefficients (light grey: $0.01 \leq p \leq 0.05$; dark grey: $p \leq 0.01$).

A redundancy analysis was performed separately for the growth and assimilate allocation parameters, with different results shown. For the lake reed stand, the water levels in spring and summer appeared to be the most important parameter (Table 4 and Figures 4 and 5). The variability of the growth and assimilate allocation parameters was mostly explained by the combined water levels for June and July, which explained 47% of the variability for growth and 52% of the variability for assimilate allocation. The water levels in the other months were less important for the lake reed stand, although when combined, they explained an additional 18% and 24% of the variability of the growth and assimilate allocation parameters, respectively. High water levels in July and August negatively affected the number of leaves and internodes for the lake reed stand. The RDA plots revealed large differences across the study years (Figures 4 and 5).

For the air temperatures for the lake reed stand, the most influential month was May, which explained 29% of the variability of the growth parameters and 19% of the variability of the assimilate allocation parameters (Table 4). Air temperatures explained a total of 86% of the variability of the growth and a total of 76% of the variability of the assimilate allocation.

Table 4. Redundancy analysis for the significant explained variance for the growth and assimilate allocation parameters for the lake reed stand (Zadnji Kraj), as defined by the monthly water levels and air temperatures.

Parameter Class	Monthly Parameter	Month	Explained Variance (%)	p
Growth	Water level	June	22	0.001
		July	25	0.001
		August	8	0.002
		April	5	0.003
		February	5	0.003
	Air temperature	May	29	0.001
		August	12	0.001
		July	10	0.001
		January	8	0.001
		March	5	<0.01
		December	5	<0.01
		September	4	<0.01
		October	4	<0.01
		April	3	<0.01
		July	3	<0.01
		February	3	<0.01
Assimilate allocation	Water level	July	23	0.001
		June	29	0.001
		April	10	0.001
		May	6	0.004
		October	4	<0.01
		September	4	<0.01
	Air temperature	May	19	0.001
		December	13	0.001
		October	8	0.001
		January	7	0.002
		March	7	0.002
		April	5	0.001
		February	5	0.001
		September	6	<0.01
		August	6	<0.01

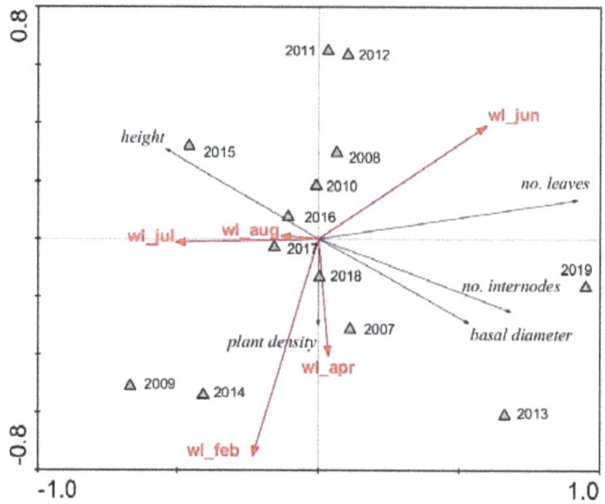

Figure 4. Redundancy analysis plot showing the strengths of the associations between monthly water levels and growth parameters of the common reed from the lake stand (Zadnji Kraj). Triangles: different years; wl: water level; no.: number.

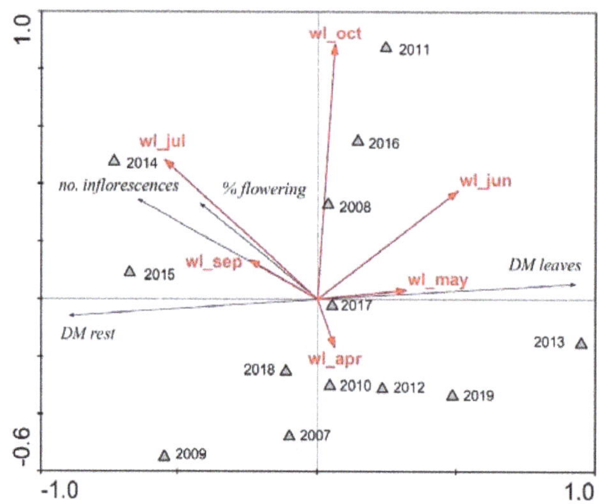

Figure 5. Redundancy analysis plot showing the strengths of the associations between monthly water levels and assimilate allocation parameters of the common reed from the lake stand (Zadnji Kraj). Triangles: different years; DM: dry mass; wl: water level; no.: number.

For the riparian reed stand, the water levels out of the vegetation season appeared to be the most important (Table 5 and Figures 6 and 7). The variability of the growth parameters was mostly explained by the combined water levels in February and December, which explained 34% of the variability. Meanwhile, the importance of the water levels in the other months was much lower, as only an additional 21% of sample variability was explained. The variability of the assimilate allocation parameters was mostly explained by the combined water levels in January and November, which explained 41% of the variability. Again, the importance of water levels in the other months was lower, with an additional 24% of the sample variability explained. The RDA plot again revealed large differences here across the study years.

Table 5. Redundancy analysis for the significant explained variance for the growth and assimilate allocation parameters for the riparian reed stand (Gorenje Jezero), as defined by the monthly water levels and air temperatures.

Parameter Class	Monthly Parameter	Month	Explained Variance (%)	*p*
Growth	Water level	February	19	0.001
		December	15	0.001
		September	9	0.002
		October	5	0.002
		January	7	0.001
	Air temperature	October	15	0.001
		June	11	0.002
		January	8	<0.01
		July	8	<0.01
Assimilate allocation	Water level	January	22	0.001
		November	19	0.001
		July	9	0.001
		October	9	0.007
		August	6	0.006
	Air temperature	July	17	0.001
		February	11	0.004
		October	9	0.007
		April	6	0.004

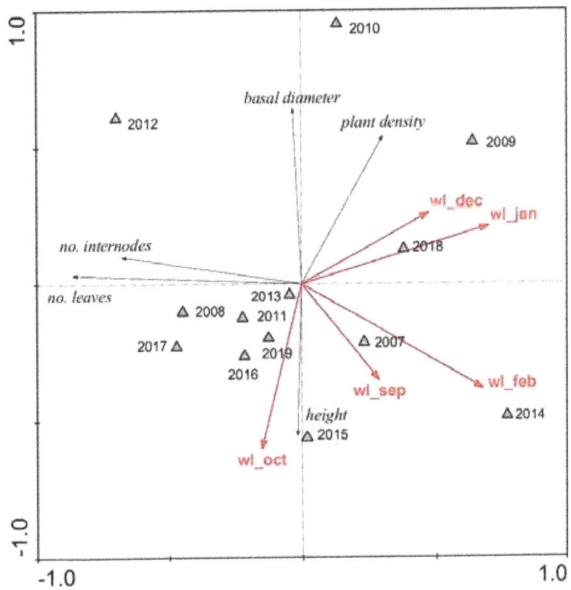

Figure 6. Redundancy analysis plot showing the strengths of the associations between monthly water levels and growth parameters of the common reed from the riparian stand (Gorenje Jezero). Triangles: different years; wl: water level; no.: number.

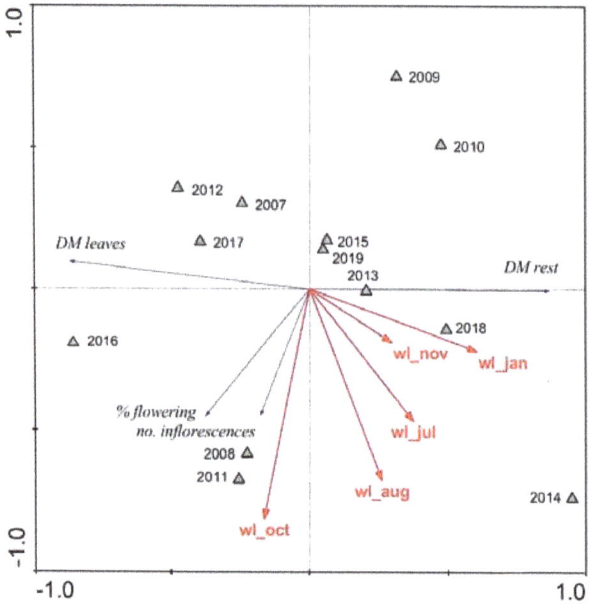

Figure 7. Redundancy analysis plot showing the strengths of the associations between monthly water levels and assimilate allocation parameters of the common reed from the riparian stand (Gorenje Jezero). Triangles: different years; DM: dry mass; wl: water level; no.: number.

For the air temperatures for the riparian reed stand, the most influential months were October and June, which explained 26% of the variability of the growth parameters, and July and February, which explained 28% of the variability of the assimilate allocation parameters (Table 5). Altogether, the air temperatures explained a total of 42% of the variability of the growth parameters and a total of 43% of the variability of the assimilate allocation parameters.

We also examined whether the reed productivity in the previous year affected the productivity parameters in the following (current) year. Here, the RDA revealed that the numbers of leaves, the number of internodes, the biomass of the leaves, and the biomass of the other parts in the previous year explained 8%, 4%, 4%, and 4%, respectively, of the productivity parameters in the following year. Therefore, these parameters combined explained 20% of the variability of the productivity parameters in the following year.

4. Discussion

The common reed is successful in habitats with stable or regular hydrology [29]. The higher variability in the measured parameters for samples from the lake reed stand in comparison to those from the riparian reed stand was a consequence of extreme water level fluctuations. For the lake location here, this could range from completely dry soil to 2 m depth of water, while the nearby River Stržen in the riparian reed stand maintains permanent high soil water table and floods up to 0.5 m. Hayball and Pearce [30] reported that in deep water, the common reed developed lower numbers of shoots, while a decrease in the water level increased the plant density and leaf numbers, which positively affected the leaf area index and consequently affected the productivity [15]. A high density of reed culms causes self-shading, which decreases the flowering rate and increases the leaf to stem mass ratio [29]. On the other hand, culm height and density significantly contribute to the competitive success of the common reed in terms of the shading out of other plant competitors in the habitat [31].

Plants that develop during moderate constant water levels grow faster than plants growing under fluctuating water levels [32,33]. This is also supported by the above-ground biomass measurements in the present study. The common reed is especially sensitive to rapid and extreme water level changes [34]. Extreme fluctuations affect reed fitness, partly due to the disturbed oxygenation of the rhizosphere [35] and possibly also due to negative effects on active mycorrhiza [36,37]. Such extreme changes are characteristic of the lake reed stand at Zadnji Kraj. In contrast, the more gradual changes that occur in the riparian reed stand at Gorenje Jezero appear to be less detrimental. In water fluctuation systems, the timing of the changes in water levels is also very important. For the lake reed stand at Zadnji Kraj, the variability of the growth and assimilate allocation parameters was explained by water levels in June and July ($p < 0.001$), which was the time of the vegetative development of plants, whereas temperatures seemed to be most important in May ($p < 0.001$). For the riparian reed stand at Gorenje Jezero, most of the variability of the growth and assimilate allocation parameters was explained by water levels out of the vegetation season and by temperatures at the end of the vegetation season, which may be related to the filling of reserves [5]. For the lake reed stand, there was positive correlation between leaf numbers and water levels in June (0.58; $p < 0.01$), and there were negative correlations with water levels in July (−0.31; $p < 0.01$) and from December to March ($p < 0.05$). On the other hand, no such significant correlations were found for the riparian reed stand, except for the negative correlation seen for December (−0.30; $p < 0.05$). Plant growth depends on the levels of available carbohydrates, which are strongly affected by environmental conditions [3]. A correlation analysis showed that in the lake reed stand, the plant height was negatively affected by the water levels in October of the previous year and in May and June of the current year, but it was positively related with air temperatures in May. In the riparian reed stand, the plant height was related to the water levels in December (−0.30; $p < 0.05$), July (0.36; $p < 0.01$), and August (0.46; $p < 0.01$). Autumnal rainfall can shorten the vegetation period and thus interrupt the filling of the reserves in the rhizomes, while high waters in late spring can prevent the development of reeds [5,32,38]. Previous studies have shown positive relationships between stem height and shoot density, which reflect productivity during the vegetation season [39]. However, no such correlation was obtained in the present study. The growth parameter that was most strongly correlated to the water level was basal diameter. For the lake reed stand, a strong negative correlation was obtained throughout the whole period of the vegetative development of reed, while for the riparian stand, this relation was strongly positive in April and June. This was possibly the consequence of the differences in maximal water levels between the two locations that might be unfavorable for lake but not for riparian stands. Thinner culms, which occurred in the years with deeper water, are more vulnerable to physical disturbances such as strong water movements and winds [40,41].

Irregular and strong water movements are usually the consequence of extreme water level fluctuations, like waves or currents, and may negatively affect the competitiveness of the common reed [29]. Water movement represents a physical force that can break reed culms and prevents the passive and humidity-induced aeration of the rhizome system at higher water levels, which also occurs via dead stems [16,42]. In water fluctuating systems, reed stands are frequently subjected to prolonged dry periods [21]. The common reed can withstand longer periods of water shortage through a reduction in leaf area and by an increased water-use efficiency [43]. Drought events are very common for the lake reed stand at Zadnji Kraj during the summer periods, with a positive correlation seen here between water levels and the numbers of leaves in June and July.

The negative effect on reed productivity may also be due to extreme temperatures, which may significantly increase the evapotranspiration rate in summer and may negatively affect reed rhizomes in winter. For example, Bodensteiner and Gabriel [44] attributed the reduction in reed stand area in the Upper Winnebago Pool Lakes in Wisconsin (USA) to the combined effects of extreme water levels and low temperatures in winter. We also obtained positive correlations with temperatures and some of the measured productivity parameters in the winter months at both locations.

In addition to differences in environmental factors like water levels, total radiation, and temperature, the fitness of plants during the previous season might also have a strong impact on plant height [3]. The present study showed little effect of the reed productivity parameters in the previous season, as these parameters explained only 20% of the variability of the productivity parameters in the current year.

This study showed that water depth represents the most important factor in common reed life. However, water depth is also an important predictor for the spread of common reed, which expands where the water levels are decreasing [45]. Water levels at the two locations affected different common reed parameters in different ways. In spite of that, water depth regulation may present a strong management tool [46], especially in human-made and human-managed aquatic systems. The increased extent of water level fluctuation in many aquatic systems may also be the consequence of different hydro-melioration measures in the landscape and thus the reduced capacity of the landscape to store water [47], as well as due to altered precipitation regimes due to global climate change. Renaturation measures to increase landscape water capacity may benefit every reed stand. Vital reed stands may increase the resilience of the landscape because by producing high biomass quantities, they provide various ecosystem services. They positively affect the local climate, purify water, present a source of biomass and a reservoir for water, and serve as a habitat and refuge for a variety of species [25,33,48–51].

5. Conclusions

This study has shown that habitats with the permanent presence of water are more suitable habitats for the common reed in comparison to those with fluctuating water levels. This presents a problem in the light of global climate change, since water level fluctuations in different water bodies are expected to become more frequent. A favorable water regime in a landscape positively affects common reed production and its ecosystems services. Thus, measures to increase landscape water capacity, including assuring favorable conditions for reed stands, are needed to mitigate future extremes.

Author Contributions: Conceptualization, A.G.; methodology, A.G. and D.A.; validation, A.G., M.G., and I.Z.; formal analysis, A.G., M.G., and D.A.; investigation, D.A.; writing—original draft preparation, A.G.; writing—review and editing, A.G., M.G., and I.Z.; visualization, A.G. and M.G.; supervision, A.G.; project administration, A.G.; funding acquisition, A.G. All authors have read and agreed to the published version of the manuscript.

Funding: This study was funded by the Slovenian Research Agency, grant numbers P1-0212 and 39096.

Acknowledgments: The authors are grateful to Christopher Berrie for his revision of the English language.

Conflicts of Interest: The authors declare no conflict of interest. The funders had no role in the design of the study; in the collection, analyses, or interpretation of data; in the writing of the manuscript; or in the decision to publish the results.

References

1. Yi, Y.; Xie, H.; Yang, Y.; Zhou, Y.; Yang, Z. Suitable habitat mathematical model of common reed (*Phragmites australis*) in shallow lakes with coupling cellular automaton and modified logistic function. *Ecol. Model.* **2020**, *419*, 108938. [CrossRef]
2. Hill, N.M.; Keddy, P.A.; Wisheu, I.C. A Hydrological model for predicting the effects of dams on the shoreline vegetation of lakes and reservoirs. *Environ. Manag.* **1998**, *22*, 723–736. [CrossRef] [PubMed]
3. Clevering, O.A.; Brix, H.; Lukavská, J. Geographic variation in growth responses in *Phragmites australis*. *Aquat. Bot.* **2001**, *69*, 89–108. [CrossRef]
4. Hocking, P.J.; Finlayson, C.M.; Chick, A.J. The biology of Australian weeds. 12. *Phragmites australis* (Cav.) Trin. ex Stuedel. *J. Aust. Institute Agric. Sci.* **1983**, *40*, 123–132.
5. Dolinar, N.; Regvar, M.; Abram, D.; Gaberščik, A. Water-level fluctuations as a driver of *Phragmites australis* primary productivity, litter decomposition, and fungal root colonisation in an intermittent wetland. *Hydrobiology* **2015**, *774*, 69–80. [CrossRef]

6. Eller, F.; Skálová, H.; Caplan, J.S.; Bhattarai, G.P.; Burger, M.K.; Cronin, J.T.; Guo, W.-Y.; Guo, X.; Hazelton, E.L.G.; Kettenring, K.M.; et al. Cosmopolitan species as models for ecophysiological responses to global change: The common reed *Phragmites australis*. *Front. Plant. Sci.* **2017**, *8*, 1833. [CrossRef]
7. Clevering, O.A.; Lissner, J. Taxonomy, chromosome numbers, clonal diversity and population dynamics of *Phragmites australis*. *Aquat. Bot.* **1999**, *64*, 185–208. [CrossRef]
8. Engloner, A.I.; Papp, M. Vertical differences in *Phragmites australis* culm anatomy along a water depth gradient. *Aquat. Bot.* **2006**, *85*, 137–146. [CrossRef]
9. Tóth, V.R. Reed stands during different water level periods: Physico-chemical properties of the sediment and growth of *Phragmites australis* of Lake Balaton. *Hydrobiology* **2016**, *778*, 193–207. [CrossRef]
10. Grašič, M.; Dobravc, M.; Golob, A.; Vogel-Mikuš, K.; Gaberščik, A. Water shortage reduces silicon uptake in barley leaves. *Agric. Water Manag.* **2019**, *217*, 47–56. [CrossRef]
11. Armstrong, W.; Justin, S.; Beckett, P.; Lythe, S. Root adaptation to soil waterlogging. *Aquat. Bot.* **1991**, *39*, 57–73. [CrossRef]
12. Asaeda, T.; Karunaratne, S. Dynamic modeling of the growth of *Phragmites australis*: Model description. *Aquat. Bot.* **2000**, *67*, 301–318. [CrossRef]
13. Li, C.; Ye, X.; Wu, M.; Shao, X. Effects of water depth and coexistence on growth characteristics of *Phragmites australis* and *Typha domingensis*. *Soc. Wetl. Sci. Bull.* **2015**, *13*, 609–615.
14. Squires, L.; Van Der Valk, A.G. Water-depth tolerances of the dominant emergent macrophytes of the Delta Marsh, Manitoba. *Can. J. Bot.* **1992**, *70*, 1860–1867. [CrossRef]
15. Tóth, V.R.; Szabó, K. Morphometric structural analysis of *Phragmites australis* stands in Lake Balaton. *Ann. de Limnol. -Int. J. Limnol.* **2012**, *48*, 241–251. [CrossRef]
16. Armstrong, J.; Armstrong, W. Light-enhanced convective throughflow increases oxygenation in rhizomes and rhizosphere of *Phragmites australis* (Cav.). *Trin. ex Steud. New Phytol.* **1990**, *114*, 121–128. [CrossRef]
17. Brändle, R.; Crawford, R.M.M. Rhizome anoxia tolerance and habitat specialization in wetland plants. In *Plant Life in Aquatic and Amphibious Habitats*; Crawford, R.M.M., Ed.; Blackwell Scientific Publications: Oxford, UK, 1987; pp. 397–410.
18. Armstrong, J.; Afreen-Zobayed, F.; Blyth, S.; Armstrong, W. *Phragmites australis*: Effects of shoot submergence on seedling growth and survival and radial oxygen loss from roots. *Aquat. Bot.* **1999**, *64*, 275–289. [CrossRef]
19. Schmieder, K.; Dienst, M.; Ostendorp, W.; Jöhnk, K. Effects of water level variations on the dynamics of the reed belts of Lake Constance. *Ecohydrol. Hydrobiol.* **2004**, *4*, 229–239.
20. Martinčič, A.; Leskovar, I. Vegetation. In *The Vanishing Lake—Monograph on Lake Cerknica*; Gaberščik, A., Ed.; Slovenian ecological society. Ljubljana, Slovenia, 2003; pp. 80–96.
21. Dolinar, N.; Šraj, N.; Gaberščik, A. Water regime changes and the function of an intermittent wetland. In *Water and Nutrient Management in Natural and Constructed Wetlands*; Vymazal, J., Ed.; Springer: Dordrecht, The Netherlands, 2011; pp. 251–262.
22. Kranjc, A. Hydrological characteristics. In *The Vanishing Lake—Monograph on Lake Cerknica*; Gaberščik, A., Ed.; Slovenian Ecological Society: Ljubljana, Slovenia, 2003; pp. 26–38.
23. Gaberščik, A.; Urbanc-Bercic, O.; Kržič, N.; Kosi, G.; Brancelj, A. The intermittent Lake Cerknica: Various faces of the same ecosystem. *Lakes Reserv. Res. Manag.* **2003**, *8*, 159–168. [CrossRef]
24. Eller, F.; Lambertini, C.; Nguyen, L.X.; Achenbach, L.; Brix, H. Interactive effects of elevated temperature and CO2 on two phylogeographically distinct clones of common reed (*Phragmites australis*). *AoB PLANTS* **2013**, *5*, pls051. [CrossRef]
25. Dolinar, N.; Rudolf, M.; Šraj, N.; Gaberščik, A. Environmental changes affect ecosystem services of the intermittent Lake Cerknica. *Ecol. Complex.* **2010**, *7*, 403–409. [CrossRef]
26. Květ, J.; Westlake, D.F.; Dykyjova, D.; Marshall, E.J.P.; Ondok, J. Primary production in wetlands. In *The Production Ecology of Wetlands*; Westlake, D.F., Květ, J., Szczepansky, A., Eds.; Cambridge University Press: Cambridge, UK, 1999; pp. 78–168.
27. Cronk, J.K.; Fennessy, M.S. *Wetland Plants: Biology and Ecology*, 1st ed.; CRC Press: Boca Raton, FL, USA, 2001.
28. Ter Braak, C.J.F.; Šmilauer, P. *CANOCO Reference Manual and CanoDraw for Windows User's Guide: Software for Canonical Community Ordination (Version 4.5)*; Microcomputer Power: Ithaca, NY, USA, 2002.

29. Packer, J.G.; Meyerson, L.A.; Skálová, H.; Pyšek, P.; Kueffer, C. Biological flora of the British isles: *Phragmites australis*. *J. Ecol.* **2017**, *105*, 1123–1162. [CrossRef]
30. Hayball, N.; Pearce, M. Influences of simulated grazing and water-depth on the growth of juvenile *Bolboschoenus caldwellii*, *Phragmites australis* and *Schoenoplectus validus* plants. *Aquat. Bot.* **2004**, *78*, 233–242. [CrossRef]
31. Onimaru, K.; Yabe, K. Comparisons of nutrient recovery and specific leaf area variation between *Carex lasiocarpa var. occultans* and *Carex thunbergii var. appendiculata* with reference to nutrient conditions and shading by *Phragmites australis*. *Ecol. Res.* **1996**, *11*, 139–147. [CrossRef]
32. Vretare, V.; Weisner, S.E.B.; Strand, J.A.; Granéli, W. Phenotypic plasticity in *Phragmites australis* as a functional response to water depth. *Aquat. Bot.* **2001**, *69*, 127–145. [CrossRef]
33. Thevs, N.; Zerbe, S.; Gahlert, F.; Mijit, M.; Succow, M. Productivity of reed (*Phragmites australis Trin. ex Steud.*) in continental-arid NW China in relation to soil, groundwater, and land-use. *J. Appl. Bot. Food Qual.* **2007**, *81*, 62–68.
34. Haslam, S.M. The performance of *Phragmites communis Trin.* in relation to temperature. *Ann. Bot.* **1975**, *39*, 883–888. [CrossRef]
35. White, S.D.; Ganf, G.G. A comparison of the morphology, gas space anatomy and potential for internal aeration in phragmites australis under variable and static water regimes. *Aquat. Bot.* **2002**, *73*, 115–127. [CrossRef]
36. Dolinar, N.; Gaberščik, A. Mycorrhizal colonization and growth of *Phragmites australis* in an intermittent wetland. *Aquat. Bot.* **2010**, *93*, 93–98. [CrossRef]
37. Dolinar, N.; Šraj, N.; Pongrac, P.; Regvar, M.; Gaberščik, A. The presence of mycorrhiza in different habitats of an intermittent aquatic ecosystem. In *Water and Nutrient Management in Natural and Constructed Wetlands*; Vymazal, J., Ed.; Springer: Dordrecht, The Netherlands, 2011; pp. 299–308.
38. Dinka, M.; Ágoston-Szabó, E.; Urbanc-Berčič, O.; Germ, M.; Šraj-Kržič, N.; Gaberščik, A. Reed stand conditions at selected wetlands in Slovenia and Hungary. In *Wastewater Treatment, Plant Dynamics and Management in Constructed and Natural Wetlands*; Vymazal, J., Ed.; Springer: Dordrecht, The Netherlands, 2008; pp. 1–12.
39. Zhang, W.; Tong, S.; Wang, X.; Liu, B.; Lv, X.; Yang, X. The interaction of water, vegetation and soil in Zhalong wetland, China. *J. Food Agric. Environ.* **2013**, *11*, 1287–1294.
40. Engloner, A.I. Structure, growth dynamics and biomass of reed (*Phragmites australis*)—GA review. *Flora* **2009**, *204*, 331–346. [CrossRef]
41. Hong, M.G.; Nam, B.E.; Kim, J.G. Differences in functional traits of leaf blade and culm of common reed in four habitat types. *J. Ecol. Environ.* **2019**, *43*, 12. [CrossRef]
42. Afreen, F.; Zobayed, S.; Armstrong, J.; Armstrong, W. Pressure gradients along whole culms and leaf sheaths, and other aspects of humidity-induced gas transport in *Phragmites australis*. *J. Exp. Bot.* **2007**, *58*, 1651–1662. [CrossRef]
43. Pagter, M.; Bragato, C.; Brix, H. Tolerance and physiological responses of *Phragmites australis* to water deficit. *Aquat. Bot.* **2005**, *81*, 285–299. [CrossRef]
44. Bodensteiner, L.R.; Gabriel, A.O. Response of mid-water common reed stands to water level variations and winter conditions in Lake Poygan, Wisconsin, USA. *Aquat. Bot.* **2003**, *76*, 49–64. [CrossRef]
45. Tulbure, M.G.; Johnston, C.A. Environmental conditions promoting non-native *Phragmites australis* expansion in Great Lakes coastal wetlands. *Wetlands* **2010**, *30*, 577–587. [CrossRef]
46. Brix, H.; Ye, S.; Laws, E.A.; Sun, D.; Li, G.; Ding, X.; Yuan, H.; Zhao, G.; Wang, J.; Pei, S. Large-scale management of common reed, *Phragmites australis*, for paper production: A case study from the Liaohe Delta, China. *Ecol. Eng.* **2014**, *73*, 760–769. [CrossRef]
47. Gaberščik, A.; Murlis, J. The role of vegetation in the water cycle. *Ecohydrol. Hydrobiol.* **2011**, *11*, 175–181. [CrossRef]
48. Kiviat, E. Ecosystem services of *Phragmites* in North America with emphasis on habitat functions. *AoB PLANTS* **2013**, *5*, plt008. [CrossRef]
49. Köbbing, J.F.; Thevs, N.; Zerbe, S. The utilisation of reed (*Phragmites australis*): A review. *Mires Peat* **2013**, *13*.

50. Gaglio, M.; Lanzoni, M.; Nobili, G.; Viviani, D.; Castaldelli, G.; Fano, E.A. Ecosystem services approach for sustainable governance in a brackish water lagoon used for aquaculture. *J. Environ. Plan. Manag.* **2019**, *62*, 1501–1524. [CrossRef]
51. Baibagyssov, A.; Thevs, N.; Nurtazin, S.; Waldhardt, R.; Beckmann, V.; Salmurzauly, R. Biomass resources of *Phragmites australis* in Kazakhstan: Historical developments, utilization, and prospects. *Resources* **2020**, *9*, 74. [CrossRef]

Publisher's Note: MDPI stays neutral with regard to jurisdictional claims in published maps and institutional affiliations.

water

MDPI

Article

Long Term Aquatic Vegetation Dynamics in Longgan Lake Using Landsat Time Series and Their Responses to Water Level Fluctuation

Wenxia Tan [1], Jindi Xing [1], Shao Yang [2,*], Gongliang Yu [3], Panpan Sun [2] and Yan Jiang [1]

1 Key Laboratory for Geographical Process Analysis & Simulation of Hubei Province/College of Urban and Environmental Sciences, Central China Normal University, Wuhan 430079, China; tanwenxia@mail.ccnu.edu.cn (W.T.); jindixing@mails.ccnu.edu.cn (J.X.); jiangyan@mail.ccnu.edu.cn (Y.J.)

2 School of Life Sciences, Central China Normal University, Wuhan 430079, China; wppxsun@hotmail.com

3 Key Laboratory of Algal Biology, Institute of Hydrobiology, Chinese Academy of Sciences, Wuhan 430072, China; yugl@ihb.ac.cn

* Correspondence: yangshao@mail.ccnu.edu.cn

Received: 19 May 2020; Accepted: 30 July 2020; Published: 2 August 2020

Abstract: Aquatic vegetation in shallow freshwater lakes are severely degraded worldwide, even though they are essential for inland ecosystem services. Detailed information about the long term variability of aquatic plants can help investigate the potential driving mechanisms and help mitigate the degradation. In this paper, based on Google Earth Engine cloud-computing platform, we made use of a 33-year (1987–2019) retrospective archive of moderate resolution Landsat TM, ETM + and OLI satellite images to estimate the extent changes in aquatic vegetation in Longgan Lake from Middle Yangtze River Basin in China using the modified enhanced vegetation index, including emerged, floating-leaved and floating macrophytes. The analysis of the long term dynamics of aquatic vegetation showed that aquatic vegetation were mainly distributed in the western part of the lake, where lake bottom elevation ranged from 11 to 12 m, with average water depth of less than 1 m in spring. The vegetation area variation for the 33-year period were divided into six stages. In years with heavy precipitation, the vegetation area decreased sharply. In the following years, the area normally restored. Aquatic vegetation area had a significant negative correlation with the spring water level and summer water level. The results showed that aquatic vegetation was negatively affected when water depth exceeded 2.5 m in May and 5 m in summer. It is recommended that water depth remain close to 1 m in spring and close to 3 m in summer for aquatic vegetation growth. Our study provide quantitative evidence that water-level fluctuations drive vegetation changes in Longgan Lake, and present a basis for sustainable lake restoration and management.

Keywords: spatial-temporal dynamics; aquatic vegetation; water level fluctuation; Longgan lake; Google Earth Engine

1. Introduction

Wetlands play an important role in ecological services and socioeconomic services. As a basic component of wetland ecosystems, aquatic vegetation in shallow freshwater provide a variety of services, including improving water quality, maintaining aquatic ecosystem balance and providing food and habitats for many aquatic animals [1–3]. According to a recent investigation, decreasing aquatic vegetation are found in around 65.2% of the world lakes [4]. Due to rapid social and economic development during the past decades, dramatic variations in the distribution area of aquatic vegetation have been occurred in many lakes [5]. In addition, lake ecosystem have changed due to water level management, nutrient input and global climate change [6,7]. In views of this, it is necessary to study the change trend of aquatic vegetation and its response to environmental factors.

Water-level fluctuation (WLF) has been well demonstrated as one of the core factors affecting biomass, diversity, composition and structure of vegetation, by causing variations in environmental factors for plant growth and germination, such as light, oxygen, temperature and nutrients etc. [8–10]. Based on the study of Great Lakes, Keddy and Reznicek concluded that during extremely high water periods, the dominant species were killed, thus creating gaps; however, many plant species and vegetation types regenerated from buried seeds during low water periods [11]. Therefore, fluctuating water levels generally increase the diversity of vegetation types and plant species. The influence of WLF on aquatic vegetation mainly manifests through variation of its amplitude and dynamic regime [12]. For example, by comparing the aquatic vegetation in two river-disconnected lakes (artificial water regime) and the river-connected lake (natural water regime), Zhang et al. [13] found that the plant species richness was highest in the disconnected lake with intermediate amplitude of WLFs, and lowest in the connected lake. The amplitude of WLFs was the most important factor in determining the distribution of lake-shore plants, followed by relative elevation and duration of submergence.

Macrophytes of different life forms response to WLF differently. By analyzing of the relationship between aquatic vegetation and WLFs in Taihu Lake from 1989 to 2010, Zhao et al. [14] found that water level from January to March had significant positive correlation with the coverage of emergent and floating-leaved vegetation, while had negative correlation with the coverage of submerged vegetation. Different macrophyte species of identical life form may also response to WLF differently. By treatment of three submerged macrophytes under different amplitude of WLF, Wang et al. [15] found *Hydrilla verticillata* exhibited more growth in static water, and *Elodea nuttallii* was inhibited by fluctuating water level treatments, while *Ceratophyllum demersum* became more abundant when water levels fluctuated.

Because aquatic vegetation in lakes and wetlands are often hard to reach, satellite images of Landsat and MODIS are employed for large scale and synoptic monitoring of lake vegetation [16–19]. Spectral indices (SIs), included various combinations of visible, near-infrared, and shortwave infrared bands, are widely used because their flexibility and capability for aquatic vegetation mapping. Villa et al. [20] and Villa et al. [21] summarized and compared several SIs for macrophyte mapping and concluded that the modified enhanced vegetation index (EVI) performed the best for three shallow freshwater bodies in different parts of Europe. The high temporal variability of aquatic vegetation in decades can offer pivotal information about the biological complexity of the lakes and wetlands. Szabó et al. [22] used Normalized Difference Vegetation Index (NDVI) for a 33-year long vegetation spread monitoring survey in an artificial lake in Hungary. Combining four vegetation indices (VI), Lopes et al. [23] studied the spatial and temporal vegetation changes within a coastal lagoon through Landsat imagery between 1984 and 2017. Google Earth Engine (GEE) [24], which is a cloud-based platform for earth science data analysis and visualization, is commonly used to study the dynamics of wetland land cover [25–27]. However, studies using GEE to investigate the long term variability of aquatic vegetation in shallow lakes are rarely found.

How water levels affect aquatic vegetation in natural lakes has been well studied. Hu et al. [28] took the use of time-series MODIS data to explore the relationship between wetland vegetation and factors linked with water-level fluctuations in Lake Poyang and Dongting. Using time-series of both MODIS and Landsat, the hydrological influence on the distribution and transition of wetland cover was studied [29]. Tan and Jiang [30], Wan et al. [31], You et al. [32], Zhang et al. [33] had also analyzed the relationship between wetland vegetation and water level for Lake Poyang and Dongting. However, Lake Poyang and Dongting are the only two lakes that are not dammed in the Yangtze river basin. All other lakes in the middle reaches of the Yangtze River are separated from the Yangtze River by dams and sluice gates. Thus, human interference is an important factor affecting lake water levels. As a national wetland reserve of China, Longgan Lake is an artificial regulated lake. However, the effects of human interference on water level to aquatic vegetation for Longgan Lake have been little explored. It is helpful to explore the relationship between aquatic vegetation coverage and human regulated water levels with long-term data for lake ecological management. As submerged

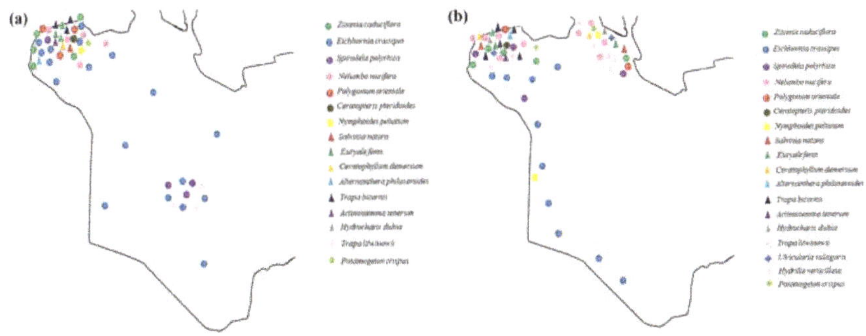

Figure 3. Distribution of macrophyte species in the western part of Longgan Lake for (**a**) September 2017 and (**b**) September 2018.

2.3. Satellite Data

All available Landsat top-of atmosphere (TOA) reflectance images for the WRS-2 footprints 121/39 and 122/39 in Google Earth Engine between 1987 and 2019 were used in this study, including Landsat 5 Thematic Mapper (TM), Landsat 7 Enhanced Thematic Mapper-plus (ETM+) and Landsat 8 Operational Land Imager (OLI) (as listed in Table 1). These datasets have a spatial resolution of 30 m. The images in late July were primarily selected to represent the vegetation cover extent of a year, because the vegetation coverage was highest during this time of a year. For years without cloud-free images in July, images in June and August were chosen instead. For Landsat 7 image in 2012, SLC-off gaps were filled with Phase 2 USGS Gap-Fill Algorithm [38].

Table 1. Temporal distribution of available Landsat images in Longgan Lake during 1987 to 2019.

Date	Sensor	Date	Sensor
27/08/1987	Landsat 5 TM	07/08/2003	Landsat 5 TM
26/06/1988	Landsat 5 TM	24/07/2004	Landsat 5 TM
15/07/1989	Landsat 5 TM	12/08/2005	Landsat 5 TM
09/07/1990	Landsat 5 TM	30/07/2006	Landsat 5 TM
26/06/1991	Landsat 5 TM	09/08/2007	Landsat 5 TM
23/07/1992	Landsat 5 TM	03/07/2008	Landsat 5 TM
10/07/1993	Landsat 5 TM	13/07/2009	Landsat 5 TM
04/07/1994	Landsat 5 TM	25/07/2010	Landsat 5 TM
05/06/1995	Landsat 5 TM	03/07/2011	Landsat 5 TM
25/07/1996	Landsat 5 TM	22/07/2012	Landsat 7 ETM +
29/08/1997	Landsat 5 TM	01/07/2013	Landsat 8 OLI_TIRS
08/07/1998	Landsat 5 TM	02/06/2014	Landsat 8 OLI_TIRS
02/07/1999	Landsat 5 TM	05/06/2015	Landsat 8 OLI_TIRS
05/07/2000	Landsat 5 TM	23/06/2016	Landsat 8 OLI_TIRS
24/07/2001	Landsat 7 ETM +	28/07/2017	Landsat 8 OLI_TIRS
11/07/2002	Landsat 7 ETM +	31/07/2018	Landsat 8 OLI_TIRS
		02/07/2019	Landsat 8 OLI_TIRS

High quality Sentinel-2 data with close acquisition dates as the corresponding Landsat images were used to validate the vegetation mapping results by the proposed method. Sentinel-2 have more than 10 bands, with spatial resolutions of 10–20 m. Moreover, a PROBA-V (Project for On-Board Autonomy-Vegetation) image of 26 June 2018 was also used in the discussion to show the aquatic vegetation cover difference between June and July 2018. Though PROBA-V has a lower spatial resolution (100 m) than that of Landsat, it was the only cloud free image that was found for June 2018 for Longgan Lake. PROBA-V is a satellite mission tasked to map land cover and vegetation growth on a daily basis. The sensor collects data in three VNIR (red, blue and near-infrared) bands and one SWIR (short-wave infrared) spectral band.

2.4. Hydrological Data

Water levels of Longgan Lake were controlled by artificial regulation using three hydrological stations (Figure 1). Since the start of the wet season (usually from May to September), the gates are closed to store water in the lakes and prevent flooding of the land. The water is released back into the Yangtze River in late autumn [39]. Daily water level data (meters above sea level, reference to Wusong datum) from May to September each year from 1987 to 2018 were collected. In this study, the spring water level was defined as the the average water level in May, and the summer water level was defined as the average water level in June, July and August. However, spring and summer water levels were missing in some years due to incomplete water level records.

The fifth generation ECMWF atmospheric reanalysis of the global climate (ERA5) from 1987 to 2018 were used to analyze the temporal trend of annual precipitation for Longgan Lake. ERA5 provides several improvements compared to ERA-Interim data with a much higher resolution in both time and space [40]. Its horizontal resolution is approximately 30 km.

2.5. Methods

2.5.1. Determination of Lake Boundaries

The spectral signatures of aquatic vegetation largely overlap with the signatures of terrestrial vegetation, which leads to the misclassification of aquatic and terrestrial vegetation patches in their transitional areas. Thus, terrestrial areas around the lake were blocked out using a mask of the lake based on the water surface in 2018. The same mask was applied to all Landsat images to obtain comparable results. Moreover, the northeastern and southern boundary of the lake have also been masked, because these area have been extensively used for aquaculture.

2.5.2. Aquatic Vegetation Mapping

Large-scale aquatic vegetation mapping usually employs a series of spectral indices (SIs), combining with threshold determination approaches [41,42]. The performance of different SIs have been assessed for aquatic vegetation mapping, and the 2-band enhanced vegetation index (EVI2) outperforms others in extracting floating and emergent vegetation [20,21,43]. EVI2 is a two-band adaptation of EVI that has been developed without a blue band. It retains sensitivity and linearity as EVI in high biomass regions [44]. EVI2 is calculated based on Equation (1),

$$\text{EVI2} = 2.5\frac{(R_{\text{NIR}} - R_{\text{R}})}{(R_{\text{NIR}} + 2.4R_{\text{R}} + 1)} \tag{1}$$

where R_{NIR} and R_{R} are the top-of-atmosphere (TOA) reflectance of the near-infrared and red bands.

EVI2 values were then used to distinguish between water and aquatic vegetation through a threshold selection method. Dynamic thresholds, instead of a static threshold, are normally recommended for remote sensing data of different phases, due to different satellite instruments used, or different atmospheric conditions or water environment for images of different dates [23,45]. For each scene, a threshold was selected based on the EVI2 histogram using the Otsu method that divide the study area into two classes (aquatic vegetation and open water) [46]. Otsu is a nonparametric unsupervised method, which find the threshold to minimize the within class variance of two classes. The histograms of EVI2 for 1997, 2001, and 2018 are shown in Figure 4 with the threshold values. For this study, the pixels with EVI2 values below and above the selected threshold were assigned as water and aquatic vegetation, respectively. The classification results and the false color composite images are also shown in Figure 4 for visual comparison.

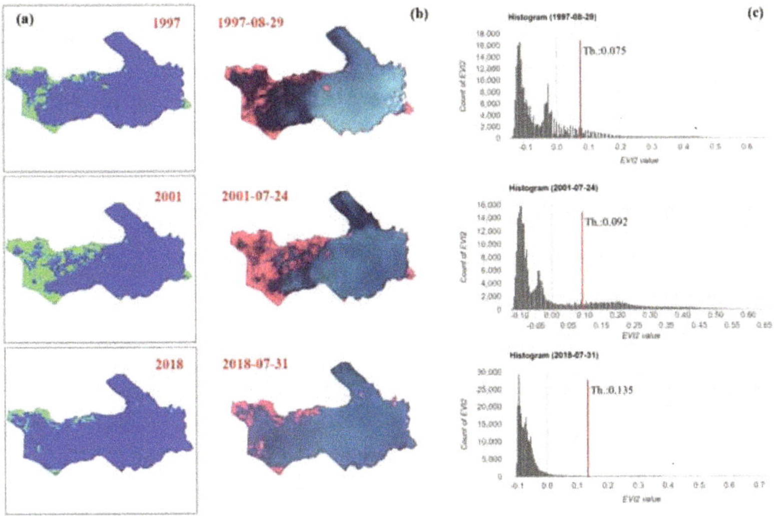

Figure 4. Classification results based on the optimized threshold for 1997, 2001, 2018. The pixels with EVI2 values below and above the selected threshold were assigned as water and aquatic vegetation, respectively. (**a**) Classification results; (**b**) The false color composite images; (**c**) EVI2 histogram and the Otsu segmentation thresholds.

2.5.3. Accuracy Assessment of Macrophytes Mapping

Extensive field survey data of aquatic vegetation in Longgan Lake were only available for September 2017 and September 2018 (Figure 3). The field survey results for a total of 48 field survey samples (pixels) were extracted and compared with the classification results of July 2017 and 2018, as shown in Figure 5 and Table 2. There were two possible reasons for the relatively low accuracy. Firstly, aquatic vegetation extent declined from summer to fall due to withering (e.g., *Trapa litwinowii* located in the center of the western lake withered from July to September). Secondly, floating macrophytes could change location quickly due to winds (e.g., *Eichhornia crassipes* located along the western shore of the lake during the survey of 2018 were not presented in the July image used for classification).

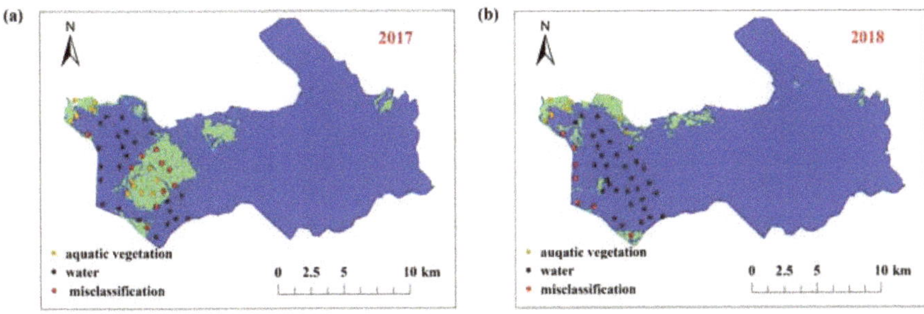

Figure 5. Comparison of field survey data and classification results for (**a**) 2017 and (**b**) 2018.

Table 2. Confusion matrix of classification result for aquatic vegetation (compared to field data).

Year	Overall Accuracy (%)	Kappa	Producer's Accuracy (%)	User's Accuracy (%)
2017	79.17	0.5833	100.00	58.33
2018	85.42	0.7267	60.00	90.00

Considering there was a two month difference between the Landsat data and the field data used, Sentinel-2 false color composite images of 26 July 2017 and 31 July 2018 were also used to validate the classification results. Aquatic vegetation boundaries were identified manually from Sentinel-2. Then, the manually delineated aquatic vegetation boundaries were compared with the classification results of Landsat using the confusion matrix. Table 3 listed the accuracy of aquatic vegetation classification. The results showed that the overall accuracies were higher than 90% and the kappa coefficients all exceeded 0.8, which demonstrated that the model performed well for distinguishing aquatic vegetation.

Table 3. Confusion matrix of classification result for aquatic vegetation (Compared to Sentinel-2).

Year	Overall Accuracy (%)	Kappa	Producer's Accuracy (%)	User's Accuracy (%)
2017	97.12	0.8682	83.04	94.62
2018	97.95	0.8183	82.75	83.10

3. Results

3.1. Seasonal and Decadal Variation of Water Level

The mean and variation of the monthly water level (meters above the sea water, reference to Wusong datum) for the study period were shown in Figure 6. It can be seen that the monthly average water level has been increasing, reaching its peak in August. Annual precipitation, average spring water level and summer water level of Longgan Lake from 1987 to 2018 were shown in Figure 7. Data gaps are caused by the lack of data. In spring, the highest water level was 13.49 m in 2010, and the lowest water level was 12.11 m in 2004. In summer, the highest water level was 16.28 m in 1992, and the lowest water level was 12.31 m in 2007. The changes of water level were generally consistent with the annual precipitation of Longgan Lake from 1987 to 2018. It can be seen that years of 1991, 1998, 2002, 2010, 2015 and 2016 have abnormally high annual precipitation and years of 2006 and 2007 have abnormally low annual precipitation.

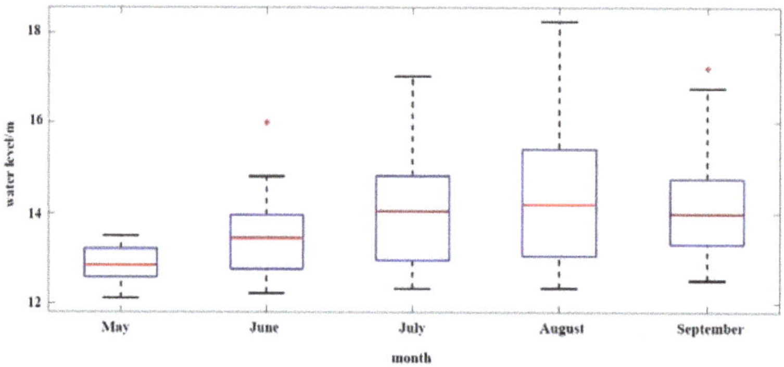

Figure 6. Box plot of the water levels from May to September for Longgan Lake (meters above sea level, reference to Wusong datum).

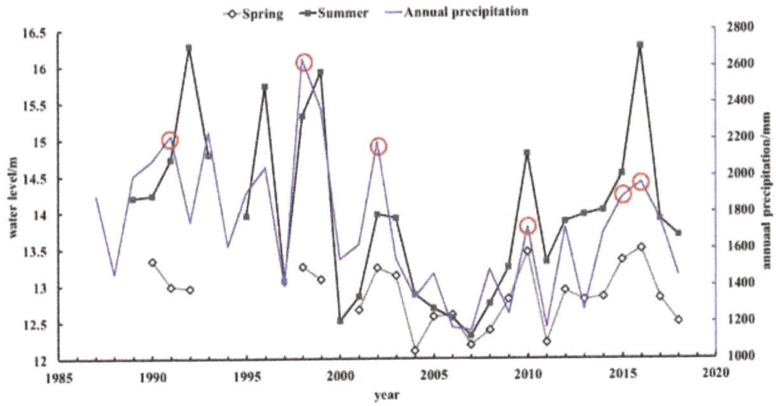

Figure 7. Annual precipitation, average spring water level and summer water level of Longgan Lake from 1987 to 2018. The red circles present years with abnormally high annual precipitation.

3.2. Inter-Annual Dynamics of Aquatic Vegetation

The spatial distribution of aquatic vegetation in Longgan Lake at the peak of vegetation coverage each year from 1987 to 2019 was shown in Figure 8. Due to the low temporal resolution of Landsat and the effects of cloud cover, it was impossible to accurately obtain the maximum aquatic vegetation coverage of each year. Even though the spatial distribution patterns of aquatic vegetation varied from year to year, aquatic vegetation were primarily distributed in the western region of the lake, where lake bottom elevation ranged from 11 to 12 m, with average water depth of less than 1 m in spring. For years that had high vegetation extent, aquatic vegetation expanded towards the center of the lake. For years with low vegetation extent, aquatic vegetation shrank towards the northwest corner.

The temporal variation of aquatic vegetation area in Longgan Lake from 1987 to 2019 were shown in Figure 9. In 2011, the vegetation area reached the maximum area of 73.99 km^2. In 2015, the vegetation area was the smallest, occupying 8.72 km^2, only about one-tenth of the aquatic vegetation area in 2011. In order to analyze the vegetation restoration process in a certain period under the influence of abnormal precipitation, the years of 1991, 1998, 2002, 2010 and 2015 were used as the dividing points. Thus, the inter-annual aquatic vegetation variation were divided into six stages. During the first stage (1987–1991), the vegetation area had remained above 35 km^2 in the first three years and then had decreased. Especially, the area was smallest in 1991, occupying only 14.87 km^2. The second stage occurred from 1992 to 1998. During this stage, the area began to increase, reaching to a maximum in 1996. Since then, the area had decreased, but at a lower rate. The third stage occurred from 1999 to 2002. During this period, the area increased firstly and then decreased. During the fourth stage (2003–2010), the area maintained at a higher value over 65 km^2. In 2010, the area dropped sharply to 12.34 km^2. During the fifth stage (2011–2015), the area increased firstly at a very high rate in 2011, and then remained above 45 km^2. In 2015, the area dropped sharply to 8.72 km^2. The sixth stage occurred from 2016 to 2019. During this stage, the area was very small except for 2019.

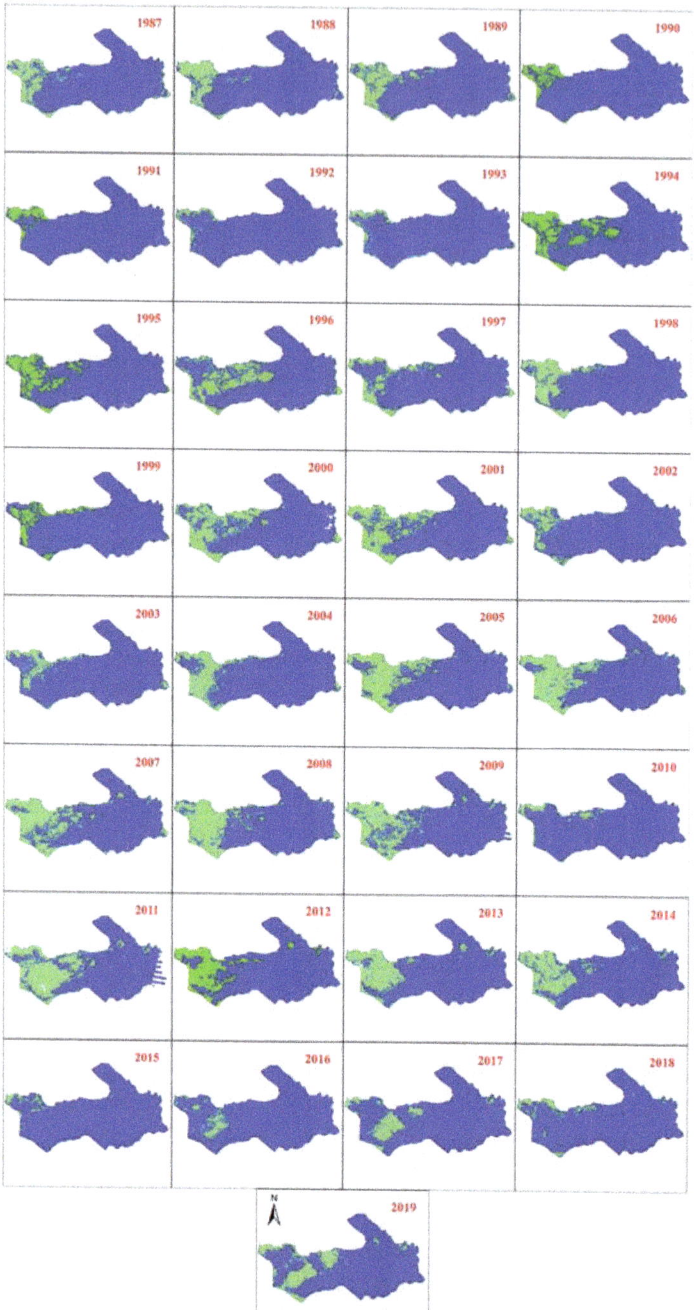

Figure 8. Spatial distribution map of aquatic vegetation in Longgan Lake from 1987 to 2019.

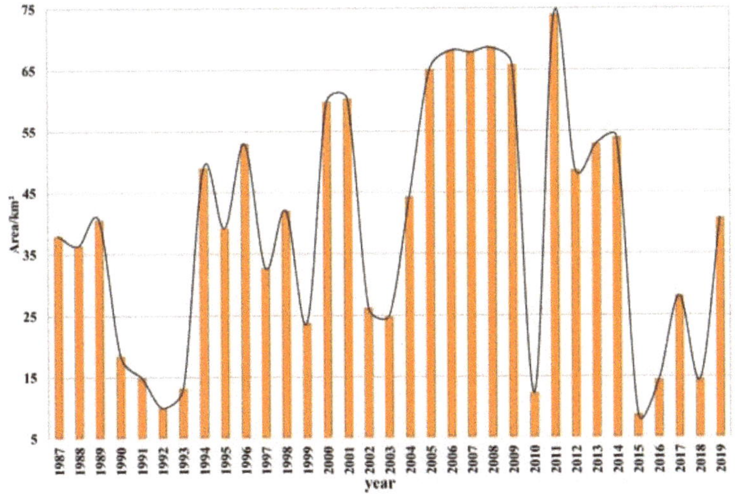

Figure 9. Aquatic vegetation area of Longgan Lake from 1987 to 2019.

3.3. Relationship between Aquatic Vegetation Area and Water-Level Change

The variation of aquatic vegetation area and water level for Longgan Lake from 1987 to 2019 were shown in Figure 10. When the water level rose above to 13 m in spring, the area of aquatic vegetation greatly reduced. When the water level exceeded 16 m in summer, the area decreased dramatically.

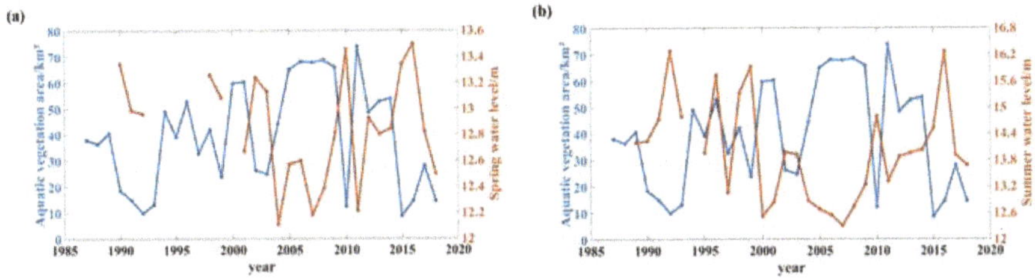

Figure 10. Decadal variations of aquatic vegetation area and water level for Longgan Lake from 1987 to 2019. (a) Aquatic vegetation area related to water level in spring; (b) Aquatic vegetation area related to water level in summer.

In order to explore the coupling relationship between the area of aquatic vegetation and water-level fluctuation rhythm, Pearson correlation analysis was used. The results revealed that aquatic vegetation area had a significant negative correlation with spring water level and summer water level, with Pearson correlation coefficients of -0.70 ($P < 0.01$) and -0.67 ($P < 0.01$), respectively. As shown in Figure 11, specifically, the area of aquatic vegetation was very large when water level was low, and vice versa. In 2011, the area of aquatic vegetation was largest, the spring water level was below 12.5 m and the summer water level was below 14 m.

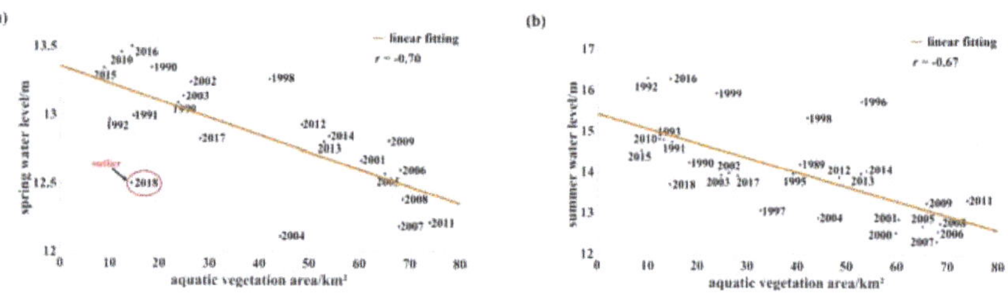

Figure 11. Correlation between aquatic vegetation distribution area and water level. (**a**) Spring water level; (**b**) Summer water level.

4. Discussion

4.1. Effects of Remote-Sensing Imagery Acquisition Time on the Maximum Aquatic Vegetation Area of a Year

According to field surveys, the main growth season for aquatic vegetation in Longgan Lake is from May to September. Thus, aquatic vegetation area normally reaches the maximum in late July. Therefore, July images were used with priority to get the maximum vegetation cover of a year. However, the maximum aquatic vegetation area extracted in this analysis might not be the true maximum cover area of a year. Firstly, images in June and August were used as supplements due to cloud coverage in July. Moreover, the locations of floating vegetation were greatly influenced by wind speed and direction, leading to large variability in their spatial distribution. As shown in Figure 11, some outliers did not follow the rule that the vegetation area was larger when the water level was lower. In 2018, the aquatic vegetation area was 14.49 km^2, and the water level in spring was 12.5 m. It showed that the water level in spring was low, but the extracted vegetation area was also relatively small compared to other years. As it can be seen in the PROBA-V image and Sentinel-2 image (Figure 12), large area of aquatic vegetation were distributed in the center of the western lake on 26 June 2018. However, these aquatic vegetation disappeared in the July 2018 images. Based on the field survey of 2017, the center of the western lake was largely covered by *Trapa litwinowii* + *Eichhornia crassipes* + *Spirodela polyrhiza* Ass. Same floating-leaved and floating plants present in June 2018, were possibly withered or blown away to other locations in July 2018. Despite these uncertainties, a good relationship between aquatic vegetation area and water level was still achieved for the three decades study period, indicating the robustness of the analysis.

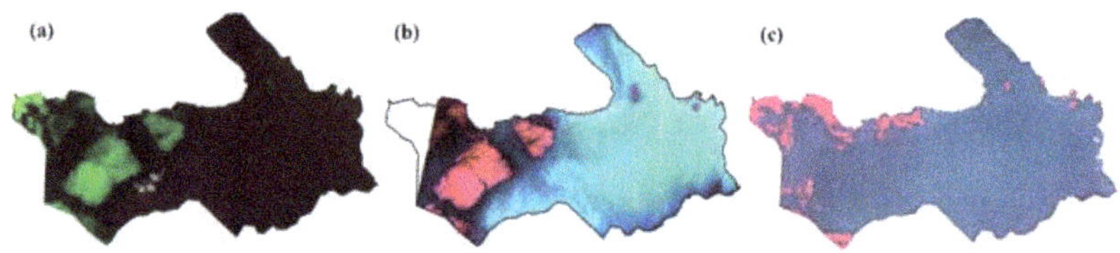

Figure 12. Comparison of PROBA-V image, Sentinel-2 image and Landsat 8 image in 2018. (**a**) PROBA-V image on 26 June 2018; (**b**) Sentinel-2 image on 26 June 2018; (**c**) Landsat 8 image on 31 July 2018.

4.2. Effects of Water-Level Fluctuation on Aquatic Vegetation Area

As shown in Figures 10 and 11, aquatic vegetation area showed a significant negative correlation with water level. The existing vegetation types in the lake have adapted to the long-term variation of water level. The expansion and retreat of aquatic vegetation are greatly related to lake bottom topography, as water depth is the determining factor for aquatic vegetation growth. The normal spring water depth in the northwestern part of the lake is less than 1 m, which is suitable for aquatic vegetation growth. For years with abnormally high spring water level, due to high water depth, aquatic vegetation retreated to areas close to the boundary of the lake where water depth was low.

When water level rose above to 13 m in spring and 16 m in summer, aquatic plants largely disappeared. To show the optimal water depth for aquatic vegetation, the distributions of spring and summer water depths for all area mapped as aquatic vegetation in all years were calculated. It can be seen in Figure 13 that in areas with aquatic vegetation, water depths did not exceed 2.5 m in May and 5 m in summer. To preserve aquatic vegetation in the western part of the lake, it is recommended that water depth remain close to 1 m in spring and close to 3 m in summer. In spring, higher water level reduced light, which was not conducive to seed germination and seedling growth [47,48]. In summer, continuous high water level could drown vegetation, resulting in reduced oxygen supplies and causing the death of aquatic plants [49]. Significantly, extreme hydrological events in summer, such as flooding, would cause massive aquatic vegetation collapse, and consequently the decline of ecosystem diversity.

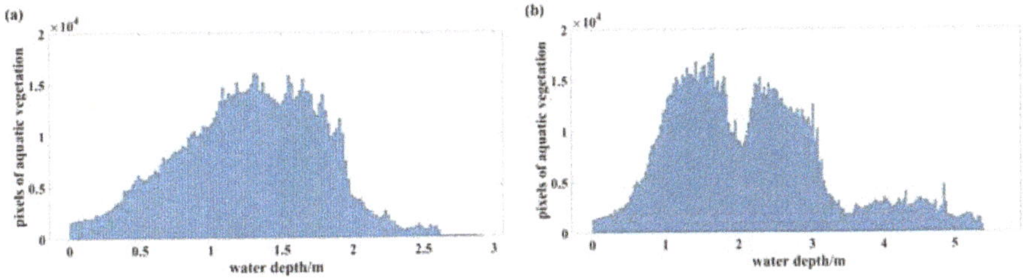

Figure 13. Water depth distribution for areas with aquatic vegetation in (**a**) Spring (average water depth in May) and (**b**) Summer (average water depth of June, July and August) for all years.

5. Conclusions

The aim of this study was to map the long term variability of aquatic vegetation extent in a human dammed lake and to investigate the correlation between aquatic vegetation and water level fluctuation of the lake. For this purpose, the modified enhanced vegetation index with Otsu threshold method was used to map the aquatic vegetation area in Longgan Lake for the 33-year period, including emerged, floating-leaved and floating macrophytes, using archived Landsat images from 1987 to 2019 on GEE platform. The classification results were validated by field survey data and other satellite data. Results showed that aquatic vegetation were mainly distributed in the western part of the lake, where lake bottom elevation ranged from 11 to 12 m, with average water depth close to 1 m in spring. The relationships between aquatic vegetation area and spring and summer water-level fluctuation were also analyzed. For years with increased vegetation cover, aquatic vegetation would expand from the northwestern to the center of the lake. Years with extremely high precipitation were related to low vegetation area. In the following years, the area normally restored. The aquatic vegetation area showed a significant negative correlation with the spring water level and summer water level. The results showed that aquatic vegetation was negatively affected when water depth exceeded 2.5 m in May and 5 m in summer. It is recommended that water depth remain close to 1 m in spring and close to 3 m in summer to assure successful aquatic vegetation growth. The results from our study provided information about the optimum water level for aquatic vegetation growth in Longgan Lake. These findings can provide the basis for wetland management and restoration. Future work will mainly focus on the classification of different aquatic vegetation types and the inter-seasonal dynamics based on multi-temporal and multi-source remote sensing imagery.

Author Contributions: S.Y. had the original idea for the study. G.Y. provided the methodology. W.T. and J.X. were responsible for analyzing the data and writing the paper. Y.J. offered background data about Longgan Lake. S.Y. and G.Y. supervised the research and provided valuable suggestions for the revision. S.Y., W.T. and Y.J. offered measured samples to validate our classification results in this paper. W.T., J.X. and Y.J. had significant contribution in coding and visualization. P.S. collected and analyzed the field data. All authors have read and agreed to the published version of the manuscript.

Funding: This work was supported by the National Natural Science Foundation of China under grants 31670464 and 41506208, partially by the Fundamental Research Funds for the Central Universities under Grant CCNU19TD002, CCNU18ZDPY03 and CCNU19TS001.

Acknowledgments: We would like to thank the staffs from the Bureau of Longgan Lake national nature reserve, Hubei, China for their assistance with fieldwork and their supports in this research. We also sincerely thank Jun Xu of Institute of Hydrobiology, Chinese Academy of Sciences , China for providing the topographic map of Longgan Lake. We also thank the anonymous reviewers for their insightful comments and suggestions, which lead to great improvements to the manuscript.

Conflicts of Interest: The authors declare no conflicts of interest.

References

1. Horppila, J.; Nurminen, L. Effects of submerged macrophytes on sediment resuspension and internal phosphorus loading in Lake Hiidenvesi (southern Finland). *Water Res.* **2003**, *37*, 4468–4474. [CrossRef]
2. Li, D.; Wang, L.; Ding, J. Ecological functions and resource utilization of aquatic plants. *Wetl. Sci.* **2011**, *9*, 290.
3. Phiri, C.; Chakona, A.; Day, J.A. Aquatic insects associated with two morphologically different submerged macrophytes, Lagarosiphon ilicifolius and Vallisneria aethiopica, in small fishless ponds. *Aquat. Ecol.* **2011**, *45*, 405–416. [CrossRef]
4. Zhang, Y.; Jeppesen, E.; Liu, X.; Qin, B.; Shi, K.; Zhou, Y.; Thomaz, S.M.; Deng, J. Global loss of aquatic vegetation in lakes. *Earth-Sci. Rev.* **2017**, *173*, 259–265. [CrossRef]
5. Guo, L. Doing battle with the green monster of Taihu Lake. *Science* **2007**, *317*, 1166. [CrossRef] [PubMed]
6. Sand-Jensen, K.; Riis, T.; Vestergaard, O.; Larsen, S.E. Macrophyte decline in Danish lakes and streams over the past 100 years. *J. Ecol.* **2001**, *88*, 1030–1040. [CrossRef]
7. Short, F.T.; Kosten, S.; Morgan, P.A.; Malone, S.; Moore, G.E. Impacts of climate change on submerged and emergent wetland plants. *Aquat. Bot.* **2016**, *135*, 3–17. [CrossRef]
8. Nõges, T.; Nõges, P. The effect of extreme water level decrease on hydrochemistry and phytoplankton in a shallow eutrophic lake. *Hydrobiologia* **1999**, *408*, 277–283. [CrossRef]
9. Geest, G.V.; Wolters, H.; Roozen, F.; Coops, H.; Roijackers, R.; Buijse, A.; Scheffer, M. Water-level fluctuations affect macrophyte richness in floodplain lakes. *Hydrobiologia* **2005**, *539*, 239–248. [CrossRef]
10. Yang, J.; Li, E.; Cai, X.; Wang, Z.; Wang, X. Research progress in response of plants in wetlands to water level change. *Wetl. Sci.* **2014**, *12*, 807–813.
11. Keddy, P.; Reznicek, A. Great Lakes vegetation dynamics: The role of fluctuating water levels and buried seeds. *J. Great Lakes Res.* **1986**, *12*, 25–36. [CrossRef]
12. Wilcox, D.A.; Nichols, S.J. The effects of water-level fluctuations on vegetation in a Lake Huron wetland. *Wetlands* **2008**, *28*, 487–501. [CrossRef]
13. Zhang, X.; Liu, X.; Wang, H. Effects of water level fluctuations on lakeshore vegetation of three subtropical floodplain lakes, China. *Hydrobiologia* **2015**, *747*, 43–52. [CrossRef]
14. Zhao, D.; Jiang, H.; Cai, Y.; An, S. Artificial regulation of water level and its effect on aquatic macrophyte distribution in Taihu Lake. *PLoS ONE* **2012**, *7*, e44836. [CrossRef] [PubMed]
15. Wang, M.; Liu, Z.; Luo, F.; Lei, G.; Li, H. Do amplitudes of water level fluctuations affect the growth and community structure of submerged macrophytes? *PLoS ONE* **2016**, *11*, e0146528. [CrossRef] [PubMed]
16. Oyama, Y.; Matsushita, B.; Fukushima, T. Distinguishing surface cyanobacterial blooms and aquatic macrophytes using Landsat/TM and ETM + shortwave infrared bands. *Remote Sens. Environ.* **2015**, *157*, 35–47. [CrossRef]
17. Liu, X.; Zhang, Y.; Shi, K.; Zhou, Y.; Tang, X.; Zhu, G.; Qin, B. Mapping aquatic vegetation in a large, shallow eutrophic lake: A frequency-based approach using multiple years of MODIS data. *Remote Sens.* **2015**, *7*, 10295–10320. [CrossRef]
18. Luo, J.; Li, X.; Ma, R.; Li, F.; Duan, H.; Hu, W.; Qin, B.; Huang, W. Applying remote sensing techniques to monitoring seasonal and interannual changes of aquatic vegetation in Taihu Lake, China. *Ecol. Indic.* **2016**, *60*, 503–513. [CrossRef]
19. Liang, Q.; Zhang, Y.; Ma, R.; Loiselle, S.; Li, J.; Hu, M. A MODIS-based novel method to distinguish surface cyanobacterial scums and aquatic macrophytes in Lake Taihu. *Remote Sens.* **2017**, *9*, 133. [CrossRef]

20. Villa, P.; Pinardi, M.; Tóth, V.; Hunter, P.; Bolpagni, R.; Bresciani, M. Remote sensing of macrophyte morphological traits: Implications for the management of shallow lakes. *J. Limnol.* **2017**, *76*, 109–126. [CrossRef]
21. Villa, P.; Pinardi, M.; Bolpagni, R.; Gillier, J.M.; Zinke, P.; Nedelcuţ, F.; Bresciani, M. Assessing macrophyte seasonal dynamics using dense time series of medium resolution satellite data. *Remote Sens. Environ.* **2018**, *216*, 230–244. [CrossRef]
22. Szabó, L.; Deák, B.; Bíró, T.; Dyke, G.J.; Szabó, S. NDVI as a Proxy for Estimating Sedimentation and Vegetation Spread in Artificial Lakes—Monitoring of Spatial and Temporal Changes by Using Satellite Images Overarching Three Decades. *Remote Sens.* **2020**, *12*, 1468. [CrossRef]
23. Lopes, C.L.; Mendes, R.; Caçador, I.; Dias, J.M. Evaluation of long-term estuarine vegetation changes through Landsat imagery. *Sci. Total Environ.* **2019**, *653*, 512–522. [CrossRef] [PubMed]
24. Gorelick, N.; Hancher, M.; Dixon, M.; Ilyushchenko, S.; Thau, D.; Moore, R. Google Earth Engine: Planetary-scale geospatial analysis for everyone. *Remote Sens. Environ.* **2017**, *202*, 18–27. [CrossRef]
25. Wang, X.; Xiao, X.; Zou, Z.; Hou, L.; Qin, Y.; Dong, J.; Doughty, R.B.; Chen, B.; Zhang, X.; Chen, Y.; et al. Mapping coastal wetlands of China using time series Landsat images in 2018 and Google Earth Engine. *ISPRS J. Photogramm. Remote Sens.* **2020**, *163*, 312–326. [CrossRef] [PubMed]
26. Wang, Y.; Ma, J.; Xiao, X.; Wang, X.; Dai, S.; Zhao, B. Long-term dynamic of Poyang Lake surface water: A mapping work based on the google earth engine cloud platform. *Remote Sens.* **2019**, *11*, 313. [CrossRef]
27. Inman, V.L.; Lyons, M.B. Automated Inundation Mapping Over Large Areas Using Landsat Data and Google Earth Engine. *Remote Sens.* **2020**, *12*, 1348. [CrossRef]
28. Hu, Y.; Huang, J.; Du, Y.; Han, P.; Wang, J.; Huang, W. Monitoring wetland vegetation pattern response to water-level change resulting from the Three Gorges Project in the two largest freshwater lakes of China. *Ecol. Eng.* **2015**, *74*, 274–285. [CrossRef]
29. Liang, D.; Lu, J.; Chen, X.; Liu, C.; Lin, J. An investigation of the hydrological influence on the distribution and transition of wetland cover in a complex lake floodplain system using time-series remote sensing and hydrodynamic simulation. *J. Hydrol.* **2020**, *587*, 125038. [CrossRef]
30. Tan, Z.; Jiang, J. Spatial-Temporal Dynamics of Wetland Vegetation Related to Water Level Fluctuations in Poyang Lake, China. *Water* **2016**, *8*, 397. [CrossRef]
31. Wan, R.; Xue, D.; David, S. Vegetation Response to Hydrological Changes in Poyang Lake, China. *Wetlands* **2019**, *39*, 99–112. [CrossRef]
32. You, H.; Fan, H.; Xu, L.; Wu, Y.; Wang, X.; Liu, L.; Yao, Z.; Yan, B. Effects of Water Regime on Spring Wetland Landscape Evolution in Poyang Lake between 2000 and 2010. *Water* **2017**, *9*, 467. [CrossRef]
33. Zhang, C.; Yuan, Y.; Zeng, G.; Liang, J.; Guo, S.; Huang, L.; Hua, S.; Wu, H.; Zhu, Y.; An, H.; et al. Influence of hydrological regime and climatic factor on waterbird abundance in Dongting Lake Wetland, China: Implications for biological conservation. *Ecol. Eng.* **2016**, *90*, 473–481. [CrossRef]
34. Zhang, E.; Cao, Y.; Langdon, P.; Jones, R.; Yang, X.; Shen, J. Alternate trajectories in historic trophic change from two lakes in the same catchment, Huayang Basin, middle reach of Yangtze River, China. *J. Paleolimnol.* **2012**, *48*, 367–381. [CrossRef]
35. Hu, H.; Kang, H.; Gong, G.; Zhu, M.; Zheng, W.; Wu, F.; He, D.; Li, Z.; Geng, D. Biodiversity of winter waterbirds in Hubei, China. *Resour. Environ. Yangtze Basin* **2005**, *14*, 422–428.
36. Zhang, S.; Dou, H.; Jiang, J. Aquatic vegetation in Longgan Lake. *J. Lake Sci.* **1996**, *8*, 161.
37. Tomaselli, V.; Di Pietro, R.; Sciandrello, S. Plant communities structure and composition in three coastal wetlands in southern Apulia (Italy). *Biologia* **2011**, *66*, 1027. [CrossRef]
38. Phase 2 USGS Gap-Fill Algorithm. Available online: https://landsat.usgs.gov/sites/default/files/documents/L7SLCGapFilledMethod.pdf (accessed on 7 October 2018).
39. Yuan, L.; Liu, G.; Li, W.; Li, E. Seed bank variation along a water depth gradient in a subtropical lakeshore marsh, Longgan Lake, China. *Plant Ecol.* **2007**, *189*, 127–137. [CrossRef]
40. Hersbach, H.; Bell, B.; Berrisford, P.; Hirahara, S.; Horányi, A.; Muñoz-Sabater, J.; Nicolas, J.; Peubey, C.; Radu, R.; Schepers, D.; et al. The ERA5 global reanalysis. *Q. J. R. Meteorol. Soc.* **2020**. [CrossRef]
41. Hu, C. A novel ocean color index to detect floating algae in the global oceans. *Remote Sens. Environ.* **2009**, *113*, 2118–2129. [CrossRef]
42. Zhao, D.; Lv, M.; Jiang, H.; Cai, Y.; Xu, D.; An, S. Spatio-temporal variability of aquatic vegetation in Taihu Lake over the past 30 years. *PLoS ONE* **2013**, *8*, e66365. [CrossRef] [PubMed]

43. Villa, P.; Bresciani, M.; Bolpagni, R.; Pinardi, M.; Giardino, C. A rule-based approach for mapping macrophyte communities using multi-temporal aquatic vegetation indices. *Remote Sens. Environ.* **2015**, *171*, 218–233. [CrossRef]
44. Jiang, Z.; Huete, A.R.; Didan, K.; Miura, T. Development of a two-band enhanced vegetation index without a blue band. *Remote Sens. Environ.* **2008**, *112*, 3833–3845. [CrossRef]
45. Ji, L.; Zhang, L.; Wylie, B. Analysis of dynamic thresholds for the normalized difference water index. *Photogramm. Eng. Remote Sens.* **2009**, *75*, 1307–1317. [CrossRef]
46. Otsu, N. A threshold selection method from gray-level histograms. *IEEE Trans. Syst. Man Cybern.* **1979**, *9*, 62–66. [CrossRef]
47. Zhang, X.; Liu, X.; Wang, H. Developing water level regulation strategies for macrophytes restoration of a large river-disconnected lake, China. *Ecol. Eng.* **2014**, *68*, 25–31. [CrossRef]
48. Zou, L.; Nie, Z.; Yao, X.; Shi, J. Effects of light on submerged macrophytes in eutrophic water: Research progress. *Chin. J. Appl. Ecol.* **2013**, *24*, 2073–2080.
49. Yuan, S.; Zhang, X.; Liu, X.; Wang, H. Ecological water level management strategy for aquatic vegetation in the mid-lower yangtze shallow lakes. *Acta Hydrobiol. Sin.* **2019**, *43*, 104–109.

MDPI
St. Alban-Anlage 66
4052 Basel
Switzerland
Tel. +41 61 683 77 34
Fax +41 61 302 89 18
www.mdpi.com

Water Editorial Office
E-mail: water@mdpi.com
www.mdpi.com/journal/water

www.ingramcontent.com/pod-product-compliance
Lightning Source LLC
Chambersburg PA
CBHW040814250926
51850CB00007BA/1164
* 9 7 8 3 0 3 6 5 2 9 6 8 4 *